INTRODUCTION TO COMPARATIVE PHYSIOLOGY

INTRODUCTION COMPARATIVE

CONTRIBUTORS

P. J. BENTLEY
Mount Sinai School of Medicine

ROBERT M. DOWBEN
University of Texas

LEON GOLDSTEIN
Brown University

T. RICHARD HOUPT
Cornell University

JOHN KANWISHER
Woods Hole Oceanographic Institute

ANTHONY R. LEECH
University of Oxford

CLAUDE J. M. LENFANT
National Institutes of Health

PAUL J. PALATT
Brown University

BERT SHAPIRO
Harvard University

STEPHEN C. WOOD
University of New Mexico

TO
PHYSIOLOGY

LEON GOLDSTEIN, editor
Brown University

HOLT, RINEHART AND WINSTON New York Chicago San Francisco Atlanta
Dallas Montreal Toronto London Sydney

Library of Congress Cataloging in Publication Data

Main entry under title:

Introduction to comparative physiology.
    Includes index.
    Bibliography: p.
    1. Physiology, Comparative.   I. Goldstein,
Leon, 1933—   II. Bentley, P. J.
QP33.I57   596'.01   76-26009
ISBN 0-03-012411-5

Printed in the United States of America
1234567890 038 0987654

PREFACE

This book was written to be read and enjoyed. Unfortunately, it has been my experience that undergraduates do not enjoy reading most of the texts assigned to them. It is not that the books are uninteresting—to the contrary, there are many interesting books on comparative physiology on the market—but the students complain that they get bogged down in details, a problem that is difficult to overcome. Ideally an author would like to present the subject material in such a manner as to both capture the readers' attention and imagination and provide sufficient information to give them an understanding of how that information was obtained. This ideal is rarely achieved in textbook writing, perhaps because authors have forgotten the student's point of view.

The aim of this text is to present the principles of comparative physiology as clearly and concisely as possible. As editor, I asked the contributors to write their sections at the level and style of a good *Scientific American* article. It is my belief that if you can capture the readers' imagination with an interesting presentation, they will want to learn more about the subject and at this point will be ready to go to the original literature. To assist at this juncture each chapter has sufficient references at the end to direct the readers to any research area they may wish to investigate.

To achieve both brevity and clarity, a certain amount of selectivity had to be exercised in choosing subject material to be included in the book. In striving for brevity the authors were forced to limit discussion of currently evolving theories and developments in their fields. To compensate they have supplied the reader with an extensive reading list at the end of each chapter including several references to current studies.

This book reflects the interests of the contributors; thus it is vertebrate-oriented. Discussion of some areas, such as neurophysiology, would be impossible, however, without mentioning the great amount of knowledge gained from the study of invertebrate systems.

Although each chapter was written by a different author, we did try to achieve some degree of uniformity. Each chapter is preceded by a short general introduction describing the theme of the chapter and its relation to other chapters. The first part of most chapters presents the general principles of the subject, followed by a discussion of how these principles operate in different animals—variations on a theme, so to speak. Where possible the subject material is presented from an organ-systems approach; that is, how the nervous, muscular, cardiovascular, respiratory, osmoregulatory, digestive, and endocrine systems operate to maintain the normal physiology of the various animals discussed. This approach was not appropriate for the chapters on physical principles, membrane physiology, and temperature regulation in which the subject material is covered in a more general manner. Finally, the chapters are cross-referenced to coordinate related material in different parts of the book.

There are some features of this book that are unique in a comparative physiology text. First, the book begins with a chapter discussing physiological systems as physical systems. Only by identifying the basic physical mechanisms underlying physiological processes can we begin to quantify physiological events. Similarly, almost all physiological systems involve membrane phenomena. Thus we felt that a chapter dealing in a comprehensive manner with membrane permeability and transport across cell membranes should be included in the text. It is our hope that inclusion of these two chapters will make the text both more educational and useful.

I wish to acknowledge the helpful editorial assistance that I received from several colleagues during preparation of the book. Dr. Howard A. Schneiderman reviewed the entire manuscript during various stages of development. Drs. Kent M. Chapman, Helen F. Cserr, John Fain, Donald C. Jackson, James T. McIlwain, and Martin Morad critically read various sections and offered many constructive suggestions.

Finally, I want to thank my wife, Barbara J. Goldstein, for moral support throughout this adventure, especially during those times when the end seemed far away, indeed.

Leon Goldstein

CONTENTS

1

PHYSICAL PRINCIPLES

Paul J. Palatt

For many readers this book will be their first encounter with physiology. Chapter 1 is therefore devoted to a general discussion of the major principles underlying quantitative analyses of physiological systems. Most of the techniques to be described were not available before the beginning of the twentieth century, and many are only now coming to the forefront. The importance of such techniques to the development of physiology as a conceptual and quantitative science cannot be overemphasized. However, important as these quantitative tools are, knowledge of them alone does not always lead to an understanding of physiological processes. In many areas of physiology the gaps in information (that is, data) preclude a detailed description of the events being studied. For instance, whereas some areas, such as contractility, can be described reasonably well at the molecular level, other areas, such as information processing in the nervous system, can be treated meaningfully only at the system, or integrated, level. These differences in levels of understanding and availability of data are reflected in the diversity of treatments of different subjects throughout the book. There are, however, common analytical elements that underlie all the treatments. These elements are presented in Chapter 1.

1 • INTRODUCTION

> *Just as man, seen in terms of paleontology, merges anatomi-*
> *cally with the mass of mammals that preceded him, so,*
> *probing backwards, we see the cell merging qualitatively and*
> *quantitatively with the world of chemical structures. Fol-*
> *lowed in a backward direction, it visibly converges toward*
> *the molecule . . .*
> —Pierre Teilhard de Chardin

Higher forms of life are composed of a multitude of subunits, which are the inheritance of sharks and shrews, elephants and algae. The science of physiology is basically an outgrowth of attempts to understand how these subunits interact within a given organism. By studying similarities and differences that exist among all forms of life, the physiologist attempts to gain a quantitative understanding of the functioning of individual living systems at all levels.

It is precisely this synthetic and universal aspect of physiology that makes it unique among the sciences. Given a hunk of unknown metal, it is a good bet that the physicist will measure its response to some field; the engineer will smash it, or stretch it, or heat it, or machine it to test its material properties; the chemist will try to dissolve it in some solvent; the mathematician will try to write an equation on or about it; and the biological scientist will probably ignore it in favor of what he considers a more interesting system, one that appears to act in a purposeful manner when perturbed by an experimenter. The physiologist of today, however, is far less likely to ignore the piece of metal than the physiologist of 50 years ago, and, what is more important, he is far less likely to ignore the studies of the physicist or engineer.

Science advances at a pace limited only by the analytical techniques available and the imagination of the scientist. The term "analytical techniques" includes not only experimental craft and ingenuity but also principles of approximation and abstraction. Today the physiological sciences are in the midst of an enormous number of changes. A major impetus for many of these changes has been the absorption and employment of quantitative concepts and methods of analysis that have classically been associated only with the physical sciences.

In particular, the consideration of organisms as physical systems has led to an increased understanding of physiological mechanisms at all levels. It has also led to a number of difficulties. The approach of the physicist and engineer is largely synthetic. Theirs is a problem of design using available hardware and concepts to meet certain performance specifications. The physiologist is faced with a different problem. The systems in which he is interested are already designed. He must discover how they work. Because physiological systems are extremely complex and the basic principles governing their functions are not immediately evident, the physiologist must resort more to analytic than to synthetic thinking to understand the complex phenomena he studies.

A related problem is that the physical scientist is often more concerned with generality than with actuality. He studies ideal systems in an attempt to understand the principles governing real systems. The physiologist's major concern must, from the outset, be less abstract. He has to deal with real, extremely complex systems, portions of which cannot be isolated for study without destroying their essential properties. As Wolfgang Köhler has pointed out:

> If organisms were more similar to the systems which physics investigates, a great many methods of the physicists could be introduced in our science without much change. But in actual fact the similarity is not very great. One of the advantages which make the physicist's work so much easier is the simplicity of his system. His systems are simple because, to a degree, the experimenter himself determines their properties. I am far from believing that organic processes are of a supernatural kind. On the contrary, the most startling difference between the organism and a simple physical system is the enormous number of physical and chemical processes, which, in complicated interrelations, occur at a given time in the organism. We are utterly unable to create simpler organic systems for elementary study. An amoeba is a more complicated system than all the systems of the inanimate world. We also know that in studying, for instance, the properties of a nerve-muscle preparation, we are not investigating "a part" of natural behavior. The functional characteristics of such a preparation differ from the characteristics which the same nerve and muscle exhibit when functioning in normal behavior. Some Behaviorists have rightly said that it is the whole organism which we have to study. Unfortunately, in the whole organism one can seldom follow the change of one particular variable, as though it alone were affected by a certain change in outer conditions. The change of one factor usually involves concomitant changes in many others, and the latter changes again affect the former. Now, isolation of functional relationships and reduction of variables which take part in an event are the great artifices by which exact investigations are facilitated in physics. Since this technique is not applicable in psychology, since we have to take the organism more or less as it is, any kind of observation which refers to the behavior of our subjects as complex acting units will be right in our case.[1]

Although Köhler was discussing psychology as an experimental science, his remarks apply equally well to physiology. The methods of physics

---

[1] From *Gestalt Psychology* by Wolfgang Köhler. Copyright © 1947 by Liveright Publishing Corp. Reprinted with the permission of Liveright Publishing Corporation.

cannot be used indiscriminately to study biological systems. One ultimate goal of physics is to explain the properties of both matter and energy within the framework of a single unified theory reducible, in the end, to structures that are molecular and statistical in origin. Since many of the concepts of classical physics, especially at the molecular level, are based on equilibrium considerations, one must be careful when applying physical principles, even seemingly simple and obvious ones, to living systems, which are in no sense in equilibrium, either internally or with their surroundings. Another problem arises when one tries to use physical concepts normally applied to isolated systems to describe living organisms. Living creatures are not isolated from their environments and cannot always be treated by the quantitative methods that the physicist normally uses to study isolated systems.

That the quantitative methods of physics and chemistry must be used with caution when dealing with living systems is not so important; that they can be used at all is extremely important. The modern physiologist is a curious blend of artisan and scientist. He picks and chooses among experimental techniques used by physical scientists and then tailors them to fit his needs. When properly carried out, his experiments and analyses are models of scientific inquiry. However, when his experiments are poorly designed or his data improperly interpreted, he may in fact still obtain "reasonable" results, that is, results differing from the proper ones only in very subtle ways. This is a major difference between physiology and physics, and it is the major danger that the experimental physiologist faces.

Partly to overcome this difficulty and partly out of necessity, physiology is evolving its own physicochemical and mathematical approaches to analyses. These newer techniques, many of them microscopic in nature, allow the physiologist to investigate his subject matter at various levels of detail and to cross-check results. Unfortunately, these techniques also lead to certain conceptual difficulties.

One obstacle that presents itself in the study of both physiological and physical systems is that the underlying mechanisms of interest seem always to be located between the microscopic and macroscopic levels. The macroscopic world is directly accessible to observation, and its mathematical representations are usually continuous and differentiable. Its physical correlates are those things we can either see with the naked eye or can touch. The microscopic world is observable only at the cost of disturbing the object of observation. Its mathematical isomorphisms are apparently discrete and discontinuous. The microscopic world, which exists below our range of vision or direct contact, is composed of molecules and atoms. It is the average behavior of these very small microscopic units that constitutes macroscopic events.

On a large scale, events seem to obey definite physical principles, the laws of classical physics. At present it is on this level that most physiological studies are performed and interpreted. On a small scale, things happen or do not happen only on the average. Quantum physics is concerned with that averaging process; physiology should also be if it is to answer many of the questions that it poses.

Somewhere between the large and the very small lie the yet undiscovered connections between quantum and classical physics. Very probably in that same inaccessible region lie the principles that separate the living from the nonliving. However, understanding or even merely identifying these principles is extremely difficult. Since all measurements on a microscopic scale are inexact and statistical in nature, we can talk about molecular events only "on the average." We must try to discover the operational principles that underlie observable physiological processes by means of conjecture, abstractions, equations, and approximations to the intractable mathematics that inevitably arise when one deals with any complex microscopic system. To discover those principles underlying the behavior of living systems is the major challenge of physiology today and its charge for the future.

2 • THE ORGANISM AS A COMPLEX SYSTEM

> . . . and everything talks to everything else with gurgles and
> squishes and other chemical factory sounds.
> —Paul J. Palatt

In principle, no sharp demarcation exists on a molecular level between living and nonliving systems. In the microscopic limit, both are composed of the same inanimate subunits. The unique properties exhibited by living systems would therefore appear to be a function of complex interactions among molecular subunits rather than of intrinsic properties of the subunits themselves.

In general we can define a system as nothing more than two or more parts connected together; it may be a group of equations, a collection of inanimate objects, or a living organism. All living systems are complex. An important property of any complex system is the ability to perform functions that are greater in both number and kind than the summation of the functions of the parts of the system. This means that it is impossible to predict the behavior of any complex system from separate analyses of the behaviors of its component parts, no matter how detailed the analyses. For instance, one could not predict the myriad functions of a digital computer from a set of hardware specifications any more than one could predict what a person is capable of saying by understanding the physical and chemical bases for the muscle actions that effect speech. In this context system structure is extremely important. What a computer is or how it works has little relationship to a bagful of its components. In an analogous fashion, many of the functions of a membrane are not present, or even hinted at, when all the chemical constituents of a membrane are brought together under the proper physiological conditions.

This leads to obvious difficulties in analysis. Any living system must, of necessity, be analyzed piecemeal if the chemical and physical properties of its constituent parts are to be understood. Such an analysis ignores structure, organization, and interactions, all of which are present in the intact animal, and all of which are of inestimable functional importance.

But we must begin somewhere, and physiology begins by assuming that (1) living systems can be explained in terms of the same chemical and physical laws that describe inanimate matter, and (2) by careful analyses and experimentation we can define the relationships among the relevant subparts of the organism that are of importance. Whether these assumptions are valid is a matter for conjecture, but the present science of physiology rests heavily on both of them.

3 • REGULATION AND CONTROL: HOMEOSTASIS

All the vital mechanisms, however varied they may be, have only one object, that of preserving the conditions of life in the internal environment.
—Claude Bernard

Life as we know it is possible only within a certain limited range of environmental conditions. In general the higher the form of life the more closely regulated must be the organism's internal and external environments.

No living thing exists in isolation. Throughout an organism's existence, matter, energy, and information pass both through the organism and between parts of it at rates that are not always predictable or even measurable. These flows provide perturbations to the physiological state of the animal that must be controlled if the animal is to remain alive. If the disturbances come from the external environment, the animal may respond at all levels from the cellular to the organismic. If the disturbances are internal to the organism, responses are usually at the cellular or organ level. In either case, stability of the internal environment is maintained by a constant adjustment of physiological parameters.

The concept of constancy of the internal environment was first put forth by Claude Bernard over a century ago. In 1929 Walter B. Cannon further developed this concept and gave it the name *homeostasis*. At the most basic levels homeostasis is accomplished by mass and energy transport controlled via feedback mechanisms.

4 • FEEDBACK

> . . . cause and effect, and secondary effect acting on primary cause. I once almost understood how light from a star long dead might have affected the future of dinosaurs.
> —Paul J. Palatt

When we try to describe biological control adequately, we are faced with a number of problems. Most biological systems that exhibit control are extremely complicated with respect to both mechanism and purpose. Loosely, we can consider any living thing to be a system that operates at a number of distinct levels to perform certain tasks or actions. The activities at all levels must be coordinated and controlled if the organism is to function efficiently or, in many cases, if it is to function at all. However, there are interactions not only between given levels of a system but also between different systems, each of which is regulated. To determine completely the individual effects of each system is often impossible.

To illustrate the principles of control theory apart from the complexities of real biological systems, we will consider a very simple example, similar to one given by Simon. Although this is not a true physiological example, it illustrates the major principles of physiological control.

Consider an individual resting in his bathtub (Figure 1.1). He notes the following: the bathtub is being filled at a rate R_{in} through the tap and being drained at a rate R_{out} through the drain (the plug is lost). While bathing, he wishes to maintain the level of water in the tub at some fixed level, L_b, despite changes in R_{out} that will occur because of the accumulation of dirt over the drain hole.

"Clearly," he thinks, "the easiest way to maintain the proper level would be to fill the tub to L_b and then somehow adjust R_{in} so that it is always equal to R_{out}. I can do this by connecting the outflow to the inflow via a pump, so that all of the liquid leaving through the drain is returned to the tub through the tap. However, that would not be a very practical solution because all of the water entering the tub during my bath would then be dirty.

"To be more realistic, I should consider what would happen if R_{out} were to change in relation to R_{in}. If R_{out} decreases below R_{in}, the level in the tub will continually rise until overflow occurs. Conversely, if R_{out} becomes greater than R_{in}, the tub will eventually empty. Only if R_{out} exactly equals R_{in} will I maintain the desired level, L_b.

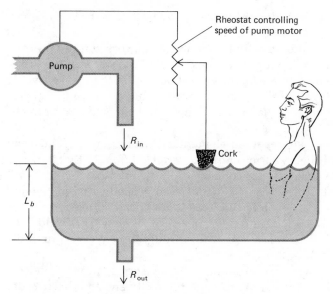

FIGURE 1.1 A simple control system in which the level of water in a bathtub controls the rate of flow into the tub.

"I can avoid the difficulty of having to maintain perfect adjustment, however, by punching a series of overflow holes just above the level L_b and then adjusting R_{in} so that it is always greater than the maximum possible value of R_{out}, that is, the value at the beginning of my bath. So long as R_{in} is greater than R_{out}, overflow will occur, and the level of the water will remain just below the level of the holes. However, this would not only waste an inordinate amount of water, but it would be difficult for anyone larger or smaller than I am to take a bath comfortably. Besides, it would ruin my dining room ceiling.

"An ideal solution would be to make R_{in} dependent on the instantaneous level in the tub. Why can't I merely place a cork in the tub and design a machine that will turn the tap water on or off depending upon the position of the cork with respect to the sides of the bathtub. I can connect the cork to a switch that controls a pump, which in turn determines R_{in}. Whenever the level falls below L_b, the switch will close, turning the pump on and restoring the proper level. Whenever the level rises above L_b, the switch will open, turning the pump off and permitting R_{out} to return the level to normal. This system has the serious disadvantage, however, that the tap is either completely on or completely off. If the response time of the pump is slow and R_{in} is high, I may find myself up to my neck in trouble;

if it is slow and the drain suddenly clears, I may find myself ankle deep in misery.

"Why don't I build a machine to monitor the cork continually, adjusting the taps in the following way: if the level of the water in the tub is far from L_b, the amount of adjustment should be greater than if the level is near L_b. This should work well within the normal range of R_{in} and R_{out} and should alleviate, at least in large part, the problem of a slow system response time."

This example illustrates in an elementary fashion most of the basic principles of control theory. The *controlled variable* refers to that variable within the system whose value is controlled. In the example it is the level of water in the bathtub. The ideal value of the controlled variable, the one that the system "aims at," is referred to as the *set point* (L_b). The value of the controlled variable is monitored by means of a *sensor* (cork) and controlled by means of a *controller* operating through an *effector* (pump effecting changes in R_{in}).

Obviously communication among the sensor, controller, and effector is necessary if the system is to operate effectively. In other words, information flow between the system's components is a necessary condition for regulation and control. The key concepts underlying control and regulation are those of negative feedback and stability. Feedback permits the controller to compare the monitored value of the controlled variable with the set point. In this way the system is made "self-conscious." In most living homeostatic systems there is no clearly defined set point, but rather there exists a normal range of operation, which is determined by the nature of the chemical and physical structures in the system. The system will operate effectively within this range only if its components remain intact and if it has an adequate supply of energy and nutrients. Within the normal range of operation, the system takes action to minimize the difference between the actual value of the controlled variable and the set point. For example, if the quantity represented by the controlled variable minus the set point, usually called the *error signal*, is positive, then action is taken to decrease the value of the controlled variable (that is, the pump is turned off). This type of control is termed *negative feedback* (see Figure 1.2) because the action taken tends to drive the controlled variable in the opposite direction of, or negative to, the direction of its disturbance. Systems that function in this manner are called regulators. Because deviations from the set point tend to be corrected rather than propagated, negative feedback control tends to be stable. It is this type of control that appears to be prevalent in biological systems.

As an example of negative feedback control, consider one of several systems that regulate arterial pressure in man (these systems are discussed

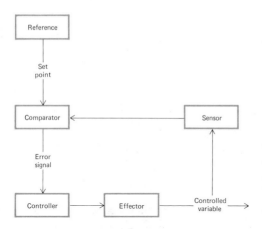

FIGURE 1.2 Components of a negative feedback control system. The boxes represent functionally distinct elements in the system. The lines connecting them represent either information flow (via neural or molecular transmission) or measurable variables.

in detail in Chapter 5, Section 4.2). The sensors for this system are nerve endings located in the walls of the aorta, in the region of the bifurcation of the carotids, and possibly in the walls of other large arteries. These nerve endings, called *baroreceptors,* exhibit an increase in activity with the mechanical stretching of vessel walls that accompanies an increase in arterial pressure.

Arterial pressure at any instant is a function of both the quantity of blood pumped by the heart per unit time and the total resistance to the flow of blood through the vessels. This so-called "peripheral resistance" depends largely on the size of the lumen of the blood vessels and is increased by the contraction of muscles located in the vessel walls. These muscles, along with those of the heart, are the major effectors in the arterial pressure control system.

The controller for the system is located in the brain stem. It responds to the level of activity in the baroreceptors by adjusting the level of sympathetic and parasympathetic nervous activity delivered to the heart and the level of sympathetic activity to the muscles in the vessel walls. The level of nervous activity in turn controls both the strength of ventricular contraction and the heart rate, as well as the amount of vasoconstriction. In the absence of baroreceptor signals, vasoconstriction and heart activity are maintained at relatively high levels. Increasing activity in the baroreceptors causes an inhibition of vasoconstriction and a decrease in heart activity. A normal range of arterial pressure is thus maintained (see Figure 1.3).

Another type of feedback control is possible. *Positive feedback* occurs if the change effected by the controller is in the same direction as the deviation from the set point. Positive feedback is said to be unstable

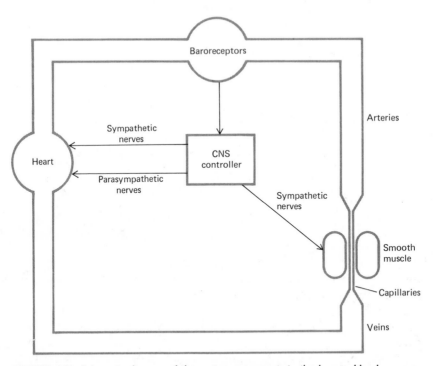

FIGURE 1.3 Schematic diagram of the major components in the human blood pressure control system.

because it leads to a "vicious cycle" which propagates perturbations in the system. For this reason examples of positive feedback in normal physiological situations are rare. Probably the most common example of positive feedback in biology is the runaway increase in sodium permeability that accompanies the active phase of an action potential in an excitable cell (see Chapter 3, Section 6, for details). Within humans, positive feedback on an organismic level often occurs during fear and anxiety reactions or in response to physiological trauma, especially pain. On a larger scale, so long as there are sufficient nutrients, proper environmental conditions, and a lack of overcrowding, the phenomenon of population growth is also an example of positive feedback.

Examples of negative feedback in living systems are much more common. Detailed discussions of physiological negative feedback systems have been presented by a number of authors (see references at the end of this chapter). Negative feedback, in the broadest sense, is what returns the neuron to a quiescent state following an action potential or what limits population size on a long-term basis. Examples of negative feedback within the organism vary from the involuntary control of the chemical environment of cells (homeostasis) to complex voluntary acts such as catching a ball. This last case illustrates another type of feedback system called a *servomechanism*, whose purpose is to follow, or track, an input. One of the classic examples of a biological servomechanism is the eye-tracking capability of higher animals.

Negative feedback is also important in behavior. It plays a crucial role in the maintenance of posture and in the performance of all purposeful acts. Memory, conditioning, and learning all act on a cognitive level in humans to alter the functions of physiological feedback loops over a period of time. Not only can physical skills be learned, but recent evidence also indicates that people are capable of exercising control over their so-called involuntary functions. The physiological changes that accompany an alteration in behavior may be due to either variations in set point or changes in the strategy of control. If the output of a system is maintained according to a specific control rule (although the set point may vary), the system is said to exhibit *nonadaptive* control. However, if the rule for control changes as the system functions, then the control is said to be *adaptive*. Adaptive control can result from changes in the controller or from an adjustment of connections on either the sensor or effector end of the system.

A rule for control refers to the way in which the controller acts on the error signal to effect changes in the controlled variable. A rule is partly determined by the physical parameters of the system (stiffness, time delays, and so on) and partly by the *mode* of control used. There are several common modes of control found in biological systems. One is *on-off* control, which is all-or-none in nature. When a certain value of the error signal is exceeded, the controller is on; otherwise it is off. This is the familiar type of control used with refrigerators and furnaces in the home, and it corresponds to the first cork-switch-pump system considered in the example. A second type of control results in corrective action *proportional* to the error signal. This type of control, also considered in the example, is probably the most obvious from an intuitive standpoint. A third kind, *integral* control, utilizes the total error over time as an additional input to the controller to minimize the steady-state error. In a fourth mode, *derivative* control, the rate of change of error is used to anticipate the corrective action required.

Because any real system possesses inertia and compliance, and because there is always a time lag associated with the transmission of information to and from the controller, the response of any physical system never tracks a stimulus precisely. For instance, consider the arterial pressure control system discussed previously. Maximal transmission of baroreceptor signals can take as long as 15 sec. Thus action taken by the controller to alter autonomic nervous activity can lag behind actual changes in values of arterial pressure by as much as 15 sec. Barring additional controls, the system would obviously pass through its set point, rather than stop precisely at it. The resulting phenomenon, called *overshoot*, would require that corrective action be taken in the opposite direction. The system would therefore oscillate about the set point.

The reason that most physiological systems fail to demonstrate oscillation is that the transmission times for control signals are rapid compared to effector response times. In addition, the systems are usually *damped;* that is, some portion of the action initiated by the controller is proportional to the rate of change of the effector's action and opposed to it. (For detailed discussions of oscillation, damping, and other concepts associated with physical and physiological control, see references.) The important point to note is that in any physical system perfect control under transient conditions is impossible, even with a combination of the more complex modes of control just described.

5 • ENTROPY, INFORMATION, ORDER, AND CHAOS

The energy of the universe is constant; the entropy of the universe increases toward a maximum.
—Clausius

In the final analysis, all that any living creature is capable of doing can be reduced to actions on its environments—external, internal, or both. These actions result from mass and energy transport on the part of effectors. In higher organisms effectors are muscles, glands, or tissues capable of transport; in lower forms of life effectors may be as simple as cilia or molecules of contractile protein. The organism monitors, either consciously or unconsciously, the output of its effectors by means of sensors and controls the output of its effectors by means of controllers. In higher animals, pathways into the organism through external sensors, such as eyes, ears, and cutaneous nerves, are generally higher energy pathways than those through the controllers to the effectors. The fact that the control portion of the system functions at relatively low energy levels allows for both precise and efficient regulation.

The amazing fact is not that biological systems regulate but that they regulate so efficiently. The control signal requires only a small amount of energy compared with the total energy in the system. The sign of the error signal is far more important in determining system function than is its magnitude; the controller recognizes and acts on information rather than on energy alone. Thus we must be concerned with two aspects of signals: energy and information; although apparently different, they can never be treated separately. Every signal has some material embodiment or transmission medium associated with it. Within living systems signals may be carried by molecules or by neural activity. In the latter case it is the complex chemical nature of the electrical activity that prohibits a separation of information flow from energy and mass transport (see Chapter 3, Section 6).

The concepts of mass and energy are firmly rooted in classical physics and everyday experience. Information and its thermodynamic correlate, entropy, are somewhat less intuitive. In principle the entropy concept is strictly applicable only to closed (isolated) systems. More generally we can define the entropy principle as the tendency for all systems, living and nonliving, to evolve toward a disordered state, and we can define entropy itself as the "amount" of disorder in the system. Note that the word "evolve" implies the existence of a time sequence; indeed, it is the entropy principle that both defines a forward time direction in physics and underlies the physiological processes of aging and death.

Information can be viewed as a measure of the amount of order in a system; it is the opposite of entropy. In fact the mathematical expression for entropy has a form that is the negative of the form of the mathematical expression for information. Over any period of time, unlike mass and energy, neither information nor entropy is conserved; the entropy of any isolated system increases, and the information content decreases with time. Qualitatively it is easy to see that information about a system reduces the uncertainty (that is, chaos) associated with the system. Obviously this reduction is a necessary prerequisite for regulation and control.

Entropy and information can be discussed only in terms of statistical distributions of large numbers of particles; both are macroscopic correlates of the average molecular behavior of a system. A quantitative approach to control and regulation is thus possible only with the entire organism, or with large portions of it, and not with molecular subsections of it. This raises interesting questions concerning mechanisms of control at the cellular level.

Averaging also raises questions about the widespread use of classical thermodynamic variables to describe physiological systems at the cellular level. For instance, pressure is often considered to be a driving force for mass transport across membranes. What the term "pressure" might refer to in a volume of cellular dimensions is not very clear. Concentration is another macroscopic concept whose use is sometimes questionable. Let us take an example. Consider a spherical mitochondrion having a diameter of 0.2μ. A simple calculation shows that at a pH of 7 there is approximately 0.25 molecule of hydrogen ion inside the mitochondrion (whatever that means). By moving fewer than three molecules of H^+ into or out of the organelle, the internal hydrogen ion concentration can be altered tenfold. What, in fact, do people mean when they discuss the chemical composition of a system as small as a mitochondrion? Even a bacterium with a diameter of $1\ \mu$ contains only about 32 H^+ ions at pH 7. Random molecular fluctuations would change its pH significantly.

The difficulties that arise when quantitative concepts of macroscopic physics are applied to physiological analyses on a microscopic level are obvious. These difficulties raise vital questions that are at the heart of contemporary physiology. The bases of all science lie in measurement and operational definitions. One problem physiology is presently faced with is how to best use concepts and definitions from the physical sciences to describe living systems. In particular, how can information and control in a living system be quantified? At present we can talk only about general aspects of entropy and information, such as maximum or minimum possible values, or a qualitative tendency toward organization or chaos. This severely limits any attempts to formalize a physicochemical approach to the study of biological organization and regulation or to the study of molecular physiology in general. It has obviously become a major burden of physiology to develop its own quantitative techniques, to verify their validity as analytical tools, and to extend their usefulness to the microscopic level. That burden will increase, not decrease, in the future.

6 • PHYSIOLOGY AND MATHEMATICS

> The observation of phenomena cannot tell us anything more than that the mathematical equations are correct: the same equations might equally well represent the behavior of some other material system. For example, the vibrations of a membrane which has the shape of an ellipse can be calculated by means of a differential equation known as Mathieu's equation: but this same equation is also arrived at when we study the dynamics of a circus performer who holds an assistant balanced on a pole while he himself stands on a spherical ball rolling on the ground.
>
> If we now imagine an observer who discovers that the future course of a certain phenomenon can be predicted by Mathieu's equation, but who is unable for some reason to perceive the system which generated the phenomenon, then evidently he would be unable to tell whether the system in question is an elliptic membrane or a variety artist.
> —Sir Edmond T. Whittaker

As physical scientists and mathematicians have become more involved in the study of physiology, and as the use of computers has become more prevalent throughout the sciences, mathematical modeling as a means

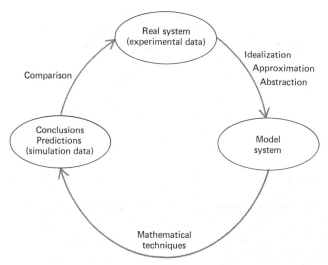

FIGURE 1.4 Summary of the interactions involved in the mathematical modeling process.

of studying living systems has become more common. The mathematical modeling process (see Figure 1.4) consists of two interacting procedures: (1) performance of experiments and the formalization of properties and relationships observed during the experiments by means of mathematical expressions called the model system, and (2) comparison of behaviors of the model system and the real system. Often this involves the performance of additional experiments that can be used to confirm, modify, or reject specific features of the model.

A model system is thus nothing more than a set of mathematical relations, together with boundary conditions, that are representative in some sense of the relationships between physical variables in the real system. For macroscopic systems, the prevalent forms of the mathematical relationships appear to be differential and partial differential equations. Values of the dependent variables in the equations are determined by specifying initial conditions for the system; different initial conditions lead to different values for the dependent variables.

For physiological systems, a majority of the dependent variables seem to settle at certain prescribed values, no matter what the initial conditions or perturbations to the system. This is a consequence of the principles of biological control discussed previously. A major problem in modeling biological systems is to ensure that the mathematical model properly accounts for the observed regulatory behavior. One possible approach is to use the mathematical formalism of control theory as developed by engineers and mathematicians for physical systems. But this provides little information about the underlying mechanisms of control or the structure of the system. Another possibility is to use the same equations as the physicist uses to describe physical systems without control, but to adjust the boundary conditions (mathematical expressions describing what happens at the physical boundaries of the real system) to account for the observed regulatory principles of living systems. This approach hopefully tells us something about the system's physics and its functional characteristics.

Obviously for biological systems the boundary conditions are extremely important. Polyani, in fact, argues that they are more important than the actual physics of the system. Since all matter must obey certain physical laws, and therefore certain equations, Polyani maintains that the uniqueness of living systems is the result of the operation of biological principles that control the boundary conditions within which the forces of physics and chemistry function. What these principles might be is difficult to say, but at least this approach suggests some formalism within which investigations can be made.

Although it is true that mathematical models are nothing more than expressions of relationships between variables, how we look at these relationships is a function of what we wish to explain. This problem has been discussed in detail by Bradley, among others. As he points out, both experimentalists and theoreticians try to "understand" or "explain" empirical phenomena by formulating a model described in terms of the language of a more microscopic, or finer-grained, level than the level at which the observations or experiments on the real system were made. Models defined in terms of the language at the same gross level as the data are generally considered to be merely descriptions or summaries, not explanations, of the data. Understanding is conveyed only when we explain the unknown in terms of known, simpler or more basic, concepts; usually this means in terms of more microscopic phenomena. For instance, we may describe a mammalian blood flow in terms of fluid dynamics and statics (Chapter 5), diffusion of O_2 in the body in terms of Fick's law (Chapter 2), or metabolism in terms of thermodynamics and kinetics. In each case we are attempting to explain biological phenomena in terms of known physical phenomena.

Every investigator first makes qualitative assumptions about how the real system works. From these assumptions he constructs some kind of model to explain his data. If the model "fits" the data to some arbitrary degree of accuracy, the investigator assumes that he understands the phenomenon under consideration. But to an investigator working at the level in whose terms the model was formulated, the model is not an explanation but a description, which in turn needs to be explained. This is, necessarily, a serial process.

Bradley presents an interesting illustration to emphasize the rather arbitrary division between description and explanation. The following example is based on his discussion. Consider how a number of different investigators might explain, or model, a specific biological phenomenon: the formation of two daughter cells from a single parent cell. Depending on who is trying to explain the fission, the comments may vary from a layman's naive wonder that the process occurs at all to the biologist's claim that cell division is universal in nature, occurring from the equator to the poles. The explanations encompass the cytologist's acid commentary on DNA replication and the biochemist's more basic considerations of nucleotide pair formation; the molecular chemist sticks with hydrogen bonding. On a deeper level, the quantum chemist discusses intermolecular potential functions. The quantum physicist integrates all "descriptions" with his wave equation, which the theoretical physicist then reduces to a reflection of the four-dimensional qualities of space-time. Each investigator's explanation is a description to any investigator collecting data at a more microscopic level.

The biologist claims that the models of the physicist are too microscopic—that they ignore interactions between components and are therefore incomplete. The physicist claims that the models of the biologist are merely collections of data and explain nothing. It would appear that all modeling should be done with a pencil in one hand and a grain of salt in the other. Why in fact should the physiologist model at all?

One major reason for modeling is simplification. Most physiological systems are far too complex to understand in all their details at any level and far too intricate to break up into subunits without destroying the integrity of the system. Modeling allows us, in principle, to separate the subunits and to study their interactions, that is, to investigate certain aspects of a complex system independently of other aspects. In addition, because of the model's relatively simple structure, we know more about the model than about the real system. This often helps us recognize important relationships that exist within the real system but are normally masked by complexities and interactions.

Moreover, a model system can usually be manipulated more easily, more quickly, and to a greater extent than can the real system. Perturbations can be introduced into the model and removed at will. This is often not possible in the real system without killing the organism. From these analyses, if the model has been properly abstracted, information can be obtained about both the sensitivity and stability of the system in the face of applied disturbances and about the functioning of any control mechanisms that might be present.

Building, or even just trying to build, a model system also serves an organizational purpose. Besides summarizing what we know about the physiology, it tells us what we do not know about the physics and chemis-

try of the real system. Critical concepts that are usually glossed over or ig-
nored in a verbal description of a system are brought to light. To describe
something mathematically, vague concepts have to be replaced by definite
statements about how a thing works. In this sense we are led to the design
of critical experiments and to better experimental design in general. Thus
a model is suggestive.

A model is also precise, not only in the numerical sense but in the
logical sense. It facilitates prediction and hypothesis testing; it allows de-
duction from premises. It necessitates that our thinking move away from
the general and verbal and toward the specific and quantitative. This is
perhaps its most useful function in physiology, for imprecise theories are
essentially useless. The most useful scientific theories are the least spec-
tacular; they are quite specific in nature and testable by experiment. Because they are useful in helping us to understand how some aspect of the
universe works, there exists the possibility that they can be proved
wrong. The more general theories are not so easy to disprove. In this sense
the usefulness of a theory is proportional to the information that can be
generated from its testable hypotheses. Mathematical modeling, by generating useful hypotheses, serves physiology no less than it serves the
physical sciences.

7 • PHYSIOLOGY AND PHYSICS

There is measure in everything.
—William Shakespeare

As previously discussed, the word "living" implies continuous exchanges of mass, energy, and information between different parts of the
organism and between the organism and the external environment. The
science of physiology tries to answer questions about both the rate of
these exchanges under widely varying conditions and the underlying
mechanisms that control these exchanges. Recent successes arising from
applications of mathematics and the physical sciences to physiological investigations have raised important questions concerning the traditional
barriers separating living and nonliving systems. But we must not forget
that barriers still exist. The fact that material and energy balances seem to
be applicable to physiological systems tells nothing about the basic molecular transformations that lead to these balances, nor does it indicate what
the interrelationships between relevant variables are on a molecular level.
To study these interrelationships, the principles of cybernetics, or systems
analysis, must be utilized.

But even these sciences cannot answer many of the questions posed
by physiology. For instance, it is possible to replace the physical functions

of man by machines, but what about the so-called mental or cognitive functions? What about language, art, and music? What about emotions? We can learn only certain things from a frequency analysis of Tchaikovsky's Sixth Symphony or from a spectral analysis of a Van Gogh sunset. Similarly, from the number of binary bits required to store a computer program it is possible to calculate the information content of the system as defined by Shannon and others, but this tells us nothing about the purpose, meaning, or function of the program.

Von Neumann has shown that it is logically possible to construct a first machine that is capable of producing a second machine more complex than itself. This, obviously, has important implications in biology, where the natural order of evolution (from the standpoint of both ontogeny and phylogeny) is from the less to the more complex. But why should such a machine want to reproduce? That is a question that only biology can answer.

The problems here are peculiarly reversed from those of classical physics. In physiology today it is above all necessary to determine accurately and in some well-defined manner the qualitative rather than the quantitative aspects of the systems under investigation. Problems of analysis are doubly compounded in the case of comparative physiology, where comparisons must be made not only on the basis of different levels of description but also among different systems, each highly complex in its own right. However, as will be shown in the chapters that follow, there are certain underlying principles and mechanisms that seem to function in living systems at all levels. The essential problem of physiology is to couple these objective and quantitative concepts with the subtle qualitative ones of teleology and purposeful behavior exhibited by living systems.

8 • ANALYTICAL TECHNIQUES OF THE PHYSIOLOGIST

These trees shall be my books.
—William Shakespeare

Thus far we have been concerned largely with the quantitative tools of the physiologist—his instruments, methodology, and terminology. There is another tool that the *comparative physiologist* makes extensive use of: nature. All forms of life existing today are the results of natural experiments begun long ago and continuing to the present. The comparative physiologist performs experiments on nature's experiments. Much as a physicist uses temperature and pressure as independent variables to investigate properties of nonliving systems, the comparative physiologist uses species as a controlled variable to investigate living systems. From similar experiments on different organisms, information about underlying similarities of physiological responses can be obtained. Such information is valuable in helping us understand evolutionary patterns, as well as normal and pathological functions, in all creatures.

All physiologists are interested in process, mechanism, structure, and function in plant and animal species. But unlike the general physiologist, who is interested mainly in similarities, the comparative physiologist is interested in both similarities and differences. To investigate similarities, the comparative physiologist often makes use of nature's foibles. Certain organisms have developed specializations that allow them to carry out a particular physiological process with extreme efficiency. Such specializations make investigation of specific species' similarities easier. For instance, certain desert rodents produce highly concentrated urine. Study of these animals' excretory functions provides insights into the processes of urine concentration in other mammals (see Chapter 6, Section 3.3). The fact that bats are specialized in echolocation permits details of biological sound production and reception to be studied more simply (Chapter 3, Section 11.6). Heat conservation in penguins is analogous to heat conservation in whales and man and to gas exchange in fish (see Chapters 5 and 10). All of these species specializations have one thing in common: they permit the isolation and investigation of the process of interest in a nearly ideal setting—one of exaggerated function. From studies of these ideal systems, the physiologist can extrapolate to other systems in which the properties of interest are not present in the extreme but in which the principles of function are the same.

The comparative physiologist is interested in differences as well as in similarities. Creatures distributed across the evolutionary spectrum have solved common problems of survival but in many different ways. Every species tends to evolve in a way that best suits it to a particular environment. Sometimes differences exist between closely related species. For example, both freshwater and marine fish transport salt, but they do so in opposite directions; freshwater fish extract salt from their surroundings, whereas marine fish transport salt into their environment (see Chapter 6 for details). The comparative physiologist must determine how each animal's specific characteristics contribute to that animal's ability to survive in a particular environment. To accomplish this analysis, time must be taken into account, not only as an explicit variable in the collection of data but also as an implicit variable in the sense that adaptation and specialization are the products of long-term selection processes. What the investigator sees is the end result of chemical, physical, and biological stresses that existed in the past, as well as pressures that exist in the present. Apparent differences in responses to similar situations may be due as much to diverse phylogenetic origins as to different needs in the present. For instance, both land and sea creatures depend upon oxygen for survival, but lungs and gills are two different adaptations for extracting oxygen from entirely different environments (see Chapter 5). Thus different organs in different species may perform the same physiological function. Conversely, homologous organs may do different things. For instance, to locate their prey, sharks use specialized muscle cells that generate electric fields; they then use unspecialized muscle to capture and devour that prey (see Chapter 3, Section 11). Many things that at first appear related are not and vice versa.

One consequence of all this is that the physiologist must be a detective as well as a scientist. He must be able to find patterns and extract conclusions from data other than those his instruments record. He must be capable of drawing fine distinctions as well as generalizations. He must be prepared to ask far more questions than he can answer and to sometimes find answers that seemingly make no sense. Above all, the physiologist must be curious and open, for the answers that he finds are, like nature, not always predictable or logical. We live in a world where natural variations sometimes lead to evolutionary dead ends and where environments may change after eliciting major species adaptations. To make sense from all this, the comparative physiologist must have a strong intuitive feel for his subject as well as technical knowledge, and he must of necessity be broadly trained in all areas of physiology. For him, structure, function, and their interrelationships—and curiosity—are the primary universals.

REFERENCES

Apter, M. J. *Cybernetics and Development*. New York: Pergamon Press, 1966.

Bayliss, L. E. *Living Control Systems*. London: English Universities Press, 1966.

Bernard, Claude. *Leçons sur les Phénomènes de la Vie Communs aux Animaux et aux Végétaux*. 2 vols. Paris: Baillière, 1878–1879.

Bradley, D. F. Multilevel Systems and Biology—Views of a Submolecular Biologist. *Systems Theory and Biology*. M. D. Mesarovic, ed. New York: Springer-Verlag, 1968.

Cannon, W. B. *The Wisdom of the Body*. New York: Norton, 1932.

Gatlin, L. *Information Theory and the Living System*. New York: Columbia University Press, 1972.

George, F. H. *Cybernetics and Biology*. London: Oliver & Boyd, 1965.

Grodins, F. S. *Control Theory and Biological Systems*. New York: Columbia University Press, 1963.

Kalmus, H. *Regulation and Control in Living Systems*. New York: Wiley, 1966.

Milhorn, H. T., Jr. *The Application of Control Theory to Physiological Systems.* Philadelphia: Saunders, 1966.

Milsum, J. H. *Biological Control Systems.* New York: McGraw-Hill, 1968.

Pierce, J. R. *Symbols, Signals, and Noise.* New York: Harper & Row (Harper Torchbooks), 1961.

Polanyi, M. Life Transcending Physics and Chemistry. *Chem. Eng. News* 45:54, 1967.

Riggs, D. S. *Control Theory and Physiological Feedback Systems.* Baltimore: Williams & Wilkins, 1970.

Schrödinger, I. *What Is Life* and *Mind and Matter.* New York: Cambridge University Press, 1967.

Shannon, C. E., and Weaver, W. *The Mathematical Theory of Communication.* Urbana: University of Illinois Press, 1949.

Simon, W. *Mathematical Techniques for Physiology and Medicine.* New York: Academic Press, 1972.

Von Neumann, J. *Theory of Self-Reproducing Automata.* Edited and completed by A. W. Burks. Urbana: University of Illinois Press, 1966.

Wiener, N. *Cybernetics, or Control and Communication in the Animal and the Machine.* 2d ed. Cambridge, Mass.: M.I.T. Press, 1961.

Yamamoto, W. S., and Brobeck, J. R. *Physiological Controls and Regulations.* Philadelphia: Saunders, 1965.

MEMBRANE
PHYSIOLOGY

Robert M. Dowben

The term "biological transport" is used to describe the movement of materials from one place to another in a living organism. In a complex animal or plant it is necessary to move nutrients and metabolites from one place to another and to change the rate of movement in response to environmental changes. Biological transport is also utilized by living organisms for the secondary effects, such as regulation of cell volume, for maintaining the ionic composition of body fluids and so on. All in all, biological transport processes represent a major kind of work performed by living organisms that requires the expenditure of a major fraction of the total energy generated by the organism.

This chapter focuses on localized transport across cellular membranes. However, the principles developed with regard to cellular membranes can be applied to other biological transport systems, some of which operate over great distances. The reader will encounter the latter types of biological transport in many other chapters of this book, particularly those on circulation, nutrition and digestion, and the nervous system.

Membranes of one sort or another form the barriers to most types of biological transport processes. Not only are membranes and their characteristics of selective permeability seminal to the operation of biological transport, membranes are also involved in many other physiological functions, such as mitochondrial oxidation, irritability and conduction in nerve and muscle, vision, and so on. Therefore our discussion of transport will begin with a consideration of membranes in living organisms.

1 • INTRODUCTION

Membranes are ubiquitous structures in living organisms, and therefore the transport process whereby various materials move across membranes are of fundamental importance for the well-being and even survival of all animals and plants. Biologists focus on the biogenesis, structure, and function of membranes (see Table 2.1) and on the nature of transport processes because they are among the most fundamental areas of biology that require understanding and explanation.

One may well ask why membranes are so ubiquitous and why transport systems have such far-reaching significance. First of all, the metabolic reactions of living cells require an environment that is different from the usual external environment. For example, the fluid inside cells is rich in potassium and low in sodium, while the reverse is true of seawater and most external environments. The high intracellular potassium concentration is not a trivial matter; rather, it is essential for life processes. Protein synthesis and the nerve impulse (excitation) are examples of essential life processes that depend upon the high potassium and low sodium concentrations found in the interior of cells. Furthermore, the efficiency of metabolic processes requires high concentrations of metabolic precursors, metabolic intermediates, and enzyme proteins in the cell sap. The cell membrane serves to enclose a domain separated from the environment in which the optimal composition required for carrying on living processes can be maintained. Cells are living units able to carry on all of the essential

---

TABLE 2.1 Functions of the Cell Membrane

1. Forms boundary for individual cells
2. Forms boundary for intracellular organelles
3. Provides structural framework for attachment of enzymes
4. Mediates cell adhesion
5. Forms boundary for gas exchange between interior and environment
6. Selective permeability and active transport of ions
7. Facilitates diffusion of sugars, amino acids, and other metabolites
8. Selective excretion, for example, of glucuronides
9. Mediates pinocytosis and phagocytosis
10. Mediates secretion
11. Mediates irritability and excitation
12. Forms specialized structures such as cilia and microvilli
13. Mediates motility
14. Mediates responses such as chemical mediation of neural transmission, hormone action, antigen response, chemotaxis
15. Initiates autolysis upon cell death

life processes, and in complex organisms composed of many cells, individual cells can carry on a quasi-independent existence under certain circumstances, for example, when cells are removed from the body and cultured in an artificial medium.

Second, membranes serve to divide the cells of higher organisms into compartments where different metabolic reactions and different physiological processes can be carried out. Frequently this compartmentalization takes the form of membrane-bounded intracellular organelles, which carry out specialized functions. For example, many hydrolytic enzymes, including enzymes that degrade RNA and protein, are contained in small membrane-bounded vesicles called *lysosomes,* which are found in the cytoplasm.

Third, membranes serve to orient or align the enzymes involved in a sequence of metabolic reactions. This orientation of enzymes on membranes (for example, those involved in oxidative phosphorylation in mitochondria) is believed to increase efficiency by permitting the product of one reaction to be immediately utilized in the next reaction and reducing the amount of free intermediates that are needed (see Chapter 7, Section 3). In the case of proteins destined for secretion by the cell, the proteins are routed through a sequence of membranous structures in the course of their synthesis and secretion.

Last, membranes carry on a number of processes that are particularly characteristic of living systems; these include motility, irritability, sensation, transmission of impulses, recognition, adhesion, and regulation of growth or cell division.

An important characteristic of transport processes through natural membranes is that they are very highly selective. Some compounds are transported easily and rapidly, whereas other closely related compounds in terms of structure or chemical composition are poorly transported or even excluded. Frequently there is a marked vectorial or directional component to transport; for many substances transport appears to occur more easily and rapidly in one direction than in the other. Very often transport across biological membranes takes place against a concentration gradient, moving from a dilute solution to a concentrated solution or, in the case of ionic substances, against an electrochemical gradient. Such movement of materials against a concentration or electrical gradient is necessary to maintain the special environment of the interior of living cells, but transport of this kind requires the expenditure of energy. The energy utilized comes from cellular metabolism. As a matter of fact, an appreciable fraction of the energy made available by cellular metabolism is used to maintain the unusual internal milieu. After death, the interior of the cell gradually equilibrates with the external environment; the electrochemical gradients characteristic of living cells are lost.

2 • CELL STRUCTURE

Let us review briefly the ultrastructure of cells, with particular emphasis on membrane components. The common use of electron microscopes during the past three decades has shown cells to possess a rich and complex ultrastructure that was previously unsuspected. Membranes can be identified as two dark (electron-dense) parallel bands, each about 30 to 40 Å thick, separating a light band about 40 Å wide. This very characteristic trilaminar "sandwich" or "railroad track" ultrastructure (Figure 2.1) is called the *unit membrane*. The appearance of the unit membrane is not absolutely the same for all membranes. The bands may be more or less electron-dense, and the thickness may vary from membrane to membrane or from cell type to cell type.

An electron micrograph of a typical animal cell is shown in Figure 2.2. Animal cells vary markedly in size and shape according to cell type and species, whereas plant cells are much more uniform. Cells are bounded by a unit membrane that separates the domain of living protoplasm inside from the external environment. The part of the cell membrane that faces the outside or a body cavity tends to be thicker than the membrane next to adjacent cells. Eukaryotic cells, by definition, possess a *nucleus*, which contains almost all of the DNA (genetic material), a great

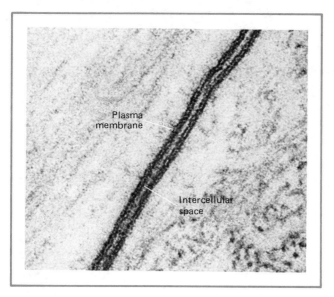

FIGURE 2.1 The intercellular space between two glial cells of an annelid, *Aphrodite*. (From D. W. Fawcett, *An Atlas of Fine Structure*. Philadelphia: Saunders, 1966.)

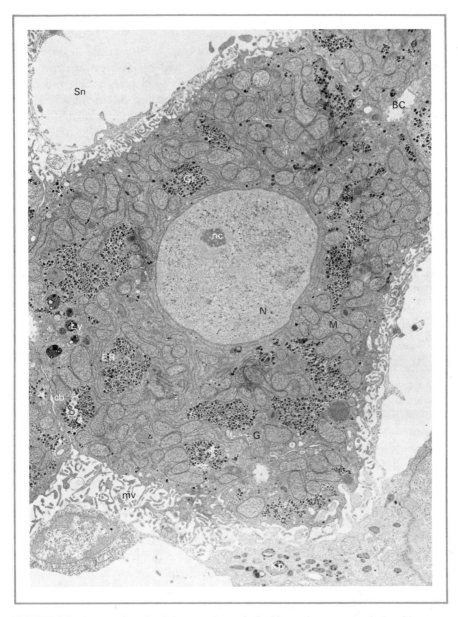

FIGURE 2.2   A parenchymal cell from rat liver. Code: N—nucleus; nc—nucleolus; M—mitochondria; G—Golgi apparatus; ER—endoplasmic reticulum; Ly—lysosomes; Gl—glycogen
granules; cb—intercellular space between two parenchymal cells; BC—bile cannaliculus; Sn—si-
nusoid; mv—microvilli; Kp—part of a Kupffer (reticuloendothelial) cell. (From R. M. Dowben, *Cell
Biology.* New York: Harper & Row, 1971.)

deal of RNA, and unique kinds of proteins. The nucleus is surrounded by a double unit membrane through which there are at intervals *nuclear pores,* 100 to 150 Å in diameter. Most of the nuclear RNA is condensed into a distinct structure, the *nucleolus.* The DNA and associated material is divided into *chromosomes,* the cells of each species containing a definite and characteristic number. The individual chromosomes are easily seen during the midst of cell division, but in nondividing interphase cells the chromosomes are poorly visualized.

In addition to the nucleus, cells contain other types of intracellular organelles. *Mitochondria* are small pleomorphic structures, usually round or sausage-shaped, containing the enzyme systems of the Krebs cycle, oxidative phosphorylation, and fatty acid metabolism (discussed in Chapter 7). They also contain small amounts of DNA and RNA and have the capacity to synthesize a few of the mitochondrial proteins. Mitochondria are surrounded by a unit membrane. The inner member of the mitochondrial membrane has numerous invaginations, the *mitochondrial cristae.* Lysosomes, mentioned earlier, contain hydrolytic enzymes. Another type of membrane-bounded organelle in which specialized enzyme systems are segregated are *microbodies.*

Cells also contain clusters of more or less parallel intracellular membranes, the *endoplasmic reticulum.* Sometimes, particularly in cells that synthesize protein for secretion from the cell, much of the surface of the endoplasmic reticulum is covered with *ribosomes* that appear as electron-dense particles, forming the *rough endoplasmic reticulum.* Ribosomes also appear free in the cytoplasm. Most, if not all, proteins secreted by the cell contain carbohydrate, which is attached to the newly formed proteins in a membranous organelle called the *Golgi apparatus.* Thence the protein is carried in membrane-bounded *secretory granules* to the cell membrane, where the membranes of the secretory granule merge with the cell membrane, discharging the entrapped protein to the outside. Endoplasmic reticulum devoid of ribosomes is called *smooth endoplasmic reticulum* and contains the enzyme systems involved in steroid biosynthesis and drug metabolism. The amount of rough endoplasmic reticulum or smooth endoplasmic reticulum varies from cell type to cell type, and in a given cell type it can change markedly in amount with changing physiological conditions.

There are several specializations of the cell membrane. Light microscopists knew that the inner surface of the intestine was covered with *villi,* projections into the lumen that serve to increase the surface area available for absorption. Examination of the luminal surface of intestinal epithelial cells by electron microscopy revealed that each villus was thrown out into numerous submicroscopic *microvilli,* thereby giving the

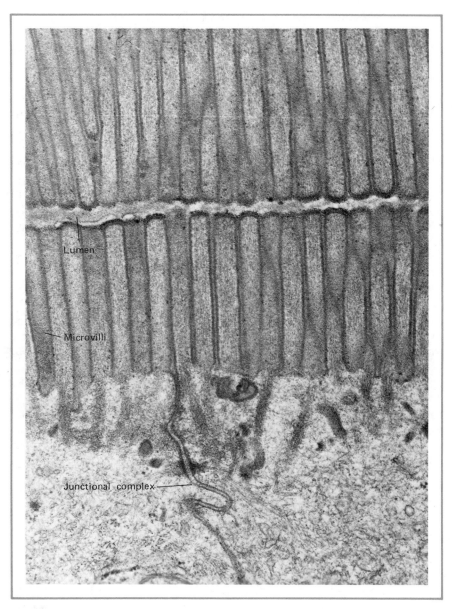

FIGURE 2.3 Cells from the duodenum of a Sprague-Dawley rat showing microvilli and a junctional complex between two adjacent cells. (From R. M. Dowben, *General Physiology: A Molecular Approach*. New York: Harper & Row, 1969.)

cell membrane an enormous surface area (Figure 2.3). Other specializations of the cell membrane will be discussed elsewhere in this book.

3 • MEMBRANE CONSTITUENTS

The major constituents of membranes are lipids and proteins; in addition, there may be small amounts of RNA. Bacterial and plant cell membranes are surrounded by a polysaccharide *cell wall*. The external cell wall is rigid; unlike the cell membrane from which it is distinct, the cell wall cannot be distended. The proportion of protein to lipid varies over a wide range for different types of membranes (Table 2.2).

Membrane lipids consist mainly of phospholipids and cholesterol. Structures of the most important membrane lipids are set out in Figure 2.4 (see also Chapter 7, Section 2.2). The phospholipids are fundamentally esters of phosphorylglycerol. The two hydroxyl groups of phosphorylglycerol are esterified by fatty acids, mainly of carbon length C_{16} to C_{22} with zero to three double bonds. An amine alcohol (ethanolamine, serine, choline) is esterified to the phosphoryl group. Related to the phospholipids are the *sphingolipids*, which contain a long-chain amino alcohol in place of glycerol. Phospholipids are characterized by a part of the molecule that is *hydrophobic* and tends to come out of aqueous solution (the fatty acid portion) and a part of the molecule that is *hydrophilic* and tends to dissolve in water (the polar phosphorylethanolamine, phosphorylserine, or phosphorylcholine portion). Molecules that possess both a hydrophobic part and a hydrophilic part are called *amphipathic*. Many of the special properties of biological membranes are due to the amphipatic lipids they contain.

Amphipathic lipids suspended in aqueous media tend to aggregate in structures called *micelles*. Micelles contain anywhere from a few dozen to thousands of amphipathic lipid molecules arranged so that the hydro-

TABLE 2.2 Protein and Lipid Composition of Animal and Bacterial Membranes

Origin of Membrane	Molar Ratio			Area Ratio (protein/ lipid)
	Amino Acid	Phospho- lipid	Choles- terol	
Myelin	264	111	75	0.43
Erythrocyte	500	31	31	2.5
Bacillus licheniformis	610	31	0	4.8
Micrococcus lysodeikticus	524	29	0	4.3
Bacillus megaterium	520	23	0	5.4
Streptococcus faecalis	441	31	0	3.4
Mycoplasma laidlawii	442	25.2	2.3	4.1

Source: E. D. Korn, *Science* 153:1496, 1966. Copyright © 1966 by the American Association for the Advancement of Science.

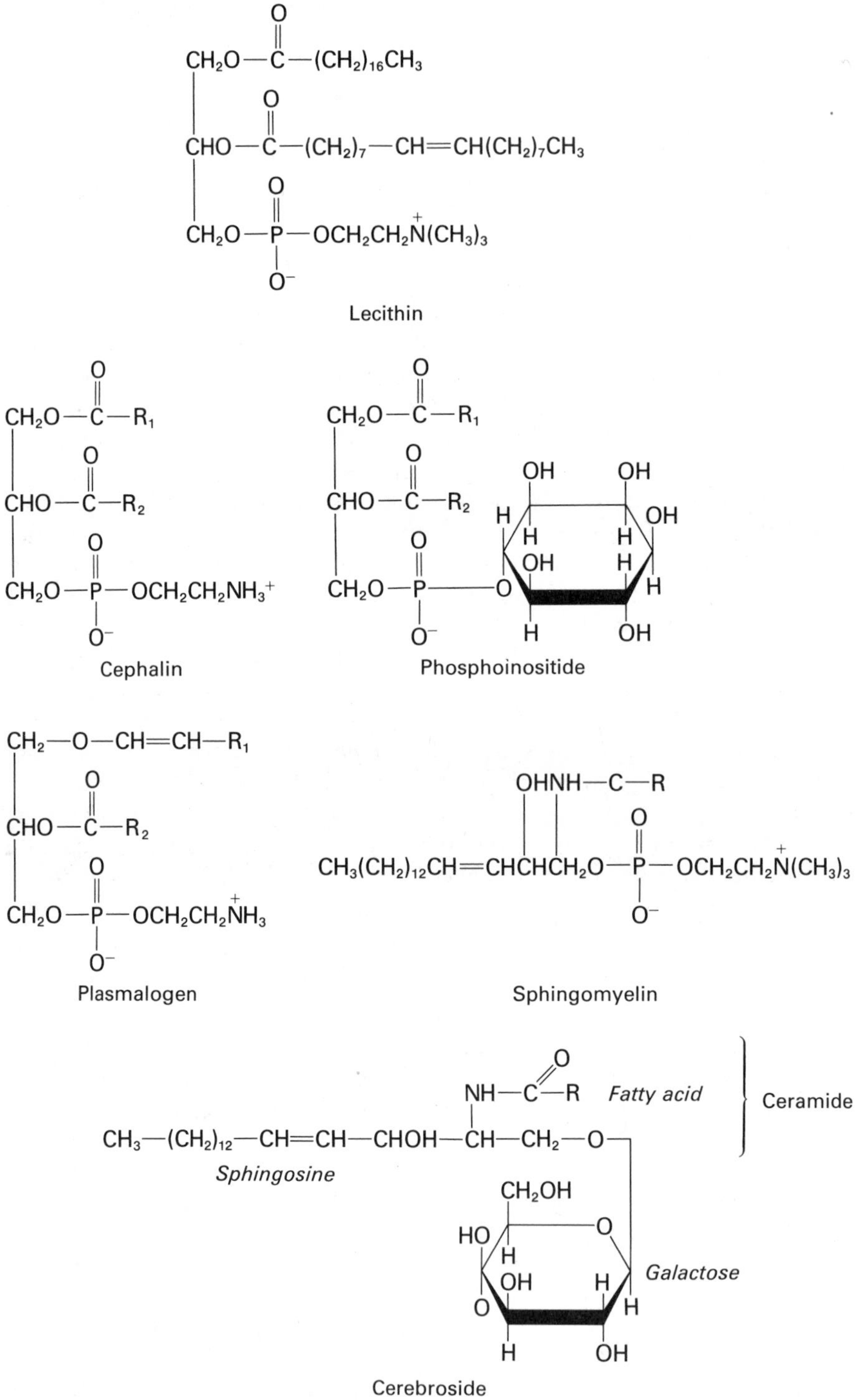

$$CH_2O-\overset{\overset{\displaystyle O}{\|}}{C}-(CH_2)_{16}CH_3$$

$$CHO-\overset{\overset{\displaystyle O}{\|}}{C}-(CH_2)_7-CH=CH(CH_2)_7CH_3$$

$$CH_2O-\overset{\overset{\displaystyle O}{\|}}{\underset{\underset{\displaystyle O^-}{|}}{P}}-OCH_2CH_2\overset{+}{N}(CH_3)_3$$

Lecithin

$$CH_2O-\overset{\overset{\displaystyle O}{\|}}{C}-R_1$$
$$CHO-\overset{\overset{\displaystyle O}{\|}}{C}-R_2$$
$$CH_2O-\overset{\overset{\displaystyle O}{\|}}{\underset{\underset{\displaystyle O^-}{|}}{P}}-OCH_2CH_2NH_3{}^+$$

Cephalin

$$CH_2O-\overset{\overset{\displaystyle O}{\|}}{C}-R_1$$
$$CHO-\overset{\overset{\displaystyle O}{\|}}{C}-R_2$$
$$CH_2O-\overset{\overset{\displaystyle O}{\|}}{\underset{\underset{\displaystyle O^-}{|}}{P}}-O$$

Phosphoinositide

$$CH_2-O-CH=CH-R_1$$
$$CHO-\overset{\overset{\displaystyle O}{\|}}{C}-R_2$$
$$CH_2O-\overset{\overset{\displaystyle O}{\|}}{\underset{\underset{\displaystyle O^-}{|}}{P}}-OCH_2CH_2\overset{+}{N}H_3$$

Plasmalogen

$$CH_3(CH_2)_{12}CH=CHCHCH_2O-\overset{\overset{\displaystyle O}{\|}}{\underset{\underset{\displaystyle O^-}{|}}{P}}-OCH_2CH_2\overset{+}{N}(CH_3)_3$$

Sphingomyelin

$$CH_3-(CH_2)_{12}-CH=CH-CHOH-CH-CH_2-O-$$

Sphingosine

NH—C—R   Fatty acid

Ceramide

CH$_2$OH

Galactose

Cerebroside

FIGURE 2.4 Structure of phospholipids.

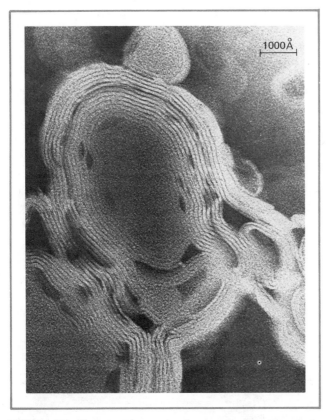

FIGURE 2.5 Electron micrograph of a myelin figure negatively stained with potassium phosphotungstate. Note the multilaminar structure. (From A. D. Bangham and R. W. Horne, *J. Mol. Biol.* 8:662, 1964.)

phobic lipid ends of the molecules are squeezed out of the aqueous phase and are oriented toward the interior of the micelle, while the hydrophilic ends face outward toward the aqueous medium. Phospholipids are approximately cylindrical in shape; they tend to form a specific kind of micelle in the form of leaflets two molecules thick, *bimolecular leaflets*, in which the fatty acid hydrophobic ends of the molecules are in apposition and the hydrophilic ends face outward. Bimolecular leaflets of phospholipids are extremely stable; as a matter of fact, manufacturers of commercial toy soap bubble solutions add a few drops of egg yolk rich in phospholipids to stabilize the soap bubbles. Phospholipids may form multilaminar structures, called *myelin figures* (Figure 2.5), made up of concentric bilayers.

Membranes also contain a number of other types of lipoid compounds in small amounts. Some of these minor lipid constituents have well-known physiological roles, such as vitamin A in the retina, carotinoids, quinones (including ubiquinone and vitamin K), and liposaccharides.

In contrast to bacterial, mitochondrial, and chloroplast membranes which contain little or no cholesterol, the plasma membranes and endoplasmic reticulum of eukaryotic cells contain approximately equimolar amounts of cholesterol, and phospholipids plus sphingolipids. The combination of phospholipid and cholesterol molecules permits tighter packing in the membrane (Figure 2.6). There appear to be major differences in properties between the cholesterol-free or cholesterol-poor membranes of mycoplasma, bacteria, blue-green algae, mitochondria, and

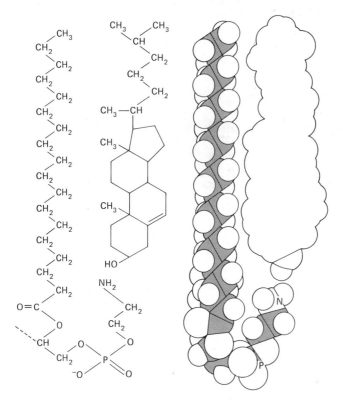

FIGURE 2.6 Conformation of a phosphatidylethanolamine molecule and a cholesterol molecule. Note that the cholesterol fits into a cavity of the phospholipid. The closer packing of molecules in cholesterol-containing membranes is believed to be responsible for their greater compactness and rigidity.

chloroplasts on the one hand, and the cholesterol-rich plasma membranes and endoplasmic reticulum of eukaryotic cells on the other. It has been suggested that mitochondria and chloroplasts evolved from primitive prokaryotic organisms; the membranes possess distinct similarities.

The fluidity of membranes depends upon the chain length and degree of unsaturation of the fatty acid residues; the shorter and more unsaturated the fatty acid residues, the lower the melting point of the membrane lipids. Even in a given species the fatty acid composition depends upon environmental conditions such as ambient temperature and diet. For example, animals in colder climates have membrane lipids that melt at lower temperatures than the membrane lipids of animals living in warm climates. Furthermore, the membrane lipids in the extremities of warm-blooded animals melt at lower temperatures than those obtained from the visceral tissues.

Less is known about the proteins in membranes than is known about the lipids. Membrane proteins are characteristically insoluble, and the lack of solubility has impeded efforts to isolate membrane proteins in pure form and to study their properties. Nevertheless, interest in the membrane proteins is very great because individual membrane proteins are most likely responsible for the enormous selectivity and distinctive transport properties of biological membranes. In addition to learning about the properties of the membrane proteins themselves, we need to learn about the protein-lipid interactions, how the proteins are inserted and arranged within membranes, and, indeed, how membranes are formed (*membrane biogenesis*) by living cells.

Although the membrane lipids can be quantitatively extracted by organic solvents and cleanly separated, purified, and identified by well-established chromatographic and spectroscopic techniques, the separation of membrane proteins requires complex and as yet imperfectly developed methods. Detergents are used to disaggregate membrane proteins, but although optically clear "solutions" are obtained, the complete molecular dissociation that characterizes a true solution is rarely achieved, and it is later difficult to completely remove the detergent from the disaggregated proteins. Many membrane proteins can be solubilized and extracted by a mildly alkaline medium in which electrolytes, particularly divalent cations, are very low or absent. Restoration of electrolytes in physiological concentrations causes reaggregation of the solubilized proteins.

For analytical purposes membrane proteins can be "solubilized" by detergents and separated by electrophoresis on polyacrylamide gels containing detergents. When sodium dodecylsulfate (SDS) is used as the detergent, differences in charge density are minimized, and the proteins are separated mainly according to their molecular weight, the electrophoretic

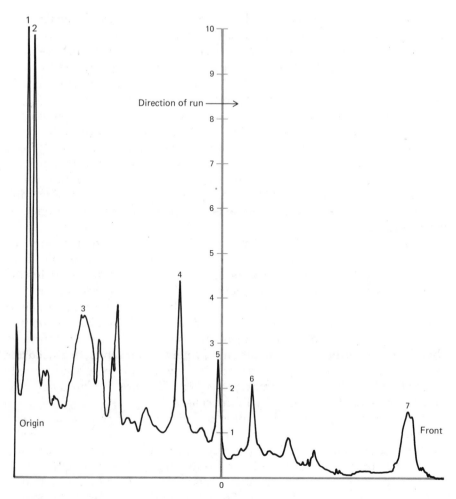

FIGURE 2.7 Scan of a polyacrilamide gradient gel of human red blood cell membrane. Purified membranes were solubilized in 2 percent sodium dodecylsulfate and separated in a slab gel electrophoresis according to molecular weight. Some of the proteins are: peak 1, spectrin (240,000 daltons); peak 2, myosinlike protein (220,000 daltons); peak 3, glycophorin (55,000 daltons); peak 4, blood group antigens; peak 5, actinlike protein (44,000 daltons); peak 6, glyceryl-3-phosphate dehydrogenase; peak 7, residual hemoglobin.

mobility varying inversely to the logarithm of molecular weight. Such SDS-polyacrylamide gel electrophoreses (Figure 2.7) have shown that membranes contain many proteins of widely varying molecular weights, from less than 20,000 to more than 300,000 daltons. Staining the gels with periodic acid–Schiff reagent reveals that many membrane proteins contain carbohydrate and are glycoproteins. The size and distribution of proteins varies from cell type to cell type and even between different types of membranes within a single cell. Thus the inner and outer membranes of rat liver mitochondria have only a single major polypeptide constituent in common. Plasma membranes are particularly rich in glycoproteins; from a variety of studies it has been shown that the glycoproteins are located on the outer surface.

Especially important are the "carrier" proteins, which are presumed to be responsible for the specific transport of glucose, specific amino acids, and other essential metabolites. A few of these "carrier" proteins have been identified; more will be said about them later. It is also presumed that certain proteins that traverse the membrane from one side to the other form the aqueous channels or "pores," which will be discussed further in a subsequent section. Some membrane proteins are undoubtedly responsible for certain enzyme activity characteristic of membranes. The more important are the following:

1. *Na^+,K^+-activated transport ATPase*. Recently purified from dog kidney membranes, shark rectal gland membranes, and from the electric organ of electric eels, this protein is responsible for maintaining the low intracellular sodium by coupling outward sodium transport to ATP hydrolysis. Though present mainly in the plasma membrane, small quantities are found in other intracellular membranes.
2. *5'-nucleotidase*. This enzyme is a glycoprotein present mainly in the plasma membrane and used as a marker for plasma membranes; it catalyzes the hydrolytic cleavage of the phosphoric acid from AMP, UMP, and other nucleotides at alkaline pH (pH optimum = 9.0).
3. *Adenylcyclase*. This enzyme, which catalyzes the formation of cyclic AMP from ATP, is ubiquitous in the cell membranes of living organisms. Depending upon the cell type, it may be activated by a specific hormone and thereby initiate the response to a hormone in the effector cell.
4. *Glucose-6-phosphatase*. This enzyme, which catalyzes the hydrolysis of glucose-6-phosphate, is present mainly in endoplasmic reticulum and is frequently used as a marker enzyme.
5. *NADH-cytochrome c reductase (or NADH diaphorase)*. This enzyme also is found mainly in endoplasmic reticulum and is used as a marker enzyme.
6. *Cholinesterase*. This enzyme is found in the motor end plate of muscle cells, electric organ cells, and other effector cells that react to acetylcholine.

Moreover, there are sequences of metabolic reactions that characteristically are carried out by enzyme systems associated with membranes; among these are:

1. Oxidative phosphorylation by mitochondrial membranes
2. Photosynthesis by chloroplast membranes
3. Fatty acid synthesis by smooth endoplasmic reticulum
4. Drug detoxification by smooth endoplasmic reticulum
5. Protein synthesis by rough endoplasmic reticulum

6. Addition of carbohydrate to form glycoproteins by the Golgi apparatus membrane system

The constituents of metazoan cells—that is, the cells of multicellular higher animals—are continually being degraded and resynthesized over a matter of days, despite the fact that the cells exist in a quiescent state for years without either dividing or dying. Rudolph Schoenheimer, who first studied this continual turnover of bodily substances using isotopically labeled compounds, named the process *dynamic equilibrium.* The lipid and protein constituents of membranes are also in dynamic equilibrium, turning over in a matter of days in cells that are alive for years but not growing or dividing. For example, studies with labeled amino acids show a continuous synthesis and degradation of rat liver membrane proteins with an average half-life of several days, whereas the liver cells of an adult rat divide about once a year in the absence of injury.

The turnover of membrane constituents could possibly take place in two ways. One would be for patches or domains of membrane to be synthesized, assembled from the individual constituents, and inserted into the membrane as a whole piece, while other patches or domains were removed from the membrane and degraded. Alternatively, individual constituent molecules could be synthesized independently and randomly inserted into the membrane, while simultaneously other constituent molecules were being randomly removed and degraded. Studies of the distribution of radioactive amino acids in membrane fragments and their removal over time give results that are not compatible with the domain alternative. A series of first-order degradations (logarithmic decay independent of protein concentration) were found with different rate constants for different proteins. Furthermore, the rates of turnover of the various proteins in rat liver endoplasmic reticulum can be changed independently of one another in response to environmental changes—for example, after the administration of phenobarbital. These data are consistent with the view that membrane protein molecules are synthesized and inserted into the membrane independently and then randomly removed and degraded.

The membrane lipids also are in dynamic equilibrium with half-lives of a few days. Cell membrane lipids may, in addition, exchange with plasma lipids carried in the bloodstream. This turnover of membrane constituents occurs continuously without disruption of membranes or interference with the transport functions.

4 • DIFFUSION

The observer who tries to understand membrane function is frequently overwhelmed by the enormous differences with which substances are transported. For example, two substances of similar molecular weight and molecular dimensions may move across membranes at vastly different rates. As mentioned above, some substances are accumulated or secreted against a large concentration or electrical gradient. To the extent that we can fit these phenomenological observations (which at times may even seem contradictory) into orderly, unified processes with rational explanations and predictive powers, we achieve a scientific understanding of membrane function.

There are four major processes by which materials move across membranes:

1. *Passive diffusion.* Movement of substances from a region of high concentration (or more properly, high electrochemical potential) to a region of low concentration as a result of kinetic (thermal) molecular motion.
2. *Active transport.* Movement of substances against a concentration gradient or electrical gradient, where the energy required for such uphill transport comes from the chemical reactions of cellular metabolism.
3. *Facilitated diffusion.* Movement of substances across membranes by virtue of interaction with a specific membrane constituent. Facilitated diffusion may be downhill from a region of high electrochemical potential, or, contrariwise, transport by facilitated diffusion may be uphill against an electrochemical gradient. Uphill transport by facilitated diffusion requires coupling to a concurrent downhill transport process.
4. *Pinocytosis and secretion.* Transport of materials across the cell membrane in discrete bulk quantities. Usually proteins or other macromolecules together with associated solvent are transported by these processes.

As just stated, passive diffusion results from the kinetic (thermal) motion of molecules. Because the process is spontaneous, it obeys the free energy law, $\Delta G \leqq 0$.[1] Examples of diffusion are common in everyday life. Consider the dissolution and spontaneous mixing after a sugar cube is dropped into a cup of tea. Initially changes in the refractive index of the liquid near the sugar cube allow the process to be observed visually. Eventually the sugar will be uniformly distributed in the tea, and every microscopical volume element will contain the same concentration of sugar. Owing to increased kinetic molecular motion as the temperature is increased, diffusion *usually* takes place more rapidly.

---

[1] ΔG, the *Gibbs free energy* change, is characteristic of each chemical reaction; it indicates the tendency of a reaction to reach equilibrium and the amount of energy that is available in doing so. If $\Delta G = 0$, the reaction is at equilibrium. If $\Delta G < 0$, the reaction will tend to occur spontaneously, although no predictions can be made about the speed with which this will happen, and ΔG is equal to energy made available by the reaction. Such reactions are called *downhill reactions.* If $\Delta G > 0$, the reaction will not only fail to occur spontaneously, but it will not occur unless an amount of energy equal to ΔG is put into the system. Reactions requiring the input of energy are called *uphill reactions.*

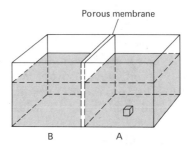

FIGURE 2.8 A tank of water divided into two compartments by a porous membrane. A sugar cube is placed in compartment A. The porous membrane is a diffusion barrier to the sucrose entering compartment B. See text for the Fick equation, which describes the diffusion of solute into B.

In a liquid, diffusion occurs much more readily than in a solid, but not as freely as in gases. The solute molecules must overcome various frictional or resistive forces including:

1. The necessity to displace solvent molecules.
2. The necessity of disrupting molecular associations between solvent molecules. Water has a loose, partially propagated internal structure owing to hydrogen bond formation between adjacent water molecules. Solute molecules must disrupt these hydrogen bonds during diffusion.
3. Overcoming the tendency of some solute molecules to interact with solvent molecules, such as the tendency of urea and sugar molecules to form hydrogen bonds with water molecules.

It is important to remember that solute molecules may encounter different resistive forces in one solvent than in another, or the resistive forces in a membrane may be less or greater than those in the solvent on either side of the membrane.

Let us now consider the situation illustrated in Figure 2.8, which consists of a tank of water separated into two compartments by a porous, sievelike barrier representing a membrane. A sugar cube dropped into compartment A will diffuse rapidly throughout compartment A. However, the diffusion of solute molecules will be retarded through the barrier. Assuming that the resistance to diffusion in the pores is uniform and continuous, the rate of diffusion of solute molecules across the barrier will be given by Fick's equation, the relation discovered by the German physiologist Adolph Fick in 1855:

$$J = \frac{\Delta n}{\Delta t} \frac{1}{A} = -DK \frac{\Delta c}{\Delta x} \qquad (2\text{-}1)$$

where J is defined as the flux of solute per unit area, Δn is the net amount of solute crossing the barrier, t is the time, A is the cross-sectional area of the barrier, Δc is the difference in concentration of solute in the two compartments in moles/liter of solvent, Δx is the thickness of the barrier, and D is the diffusion constant of the given solute in the bulk phase.

The constant K is the fraction of the barrier cross section represented by the effective pore area. We consider the effective pore area because if the pores are very small, the diffusion of solute molecules may be impeded and the effective pore area will be less than the actual pore area. If the solute does not dissolve in the membrane, K is always less than 1. If the solute is soluble in the membrane substance, Fick's equation still holds, except that K is greater than the actual fraction of pore area and may even be greater than 1.

Thus we can make the following predictions from Fick's equation:

1. The thicker the membrane the less the solute flux.
2. As the molecular size of solute approaches the diameter of the pores, the solute flux will fall. For a given membrane with a fixed pore size, therefore, the diffusion across the membrane will fall as the molecular weight of solute increases. The lumped permeability constants $(DK/\Delta x)$ for a number of solutes (that do not dissolve in the membrane) across the erythrocyte membrane are shown in Figure 2.9, and it is seen that this relation holds.
3. The greater the solubility in the membrane material the greater the solute flux. At the turn of the century the American physiologist Overton suggested that the

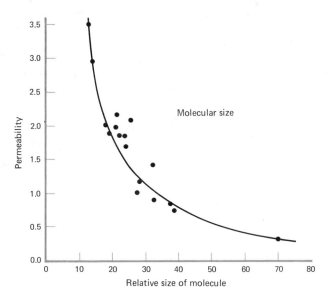

FIGURE 2.9 Relative permeability of various solutes in the large sulfur bacterium *Beggiatoa* as a function of molecular size. (Data from W. Ruhland and C. Hoffmann, *Planta* 1:1, 1925.)

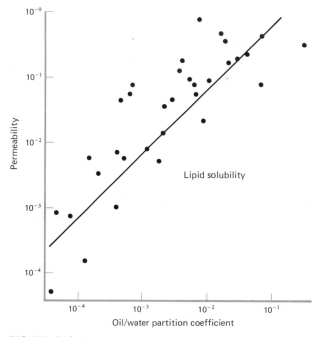

FIGURE 2.10 Permeability of various nonelectrolytes into cells of the alga *Chara ceratophylla* as a function of the partition coefficient between olive oil and water. The higher the olive oil/water partition the greater the permeability. (From R. Collander, *Physiol. Plantarum* 2:300, 1949.)

solubility in membranes could be evaluated by partition of a solute between olive oil and water in a two-phase system; the greater the solubility in olive oil the greater the permeability coefficient of the solute. Some typical data are shown in Figure 2.10. The last two sets of observations formed the basis of Collander's *lipoid-filter* theory of membrane permeability. In Chapter 5, Section 3.1, Fick's law is applied to gaseous diffusion in the process of respiration.

5 • OSMOTIC PRESSURE AND COLLIGATIVE PROPERTIES

In the foregoing analysis it was blithely assumed that only solute molecules move across the membrane. In fact solvent molecules also move back and forth through the membrane pores as a result of kinetic molecular motion. If a small quantity of radioactive tritiated water is added to compartment A in Figure 2.8, it will diffuse throughout the solvent water until mixing is complete and the radioactive water is homogeneously distributed in both compartments A and B.

Now let us assume that the membrane in Figure 2.8 contains pores of such dimensions that it is relatively permeable to solvent water but impermeable to sucrose. Biological membranes frequently are permeable to solvent but impermeable to some solutes; such membranes are commonly called *selectively permeable membranes*. If a sugar cube is now added to compartment A, the sucrose cannot diffuse through the membrane into compartment B. However, the tendency for mixing to take place is not diminished, and there will be a net flow of water from compartment B through the membrane into compartment A tending to dilute the sucrose concentration.

Let us use a somewhat different experimental setup—that shown in Figure 2.11, where compartment A is completely enclosed and filled with sucrose solution. The same tendency exists for solvent to pass through the selectively permeable membrane into compartment A because of the tendency to mix until equilibrium is reached. However, because the volume of compartment A cannot expand, the net quantity of solvent that actually moves across the membrane is infinitesimal. If the membrane is rigid, the hydrostatic pressure in compartment A increases until it offsets the tendency of solvent to move across the membrane. The increase in hydrostatic pressure is numerically equivalent to the *osmotic pressure* of the solution in compartment A. It should be emphasized that the pressure in compartment A is not caused directly by the solute molecules but because of the tendency for solvent to move from compartment B in an effort to achieve equilibrium of mixing.

The Dutch chemist van't Hoff demonstrated that if the temperature remains constant, the osmotic pressure of a dilute solution depends only upon the *concentration* of solute particles and not upon their size or molecular weight. The osmotic pressure depends upon the total concentration of solute particles which may be of two or more kinds. He also showed that the osmotic pressure was proportional to the absolute temperature if the solute concentration remained constant. In short, van't Hoff discovered an

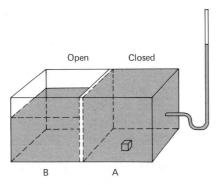

Open   Closed

B   A

FIGURE 2.11 A tank of water is divided into two compartments by a semipermeable membrane which is permeable to water but not to sucrose. Compartment A is closed and filled with water and a cube of sugar; compartment B is open and contains water. Water moves from compartment B to A until the pressure of the water head in the manometer tube equals the osmotic pressure of the sucrose solution.

equation similar to the ideal gas law that described dilute or "ideal" solutions as

$$\Pi = \frac{nRT}{V} = RTc \qquad (2\text{-}2)$$

where Π is the osmotic pressure, R the gas constant, T the absolute temperature, and $c = n/V$ the concentration of solutes in moles/liter solvent. Note that a mole of gas at 0°C must be compressed to 22.4 atm to occupy 1 liter, and a 1 molal solution at 0°C has an osmotic pressure of 22.4 atm.

Just as gases tend to deviate from the ideal gas law as the pressure increases, so do solutions tend to deviate from van't Hoff's equation as the total concentration of solutes increases. The deviations result from molecular interactions between solute molecules with each other and between solute and solvent molecules.

Osmotic pressure is one of the several properties of solutions that depend upon the total concentration of solutes. These properties are collectively called *colligative properties,* and they also include:

1. Elevation of boiling point of a solution: $\Delta T_b = K_e c$
2. Depression of freezing point of a solution: $\Delta T_f = K_f c$
3. Reduction in vapor pressure of a solution: $\Delta P = -c$
4. Reduction in surface tension of a solution: $\Delta \Gamma = -K_s c$
   where K_e, K_f, and K_s are constants which are characteristic of each solvent and independent of the nature of the solutes. For water, $K_e = 0.51$ and $K_f = 1.86$.

In practice the osmotic pressure of a solution is frequently determined from the depression of the freezing point

$$\Pi = RT \frac{\Delta T_f}{K_f} \qquad (2\text{-}3)$$

The regulation of concentration of solutes in bodily and intracellular fluids is an important requirement for life (see Chapter 6).

6 • REFLECTION COEFFICIENT

In the previous section we considered a membrane that was perfectly selectively permeable and excluded solute molecules completely. Particularly concerning low molecular weight solutes, biological membranes are imperfect selectively permeable membranes; that is, solute molecules do permeate the membrane but at very low rates. Let us suppose, for example, that the diameter of the membrane pores is small, and therefore diffusion of solute molecules is impeded.

If a solute is added to compartment A in Figure 2.8, two processes will occur initially: (1) solute molecules will diffuse through the membrane from A to B at a slow rate, and (2) net flow of solvent will occur from B to A because the osmotic pressure of the solution in A is greater than that in B. However, because solute is moving from A to B, the effective osmotic pressure as judged by the movement of solvent from B to A will be less than the osmotic pressure calculated from the solute concentration. The effective osmotic pressure is given by

$$\Pi_{eff} = \sigma RT \, \Delta c \tag{2-4}$$

where σ is Staverman's *reflection coefficient*. The reflection coefficient depends upon the particular membrane and solute.

Reflection coefficients for a number of solutes for a few different membranes are listed in Table 2.3. You will notice that the greater the permeability coefficient, the smaller the reflection coefficient. It can be shown

TABLE 2.3 Measured Values of the Reflection Coefficient for Various Solutes through Various Membranes and the Calculated Value of the Equivalent Pore Radius

Cell Type	Solute	σ	Equivalent Pore Radius, Å
Human erythrocyte	Glycerol	0.88	4.2
	Propylene glycol	0.85	
	Thiourea	0.85	
	Methylurea	0.80	
	Propionamide	0.80	
	Urea	0.62	
	Acetamide	0.58	
Frog muscle fibers	Mannitol	1.00	4.0
	Sucrose	1.00	
	Glycerol	0.86	
	Urea	0.82	
	Formamide	0.65	
Squid axon	Glycerol	0.96	4.25
	Ethylene glycol	0.72	
	Urea	0.70	
	Ethanol	0.63	
	Formamide	0.44	
Necturus kidney	Sucrose	1.00	5.6
	Erythritol	0.89	
	Glycerol	0.77	
	Urea	0.52	

Source: R. M. Dowben, *General Physiology: A Molecular Approach*. New York: Harper & Row, 1969, p. 439.

that the relation of reflection coefficient to effective pore areas of solute and solvent is given by

$$\sigma = 1 - \frac{A_s}{A_w} \tag{2-5}$$

where A_s is the effective fraction of area available for solute diffusion and A_w is the effective fraction of area available for solvent diffusion. As the solute molecule approaches the pores in size, they cannot diffuse, and the membrane becomes semipermeable to that solute, A_s/A_w approaches zero, and σ approaches 1.

All of this is particularly important in dealing with living cells. Cells are very permeable to water, but much less permeable to other substances. Mammalian red cells, for example (see Figure 2.12), will shrink when placed in a concentrated solution and, contrariwise, will swell when placed in a dilute solution owing to osmotic movement of water across the red cell membrane. We say that the concentrated solution that makes the red cell shrink is *hypertonic,* whereas the dilute solution that makes the red cell swell is *hypotonic.* A solution is said to be *isotonic* when it produces no change in cell size.

In contrast to tonicity, which refers to swelling or shrinking of cells, *osmolarity* refers to the concentration of solutes in the solution. The osmolarity of solutions is usually expressed as *milliosmoles;* 1 mOs is equivalent to a solution of 1 millimole of a nonelectrolyte per liter of water. Electrolytes, of course, dissociate into cations and anions; 1 millimole of electrolyte will produce almost 2 or more mOs. Osmolarity expresses the combined osmotic pressure of all solutes.

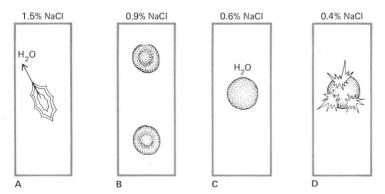

FIGURE 2.12 A diagrammatic representation of osmotic effects in animal cells without a rigid cell wall. When cells, such as red blood cells, are placed in isotonic media, they retain their size and shape (**B**). In hypertonic media (**A**) they lose water and shrink. In hypotonic media (**C**) cells take up water and swell. If the media are sufficiently hypotonic (**D**), they burst.

The osmolarity of the human red blood cell is approximately 280 mOs. When red blood cells are suspended in 280 mM mannitol, their volume will not change initially because $\sigma = 1.0$ for mannitol. However, an *isotonic* solution of glycerol must be 318 mM because $\sigma = 0.88$ for glycerol, and this concentration is required to give an effective osmolarity of 280 mOs. Similarly, a malonamide solution must be 337 mM to be isotonic for red cells because $\sigma = 0.83$ for malonamide.

7 • ELECTROLYTES

So far we have concerned ourselves only with nonelectrolyte solutes. An additional consideration must be dealt with when quantifying the transport of *electrolytes*—that is, salts such as sodium chloride and magnesium acetate that dissociate into positively and negatively charged particles called *ions* when dissolved in water. The ions are free to move independently in solution. When an electric field is applied to an electrolyte solution, the positively charged ions (*cations*) are attracted to the negative electrode (*cathode*), while the negatively charged ions (*anions*) move toward the positive electrode. However, the dissociation is not complete, and the concentration in the osmotic pressure equation must be corrected by a concentration dependent dissociation constant γ

$$\Pi = RT(\gamma c^+ + \gamma c^-) \tag{2-6}$$

where c^+ is the concentration of cations and c^- that of anions. The dissociation depends upon the particular salt and decreases as the salt concentration increases (Table 2.4).

TABLE 2.4   Percent Dissociation of Various Electrolytes as a Function of Concentration at 0°C*

	Concentration (molal)				
	0.02	*0.05*	*0.1*	*0.2*	*0.5*
NaCl	92.1%	89.7%	87.2%	84.3%	—
LiCl	92.8	91.2	89.5	88.4	82.7%
KCl	91.9	88.5	85.7	82.7	78.4
KNO$_3$	90.4	84.7	78.4	69.8	55.1
MgCl$_2$	85.4	83.8	82.9	83.9	84.8
CaCl$_2$	83.7	81.5	80.1	78.7	78.0

* Calculated from determination of cryoscopic coefficient.

TABLE 2.5   Radii and Hydration of Alkali Earth Ions

Ion	Crystal Radius, Å	Hydrated Radius, Å	Hydration (number of water molecules)
Li^+	0.68	2.51	6.0
Na^+	1.01	2.25	4.6
K^+	1.30	1.86	2.3
Rb^+	1.50	1.75	1.7
Cs^+	1.75	1.56	—

Ions in solution attract a number of water molecules that tend to adhere closely to a dissolved ion, a phenomenon known as *hydration*. The smaller the ion the more water molecules in the layer of hydration. Therefore the hydrated diameter of ions in solution increases as the diameter of the nude ion in the salt crystal decreases (Table 2.5).

Biological membranes exhibit marked selectivity between ions. For example, the resting squid nerve membrane is about 30 times more permeable to K^+ than to Na^+, whereas during the ascending phase of a nerve action potential the selectivity is reversed, and the permeability of Na^+ is 10 times greater than that of K^+. If all the alkali cations are studied, the sequence of permeabilities are found to be

$$\text{Resting squid axon:} \quad K^+ > Rb^+ > Cs^+ > Na^+ > Li^+$$
$$\text{Rising action potential:} \quad Li^+ > Na^+ > K^+ > Rb^+ > Cs^+$$

The number of possible permutations of the five alkali cations is $5! = 120$. Note that of the 120 possible selectivity sequences, only 11 have been observed in nature, either in living or nonliving systems:

$$
\begin{aligned}
\text{I}_{c1} \quad & Cs^+ > Rb^+ > K^+ > Na^+ > Li^+ \\
\text{II}_{c1} \quad & Rb^+ > Cs^+ > K^+ > Na^+ > Li^+ \\
\text{III}_{c1} \quad & Rb^+ > K^+ > Cs^+ > Na^+ > Li^+ \\
\text{IV}_{c1} \quad & K^+ > Rb^+ > Cs^+ > Na^+ > Li^+ \\
\text{V}_{c1} \quad & K^+ > Rb^+ > Na^+ > Cs^+ > Li^+ \\
\text{VI}_{c1} \quad & K^+ > Na^+ > Rb^+ > Cs^+ > Li^+ \\
\text{VII}_{c1} \quad & Na^+ > K^+ > Rb^+ > Cs^+ > Li^+ \\
\text{VIII}_{c1} \quad & Na^+ > K^+ > Rb^+ > Li^+ > Cs^+ \\
\text{IX}_{c1} \quad & Na^+ > K^+ > Li^+ > Rb^+ > Cs^+ \\
\text{X}_{c1} \quad & Na^+ > Li^+ > K^+ > Rb^+ > Cs^+ \\
\text{XI}_{c1} \quad & Li^+ > Na^+ > K^+ > Rb^+ > Cs^+
\end{aligned}
$$

It should be noted that sequence I corresponds to the order of diameters of the naked ions, whereas sequence XI corresponds to the order of the hydrated ion diameters. Now we must ask what rules govern the occurrence of a particular selectivity sequence. Remember that the constituents of biological membranes, phospholipids and proteins, are polyelectrolytes carrying multiple charges of both signs. Ions in solution near such a membrane will alternatively interact with water molecules or with charged sites of opposite sign on the membrane. Let us consider the two limiting cases. If the charged sites on the membrane are very strong and closely spaced, they will tend to overwhelm the tendency for ions to interact with water molecules. Under these circumstances the smallest nude ion will be attracted with the largest force, and selectivity sequence XI will follow. The opposite extreme is when the charges on the membrane are very weak and sparsely spaced and do not interfere with the hydration of ions; in this case the smallest hydrated ion is attracted most strongly and selectivity sequence I is operative. For intermediate strengths of charged sites, intermediate selectivity sequences are observed.

Similar selectivity sequences are found for the alkaline-earth cations

$$I_{c2} \quad Ba^{2+} > Sr^{2+} > Ca^{2+} > Mg^{2+}$$
$$II_{c2} \quad Ba^{2+} > Ca^{2+} > Sr^{2+} > Mg^{2+}$$
$$III_{c2} \quad Ca^{2+} > Ba^{2+} > Sr^{2+} > Mg^{2+}$$
$$IV_{c2} \quad Ca^{2+} > Ba^{2+} > Mg^{2+} > Sr^{2+}$$
$$V_{c2} \quad Ca^{2+} > Mg^{2+} > Ba^{2+} > Sr^{2+}$$
$$VI_{c2} \quad Ca^{2+} > Mg^{2+} > Sr^{2+} > Ba^{2+}$$
$$VII_{c2} \quad Mg^{2+} > Ca^{2+} > Sr^{2+} > Ba^{2+}$$

and for monovalent anions

$$I_a \quad I^- > Br^- > Cl^- > F^-$$
$$II_a \quad Br^- > I^- > Cl^- > F^-$$
$$III_a \quad Br^- > Cl^- > I^- > F^-$$
$$IV_a \quad Cl^- > Br^- > I^- > F^-$$
$$V_a \quad Cl^- > Br^- > F^- > I^-$$
$$VI_a \quad Cl^- > F^- > Br^- > I^-$$
$$VII_a \quad F^- > Cl^- > Br^- > I^-$$

With regard to the relative affinities of negative membrane sites for monovalent or divalent cations, the spacing between sites is determining. Closely spaced sites have a preference for divalent cations.

Among membranes derived from living systems, almost every possible pattern of selectivity has been observed. Since we are concerned mainly with general principles here, only a few more examples will be cited for purposes of illustration. Salt receptors of blowflies are stimulated in the order of sequence VI_{c1}. DNA from calf thymus binds cations in the order of sequence XI_{c1}. The permeability of rabbit gallbladder to monovalent cations shows sequence VI_{c1} at pH 7.4 'and sequence III_{c1} at pH 2.4.

The potency of divalent cations in blocking membrane negative charges and thus lowering the permeability to NaCl in rabbit gallbladder follows sequence III_{c2}. The rate of rise of the spike of membrane depolarization in barnacle muscle is influenced by divalent cations according to sequence V_{c2}. Divalent cations activate the enzyme amylase (which hydrolyzes starch) according to sequence VII_{c2}. Divalent cations bind to the contractile protein G-actin according to sequence VI_{c2}.

Bullfrog skin is permeable to anions according to the sequence III_a. For human red blood cells, the internal equilibrium concentration of anions follows sequence I_a, while efflux rates (the rates of anion outflow) follow sequence V_a. Anion conductances in cardiac muscle membrane follow sequence I_a, while in frog skeletal muscle membrane, sequence IV_a is found.

8 • MEMBRANE POTENTIALS

We have repeatedly mentioned the remarkable ability of living cells to maintain intracellular ion concentrations that are very different from those of the extracellular fluid or of the environment. The basic differences are that the intracellular fluid is high in K^+ and Mg^{2+} ions and low in Na^+ and Ca^{2+} ions compared with the extracellular fluid. A consequence of the differences of ion concentrations across the cell membrane is the presence of an electrical potential difference. The membrane potential varies in magnitude from cell type to cell type (as do the differences in ion concentration), but the interior of the cell is always slightly negative with respect to the exterior. The importance of membrane potentials in nerve function is considered in Chapter 3, Section 4.

As early as 1888 the German physical chemist Nernst discovered the relation (which bears his name) between the magnitude of the potential across the membrane and the difference in ion concentrations:

$$E = \frac{RT}{zF} \ln \frac{c_2}{c_1} = 2.302 \frac{RT}{zF} \log \frac{c_2}{c_1} \tag{2-7}$$

where E is the potential in volts, R the universal gas constant, T the absolute temperature, F the faraday (96,000 coulombs/mole), c_1 and c_2 the concentrations of ions on the two sides of the membrane, and z the sign (positive or negative) and valence of the ion. The equation was originally derived in terms of natural logarithms of the ion concentrations, but it can be restated in terms of logarithms to the base 10 by use of the conversion factor 2.302.

As we said previously, the cell membrane is not equally permeable to all species of ions. On the contrary, it is more permeable to some kinds of ions than others. The resting cell membrane is much more permeable to potassium ions than to other kinds of ions; therefore, in the case of most resting or unstimulated cells, the membrane potential is approximately given by

$$E = -2.302 \frac{RT}{F} \log \frac{[K^+]_{in}}{[K^+]_{out}} \tag{2-8}$$

When the intracellular K^+ is depleted, or when a cell is stimulated and becomes very permeable to Na^+, Equation 2-8 no longer holds.

In 1943 Goldman developed an equation that takes into account all of the ion species that influence the membrane potential by considering the conditions necessary to maintain constant field strength as the distribution of charges across the membrane changes. The derivation of the Goldman constant field equation is beyond the scope of this book, but it is a very useful relation for physiologists, particularly when considering the membrane potential of intestinal or kidney cells or excitable cells like muscle and nerve. The equation is

$$E = -\frac{RT}{zF} \ln \frac{P_K[K^+]_{in} + P_{Na}[Na^+]_{in} + P_{Cl}[Cl^-]_{out}}{P_K[K^+]_{out} + P_{Na}[Na^+]_{out} + P_{Cl}[Cl^-]_{in}} \tag{2-9}$$

where P_K, P_{Na}, and P_{Cl} are the permeability for potassium, sodium, and chloride, respectively. The resting potential of the squid giant axon, for example, can be calculated very accurately from Equation 2-9 by using $P_K:P_{Na}:P_{Cl} = 1.0:0.04:0.45$. The membrane potential during excitation and in relation to kidney function and intestinal activity will be considered later on (Chapters 6 and 8).

There are two other items that should be remembered concerning ion transport. The first is that cations will tend to move from the positive side of the membrane to the negative side over and above the tendency for the ion to move by diffusion, while for anions transport due to the electrical gradient occurs in the opposite direction. Second, when ions move across the membrane, they cause an electric current to flow. The electric current generated by ion movement is important in causing the spread of excitation in muscle and nerve, and it has been used extensively for studying ion transport.

9 • DONNAN EQUILIBRIUM

Proteins, nucleic acids, and other kinds of natural macromolecules represent a special type of electrolyte. All of these individual molecules possess multiple charges, both positive and negative, and therefore are called *polyelectrolytes*. Macromolecular polyelectrolytes usually have a preponderance of either positive or negative charges and are said to have a net positive or net negative charge. The macromolecular polyelectrolytes are large, bulky molecules which diffuse very slowly in solution. The individual charges on the surface tend to attract small, mobile ions of the opposite charge. For example, consider a protein dissolved in a solution of NaCl. The positive charges on the surface of the protein molecule will tend to attract Cl^-, while the negative surface charges will tend to attract Na^+ ions. These small ions of opposite charge (called *counterions*) are sufficiently strongly attracted to the macromolecular polyelectrolyte that they tend to move together, creating an *electrical double layer* of charges at the surface. It should be appreciated that a given Cl^- ion or Na^+ ion will not always be associated with the same charged group of a given polyelectrolyte molecule, but that the counterions will continually exchange with ions in the solution.

Let us now consider, as an example, a selectively permeable membrane separating two compartments, both of which contain a solution of a permeable salt such as NaCl. In addition, only one compartment contains a nonpermeable polyelectrolyte, such as a protein or nucleic acid. For the purposes of illustration, let us assume that the polyelectrolyte bears a net negative charge. A suspension of red blood cells in an isotonic saline solution will fit the conditions we have set out, and the conclusions that follow can be regarded as holding for such a suspension of red blood cells. When equilibrium is attained, we find that the concentrations of permeable salt are different on the two sides of the membrane (*Donnan equilibrium*).

The condition of electroneutrality requires that the total concentration of cations and anions in each compartment must be equivalent. Let us designate the polyelectrolyte X^- and call the compartment containing it the inside. At equilibrium,

$$[Na^+]_{in} = [Cl^-]_{in} + [X^-]_{in} \tag{2-10}$$

$$[Na^+]_{out} = [Cl^-]_{out} \tag{2-11}$$

It can be shown that the distribution of ions at equilibrium is given by the relation

$$r = \frac{[Na^+]_{out}}{[Na^+]_{in}} = \frac{[Cl^-]_{in}}{[Cl^-]_{out}} = \left[\frac{[Cl^-]_{in}}{[Cl^-]_{in} + [X^-]_{in}} \right]^{1/2} \tag{2-12}$$

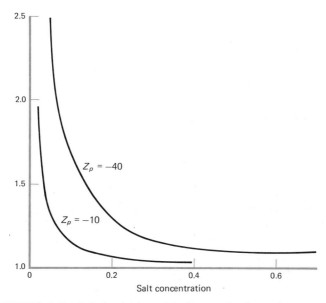

FIGURE 2.13 Calculated values of the Donnan distribution function r for varying salt concentration in the compartment that does not contain protein. The other compartment contains 1 g protein per 100 g water. The protein has a molecular weight of 40,000 daltons and may have a net charge Z_p of -10 or -40. (From H. B. Bull, *An Introduction to Physical Biochemistry*. Philadelphia: F. A. Davis, 1964.)

where r is a distribution ratio that depends upon the relative concentrations of polyelectrolyte and permeable salt. Values for r are plotted in Figure 2.13.

It will be appreciated that the difference in ion concentrations that develops gives rise to a membrane potential, the *Donnan potential*, which is given by

$$E = \frac{RT}{F} \ln \frac{[\text{Na}^+]_{\text{out}}}{[\text{Na}^+]_{\text{in}}} = \frac{RT}{F} \ln \frac{[\text{Cl}^-]_{\text{in}}}{[\text{Cl}^-]_{\text{out}}} \qquad (2\text{-}13)$$

The considerations involved in the reflection coefficient and Donnan equilibrium explain the following phenomena, which can be observed in red blood cells. If red blood cells are suspended in isosmolar sucrose solution, they will shrink. This is attributable to the fact that the red cell membrane is impermeable to sucrose ($\sigma \simeq 1.0$) but slightly permeable to the intracellular contents ($\sigma \simeq 0.94$). Thus isosmolar sucrose is hypertonic, and the cells will shrink. If solid NaCl is then added to the medium, the cells will swell. This unexpected occurrence results from a redistribution of NaCl to achieve a Donnan equilibrium, when the concentration of total solute in the medium, even though greater than before, is hypotonic with respect to the intracellular fluid of the red cell.

10 • EFFECT OF IONIZATION UPON PERMEABILITY

Many metabolites, drugs, and other biologically important substances are weak organic acids (less often they are weak organic bases) which dissociate poorly. The permeability of these substances, including absorption from the gastrointestinal tract and excretion by the kidney, usually depends upon whether or not they are ionized. The dissociation of weak acids depends upon the pH of the medium and is given by the *Henderson-Hasselbach equation*

$$pH = pK_a + \log \frac{[A^-]}{[HA]} = pK_a + \frac{\alpha}{1 - \alpha} \tag{2-14}$$

where α is the fraction of ionized organic acid and $1 - \alpha$ is the fraction that is nonionized. The pK_a is the negative logarithm of the dissociation constant and is numerically equal to the pH at which the concentrations of ionized and nonionized forms are equal.

In general, the nonionized forms of weak acids and bases permeate membranes more readily than the ionized forms. Thus weak acids are more permeable in acid solutions, and weak bases are more permeable in basic solutions. This phenomenon is particularly important in regard to the absorption and excretion of metabolites and drugs. Drugs that are weak acids—aspirin, for example—are absorbed from the stomach where the pH is low, whereas drugs that are weak bases, like barbiturates, are absorbed from the small intestine where the pH is high. Tests of the antibacterial activity of a series of sulfonamides with pK_a values less than 7 show that antibacterial activity falls off with lower pK_a, because a greater fraction of the sulfonamide is ionized and less of the drug enters the bacteria. In cases of barbiturate poisoning the patient is given alkali; because barbiturates are weak bases, a greater fraction of the drug will be in the nonionized form as the pH increases, and the drug will be more readily mobilized from the tissues into the circulation and more rapidly excreted.

11 • FACILITATED DIFFUSION

Some materials, particularly sugars, amino acids, vitamins, and other substances that are metabolized by the organism, are transported across cell membranes at rates much greater than would be expected on the basis of their molecular size or lipid solubility. Indirect evidence indicates that the transport of these materials involves an interaction with a limited number of specific membrane constituents called *carriers*. Transport that is mediated by a membrane carrier is called *facilitated diffusion*. Let us list the characteristics of facilitated diffusion and then give some illustrative examples:

1. Transport is highly stereospecific. Closely related compounds and even isomers or optical enantiomorphs may be transported at markedly different rates.
2. Except at very low concentrations, the rate of transport is not proportional to concentration and does not follow Fick's law, but rather it tends to reach a limiting value as the concentration of solute in the medium is increased.
3. The rate of transport may be decreased by the presence of some structurally similar solutes that appear to compete for participation in the facilitated diffusion process, that is, compete for carrier sites. Frequently the competition is not reciprocal; for example, glucose competitively inhibits the transport of sorbitol and fructose in red blood cells but not vice versa.
4. Facilitated diffusion may occur uphill, against an electrical or concentration gradient, if it is coupled with the simultaneous downhill transport of another substance.
5. Transport is markedly reduced by substances, such as organic mercurials, that are inhibitors of enzyme reactions. Some inhibitors will reduce some transport systems, whereas other inhibitors will act on different transport systems.
6. The unidirectional flux of the solute from one side of the membrane to the other as measured by the use of radioactively labeled material is usually much greater than the rate of net (or overall) transport.
7. The temperature coefficients of transport resemble those of chemical reactions ($Q_{10} \simeq 3$).

The transport of glucose is one of the best-studied facilitated diffusion systems. Glucose would *not* be expected to permeate cell membranes readily because it is a relatively large solute molecule (180 daltons), and, owing to the many hydroxyl groups, it is not lipid soluble. Yet glucose and certain other monosaccharides readily permeate animal cells. For most cell types, net transport is *downhill,* occurring only from a higher to a lower concentration. Because they are easily obtained, human red blood cells have been extensively used for studies of glucose transport. While most fetal mammalian red blood cells show high net glucose transport, the adult red blood cells of ungulates, rodents, carnivores, and other nonprimates have lost the ability to transport glucose readily, but this ability is retained by adult red cells of primates.

The rate of entry of sugars may be measured by changes in light scattering resulting from volume changes of suspended cells. Another method measures the time required for a given fraction of red blood cells to swell and hemolyze after being placed in a sugar solution. The extent of

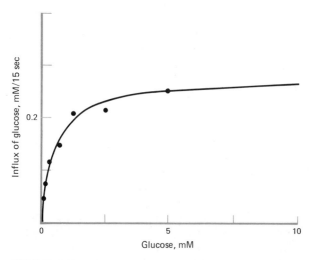

FIGURE 2.14   Influx of ^{14}C-glucose into human red blood cells as a function of glucose concentration in the incubation medium. (From W. D. Stein, *The Movement of Molecules across Cell Membranes.* New York: Academic Press, 1967.)

hemolysis can be followed in a photometer, and absorbance changes may be recorded as hemolysis progresses. Finally, the sugar content of cells can be determined chemically, enzymatically, or by use of radioactively labeled sugars at various time intervals after they are placed in a sugar solution.

As the concentration of glucose is increased, the glucose transport system becomes saturated (Figure 2.14). The process seems analogous to an enzyme-catalyzed reaction, the solute combining with carrier to form a transient solute-carrier complex. An expression similar to the Michaelis-Menten equation can describe the net flow of a solute that is transported by facilitated diffusion; that is,

$$J = \frac{c_s J_m}{K_m + c_s} \tag{2-15}$$

where c_s is the solute concentration, J_m the maximal solute flow, and K_m the solute concentration at which flow is half maximal.

The transport mechanism is highly stereospecific: D-glucose and D-galactose are transported, but L-glucose and L-galactose are not. Cells also are permeable to several related sugars: D-mannose, L-sorbose, D-fructose, D-xylose, and L-arabinose. Some sugar derivatives such as 3-O-methylglucose penetrate readily, whereas others, like mannitol, do not.

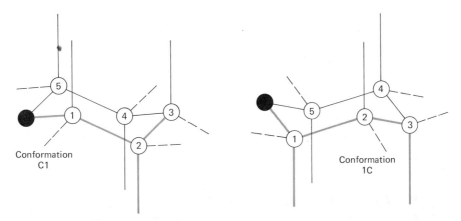

FIGURE 2.15 A diagram of the two chair conformations of the six-membered pyranose ring of simple sugars. The filled circle represents the oxygen atom in the ring, whereas the numbered circles represent the corresponding carbon atoms. The side of the ring facing the observer is heavier, equatorial bonds are shown by dashed lines, and vertical solid lines depict axial bonds. (From P. G. LeFevre and J. K. Marshall, *Am. J. Physiol.* 194:335, 1958.)

Can we discover some simple rules that explain the selectivity of the glucose transport system? Most simple sugars in solution are found in the form of a six-membered *pyranose* ring, composed of five carbon atoms and one oxygen atom (Figure 2.15). The sugar pyranose rings cannot quite exist in a perfectly planar form owing to the dimensions of the chemical bonds and bond angles; rather, they exist in two slightly bent conformations frequently called the *chair form* and the *boat form*.

There are six possible boat forms, one with an oxygen atom in each possible position; the six boat forms are readily interconvertible and can be considered to behave as a single conformational form. There are two possible chair forms: the 1C form, in which the bulky side chain is approximately perpendicular to the plane of the pyranose ring, and the C1 form, in which the bulky side chain extends approximately in the plane of the ring (Figure 2.15). The chair forms are relatively rigid and are not easily interconverted with each other or with the boat form.

It has been shown that the glucose transport system requires that the sugar be largely in the C1 form. Those sugars that exist largely in the 1C or boat forms do not seem to interact with the carrier and are not transported. There are, of course, variations from cell type to cell type and between different species. For example, D-fructose and α-methyl-D-glucose are transported almost as well as D-glucose by fetal rabbit erythrocytes but very poorly by human erythrocytes.

The various sugars that are transported reversibly compete with each other (Figure 2.16). Thus the transport of D-glucose is depressed in

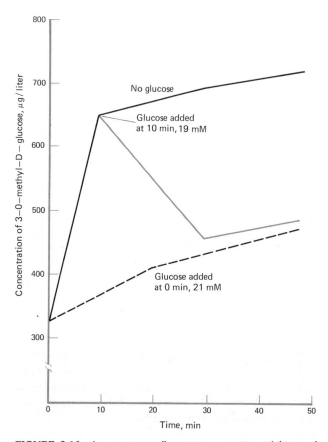

FIGURE 2.16 An experiment illustrating competitive inhibition of transport and countertransport. Hearts were perfused with 0.75 mM 3-0-methylglucose and the uptake determined (solid line). In the presence of 21 mM glucose, less 3-0-methylglucose was taken up (broken line), owing to competitive inhibition of transport. When 19 mM glucose was added to the perfusion medium after 10 min, an actual efflux of 3-0-methylglucose was observed, the level falling to approximately that of the glucose uptake curve (dashed line). This phenomenon is known as countertransport. (From H. E. Morgan, D. M. Regen, and C. R. Park, *J. Biol. Chem.* 239:370, 1964.)

the presence of D-galactose. Double reciprocal plots of $1/J$ against $1/c$ (similar to the double reciprocal plots used in quantifying enzyme kinetics) yield straight lines and show the form typical of competitive inhibition; namely, K_m remains constant and J_m is diminished. On the other hand, inhibition by the enzyme poisons p-chloromercuribenzoate or 1-fluoro-2,4-dinitrobenzene is noncompetitive and irreversible.

In skeletal muscle, cardiac muscle, and fat cells, but not in red blood cells and other tissues, glucose uptake is markedly stimulated by the hormone insulin. In part the action of insulin is due to stimulation of glucose metabolism so that the intracellular glucose concentration is kept low. In part, however, insulin directly affects the membrane and its capacity to transport glucose and related sugars. The partial proteolysis of the outer membrane surface by exposure to the digestive enzymes trypsin or chymotrypsin also increases the transport of glucose. Both insulin and trypsinization leave K_m unchanged, but J_m is greatly increased. The insulin and trypsin effects appear to be equivalent and nonadditive.

In the tissues thus far mentioned, glucose transport by facilitated diffusion is strictly a downhill process from high concentration to low concentration. In intestinal epithelium, kidney tubular epithelium, and certain other tissues, glucose transport is *uphill,* against the concentration gradient. The concentration of glucose requires energy input; this is obtained in most instances by coupling the uphill transport of glucose to the downhill transport of Na^+ ions. Other monovalent cations (K^+, for instance) do not substitute for extracellular Na^+ in supporting uphill glucose transport.

Amino acids also are transported by facilitated diffusion, and most animal cells are able to concentrate amino acids by a factor of approximately 20. In many cases the uphill transport of amino acids into cells is also coupled to the simultaneous downhill movement of Na^+ ions into cells, and extracellular K^+ cannot substitute for Na^+. There are several distinct amino acid transport systems: (1) a leucine-preferring system for neutral amino acids with hydrophobic side chains; (2) an alanine-preferring system for neutral amino acids and preferring amino acids with short, polar side chains; (3) perhaps an alanine-, serine-, and cysteine-preferring system that may be specifically inhibited by N-methylaminoisobutyric acid; (4) a transport system for β-amino acids, as distinct from systems transporting the more common α-amino acids; (5) a transport system for acidic amino acids; and (6) a transport system for basic amino acids.

12 • ACTIVE (OR METABOLICALLY COUPLED) TRANSPORT

The term *active transport* should really be limited to those transport systems in which a solute moves uphill against an electrochemical gradient by a process wherein the required energy is derived from a metabolic reaction. The metabolic reaction usually is the hydrolysis of ATP or some other energy-rich compound or an oxidation-reduction reaction.

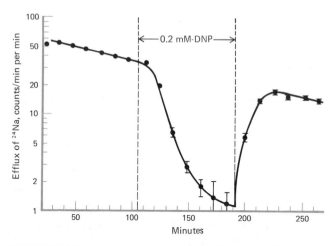

FIGURE 2.17 Active outward transport of Na^+ from a giant squid axon, showing its abolition upon treatment with dinitrophenol and its restoration when the drug was washed away with sea water. (From A. L. Hodgkin and R. D. Keynes, *J. Physiol.* 128:34, 1955.)

The different intracellular ion concentrations, high K^+ and low Na^+, are maintained by the active, uphill outward transport of Na^+ ions coupled to ATP hydrolysis. The requirement for ATP is obligatory; neither ITP, GTP, or UTP will substitute as the metabolic energy source. Metabolic inhibitors such as dinitrophenol or sodium azide inhibit active Na^+ extrusion (Figure 2.17) by interfering with the biosynthesis of ATP. The digitalis group of drugs inhibits active Na^+ transport directly without inhibiting any metabolic reactions.

There appears to be a rough stoichiometry for active Na^+ transport. Roughly, 2 to 3 moles of Na^+ ions are extruded for every mole of ATP hydrolyzed, and the entry of approximately two K^+ ions is coupled to the extrusion of three Na^+ ions.

Cell membranes contain a lipoprotein enzyme that hydrolyzes ATP in the presence of Na^+, K^+, as well as Mg^{2+}. This lipoprotein is believed to mediate the active transport of Na^+; that is, it is thought to be the "sodium pump." The requirement for Na^+ is specific, while other monovalent cations can substitute for K^+. Ouabain and other digitalislike drugs inhibit the Na^+,K^+-activated ATPase directly. The amount of enzyme found in different types of cells is roughly proportional to the amount of active transport, being high in the tubular epithelial cells of the kidney and low in red blood cells. The Na^+,K^+-activated ATPase is abundant in the salt glands of marine birds and rectal glands of certain fishes, organs that secrete salt against a concentration gradient (see Chapter 6, Section 2.1).

The enzyme consists of two polypeptide chains of molecular weights 130,000 and 40,000, which completely traverse the membrane. Thus portions of the enzyme are exposed on the inner surface and other parts on the outer surface. On the basis of antigenicity the portions of the enzyme exposed on the inner and outer surfaces are quite distinct, and they do not interchange sides. Thus the Na^+,K^+-activated ATPase is inserted into the membrane vectorially, and this may explain why active Na^+ transport is unidirectionally outward. The ATP-binding and Na^+-binding sites of the enzyme face the interior of the cell, whereas the K^+-binding and ouabain-binding sites are on the external surface of the membrane.

Under some conditions it is possible to run the sodium pump backward and resynthesize ATP from ADP. For example, this can be done using red blood cells in which glycolysis and ATP production have been poisoned with iodoacetate by incubation in a Na^+-rich, K^+-free medium.

The amino acid composition has been determined by using highly purified enzyme from the electric organ of the electric eel, the rectal gland of sharks, and the renal medulla (cell membrane of kidney tubular epithelium) of dogs. When these data were subjected to statistical analysis, it could be concluded that no (or very few) differences in the amino acid composition occur in these diverse species. Like the genetic code, the Na^+,K^+-activated ATPase occupies a key position in the physiological function of animals, and it has been retained through evolution with remarkable constancy of structure.

A great deal of research effort is being devoted to unraveling the mechanism of the sodium pump. It is thought that the enzyme exists in two different conformations, E_1 and E_2, and that ATP hydrolysis proceeds by a sequence of reactions:

$$E_1 + ATP \xrightarrow{Mg^{2+},\ Na} E_1 - P + ADP$$

$$E_1 - P \xrightarrow{Mg^{2+}} E_2 - P$$

$$E_2 - P + H_2O \xrightarrow{Mg^{2+},\ K^+} E_2 + P_i$$

$$E_2 \longrightarrow E_1$$

The phosphorylated intermediate is quite stable at low pH in the absence of K^+ (or other monovalent cation other than Na^+) and in the presence of ouabain. However, the mechanism by which the Na^+ ions actually move across the membrane is still a mystery.

Metazoan organisms have other active transport systems, although there are not as many different active transport systems as in prokaryotes. The sarcoplasmic reticulum membrane actively transports Ca^{2+} ions in a process linked to ATP hydrolysis so effectively that the intracellular Ca^{2+} concentration in resting muscle is about 10^{-8} M. The stomach mucosa of many animals actively secretes H^+ ions, lowering the pH to 1.5 or less. Mitochondrial membranes contain Ca^{2+} and other active transport systems coupled to oxidation-reduction, electron transport reactions.

13 • COMPOSITE MEMBRANES

Many physiologically important transport processes take place across a layer of epithelial cells from one body cavity to another. Examples are transport across the eipithelium of the gut or across the renal tubular epithelium. In essence such systems consist of two plasma membranes back to back and constitute a composite membrane system. The frog skin is one of the best studied composite membrane systems; it has the ability to transport sodium chloride uphill from a low concentration on the outer side of the frog skin to a high concentration on the inner side by a process that is coupled to the metabolism in the epithelial cells.

When a piece of frog skin is placed between compartments containing identical Ringer's salt solution, using an apparatus shown diagrammatically in Figure 2.18, a potential of about 50 to 100 mV will develop between the two compartments. The inner surface of the frog skin will be positive, and the outer surface will be negative. Roughly, the magnitude of the potential will be proportional to the flux of sodium chloride transported.

The presence of Na^+ ions in the medium bathing the outer surface of the frog skin is required for transport and for generating the potential. If a counterpotential of opposite sign is applied in series with the potential generated by the frog skin so that the net potential between the two compartments is zero, the electric current that flows through the connecting wire is exactly equal to the inward flux of Na^+ ions through the frog skin. In this way it is possible to utilize the apparatus depicted in Figure 2.18 to study the effect of ion composition and of various agents like metabolic inhibitors on sodium transport through the frog skin.

If a microelectrode is pushed from the outer surface of the frog skin toward the inside surface, a few millivolts of negative potential is found as the microelectrode passes through the outer layer of keratinized epithelial cells. At a depth of about 50 μ below the outer surface, a positive potential of about 60 mV suddenly develops. If the microelectrode is pushed deeper, a potential of about $+100$ mV is observed when the microelec-

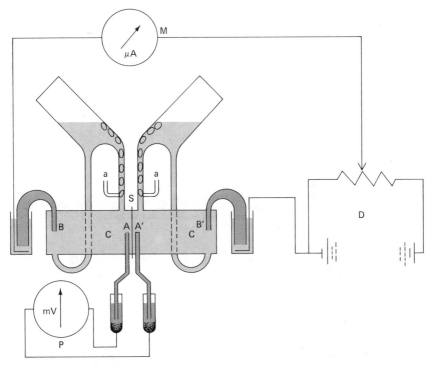

FIGURE 2.18 A diagram of the apparatus used for measuring ion fluxes and potential differences and currents across a frog skin. Chambers (C) containing Ringer's salt solution are located on either side of the frog skin (S), and they contain inlets (a) for oxygen, salt bridges to calomel electrodes (A) for measuring the potential with a potentiometer (P), and salt bridges (B) for applying a potential from an external source (D) through a current meter (M). (From H. H. Ussing and K. Zerahn, *Acta Physiol. Scand.* 23:113, 1951.)

trode reaches the basement membrane. Thus the potential jumps are observed on either side of the basilar layer of epithelial cells in the frog skin, transport taking place by a two-stage process. It is thought that Na^+ ions enter the interior of the basilar epithelial cells passively by downhill diffusion (Figure 2.19), and they are extruded through the inner surface by active transport. The net effect is to pump Na^+ ions actively from the outside to the inside of the frog skin.

The toad bladder is another composite membrane system. The toad conserves water by reabsorbing it from the urine in the bladder. Actually the toad bladder pumps sodium chloride from the bladder cavity into the extracellular fluid on the outer bladder surface by an active process coupled to metabolism, just as the frog skin does. Water movement passively accompanies the active transport of NaCl; that is, the water is carried

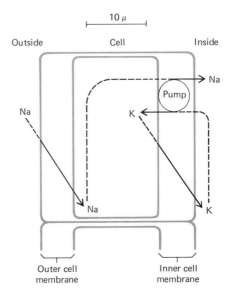

FIGURE 2.19 Ussing's view of the epithelial cell of the frog skin as an asymmetric composite membrane system. The outward-lying membrane is highly permeable to Na^+, which enters the cell by moving down an electrochemical gradient; the inward-facing membrane actively transports Na^+ out of the cell to the inside compartment. (From H. H. Ussing, *J. Gen. Physiol. (Suppl. 1)* 43:140, 1960.)

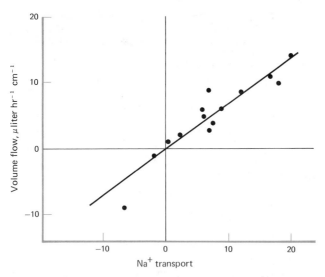

FIGURE 2.20 Relation between volume flow and net Na^+ transport across rat small intestine. Positive values indicate flow from the mucosal surface to the serosal surface. (From P. F. Curran, *J. Gen. Physiol.* 43:1140, 1960.)

(*dragged*) along with the actively transported salt. A similar flow of fluid accompanies active sodium transport across the intestinal epithelium (Figure 2.20).

14 • STRUCTURE OF MEMBRANES

No discussion of transport would be complete without some consideration of our present concepts of membrane structure and organization at the molecular level, which is admittedly still being developed and is as yet somewhat uncertain. The contemporary ideas of membrane structure originated from the classical experiments in 1925 of Gorter and Grendel, who extracted the lipids from washed red blood cells and spread them as a monolayer on water in a Langmuir trough. Gorter and Grendel found that the area of the monolayer film was almost exactly twice the total surface area calculated for the intact red blood cells used in the experiment.[2] From this result they suggested that the cell membrane consisted of a bimolecular layer of phospholipids packed with the hydrophobic ends facing each other in the center of the membrane and the hydrophilic ends facing outward on either side (Figure 2.21).

On the basis of surface tension studies of marine eggs, model systems, and other membranes, Davson and Danielli proposed their now-famous model in which they suggested that a protein layer was attached to the polar groups on either surface of the bimolecular lipid leaflet (Figure 2.22). Belief in the validity of the Davson-Danielli model was reinforced

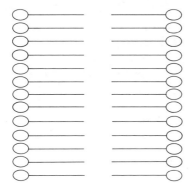

FIGURE 2.21 A model of the phospholipid bilayer membrane according to Gorter and Grendel. The polar head groups are facing outward on either side, whereas the hydrophobic side chains are oriented toward the interior of the membrane.

[2] It appears that Gorter and Grendel made two errors in their experiments: they failed to extract about one-third of the lipid from the erythrocytes with their acetone procedure, and they underestimated the red cell surface area by one-third. Happily, these two errors offset each other.

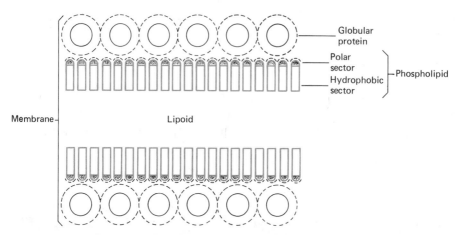

FIGURE 2.22 A schematic diagram of the Davson-Danielli model of membrane structure. The membrane is thought to consist of a lipid bilayer sandwiched between protein molecules. (From J. F. Danielli and H. A. Davson, *J. Cell. Comp. Physiol.* 5:498, 1935.)

by the finding that membranes have a characteristic trilaminar appearance in electron micrographs.

The bimolecular leaflet is a very stable structure for amphiphilic phospholipids, such as those found in biological membranes. The evidence for existence of a bimolecular leaflet in prokaryotic cell, mitochondrial, and chloroplast membranes is strong. These membranes undergo a sharp phase transition, which can be detected by microcalorimetry, that corresponds to freezing-melting of a similar mixture of bulk lipids. X-ray diffraction studies and studies using spin-label probes also support the presence of bimolecular lipid leaflets in these membranes. However, good supporting evidence for bimolecular lipid leaflets in eukaryotic cell and endoplasmic reticulum membranes is lacking at present.[3]

Spectroscopic studies indicate that generally membrane proteins are globular, and an appreciable fraction of their amino acids are in α-helical structures, rather than occurring in extended sheets as envisioned in the Davson-Danielli model. We have noted, in discussing individual membrane proteins in this chapter, that in terms of their function, some proteins face only the inner membrane surface, others only the outer sur-

[3] The division between prokaryotic cells and eukaryotic cells represents a major watershed in evolution. According to one view of the evolution of eukaryotes, mitochondria, chloroplasts, and centrioles represent latter-day derivatives of free-living prokaryotic cells that were incorporated into large ameboid cells and survived as a result of an advantageous endosymbiotic relationship. If this is true, it would not be surprising that the membranes of mitochondria and chloroplasts bore some similarity to membranes of prokaryotes, while the structure of the cell membrane, nuclear membrane, and endoplasmic reticulum of eukaryotic cells would be different.

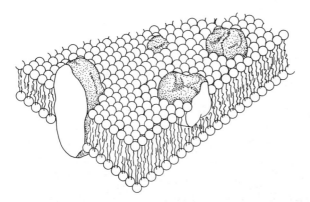

FIGURE 2.23 The lipid mosaic model of membrane structure. The membrane is thought to consist of a fluid, lipid bilayer containing globular protein molecules randomly distributed in the plane of the membrane. Some protein molecules are exposed only to the outer surface and some only to the inner surface, whereas others traverse the membrane and are exposed to both surfaces. (From S. J. Singer and G. L. Nicolson, *Science* 175:723, 1972. Copyright © 1972 by the American Association for the Advancement of Science.)

face, and still others seem to penetrate the membrane and show functional interactions on both membrane surfaces.

Singer has proposed a *fluid mosaic model* of membrane structure (Figure 2.23) which incorporates these considerations. In his model the membrane consists of a mosaic of globular protein molecules suspended in a matrix that is a bimolecular lipid leaflet with the polar lipid head groups facing outward and stabilized by contact with the aqueous medium on either side of the membrane. At physiological temperatures the lipid matrix is fluid, and the protein molecules can move freely in the plane of the membrane. Some proteins are only partly embedded and face only the inner surface, others face only the outer surface, and still others penetrate entirely through the membrane and are exposed on both sides. Membrane proteins are thought to be characterized by extensive domains of surface formed by neutral amino acids, with hydrocarbon side chains that give these domains a very hydrophobic character.

During the past decade a novel technique, called *freeze-etching*, has been used to study membranes. The specimen to be examined first is frozen, then it is fractured with a microtome knife at a shallow angle so that cleavage occurs near the surface, and finally some water is permitted to evaporate by sublimation from the surface. A replica of the surface is made by vacuum evaporation of metal. When the shadow-cast metal replica of fractured membranes is examined by electron microscopy (Figure

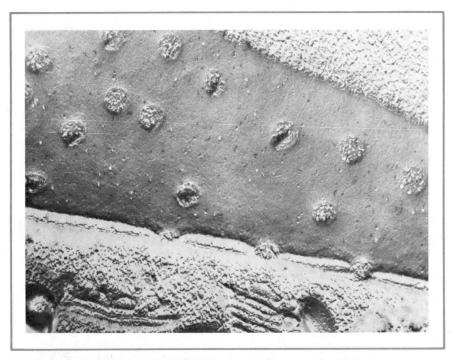

FIGURE 2.24 Electron micrograph of a replica of a frozen-etched fractured membrane showing depressions and masses in the middle section of the membrane which may correspond to globular protein units. The outer surface is at the upper right, and the inner membrane surface is at the bottom of the micrograph. (From D. Branton, *J. Ultrastruct. Res.* 11:407, 1964.)

2.24), a mosaic is seen consisting of a smooth matrix in which are embedded irregularly spaced particles about 85 Å in diameter. This picture fits the fluid mosaic model of membrane structure described above.

The major protein of the human red cell membrane is a sialoglycoprotein, named *glycophorin,* which has been isolated and partially characterized. It consists of a single polypeptide chain with three distinct segments. All of the carbohydrate side chains associated with the protein are located on the N-terminal portion and are exposed on the outer surface of the red cell membrane. This portion of the molecule contains many of the blood group antigens and sites that combine in a highly specific fashion with certain vegetable proteins, called *lectins,* which cause agglutination of red blood cells. The C-terminal portion of the molecule faces the interior of the red blood cell and extends into the cytoplasm. The middle segment of the polypeptide chain that spans the membrane is composed of nonpolar, neutral amino acids; it appears to be connected to the 85 Å intramembranous globular particles found in freeze-etching electron micrographs.

As we said previously, the fluid mosaic model of membrane structure predicts freedom of movement of protein constituents in the plane of the membrane. One type of experiment that demonstrates such translational mobility involves the use of antibodies to cellular surface antigens conjugated to ferritin so that they can be visualized by electron microscopy. In one such experiment, Rh-positive human red blood cells were treated with saturating concentrations of human antibody to the Rh antigen. Under these conditions one antibody molecule combines with a single Rh antigen. The Rh antigen-antibody complex is now reacted with ferritin-conjugated goat antibody, which reacts with each complex. Electron micrographs of the membrane surface show an irregular, random distribution of Rh sites visualized in this way.

Under other conditions the antibody reacts in a multivalent fashion, combining with two sites to form a propagated network of linked antigenic sites. For this to happen the antigenic membrane proteins must move together in the plane of the membrane to form clusters. Such clusters of membrane proteins can again be visualized by the use of ferritin-labeled antibodies.

15 • PINOCYTOSIS

In addition to permeating the cell membrane as individual molecules by diffusion, facilitated diffusion, or active transport, substances can be taken up in bulk by eukaryotic cells by a process called *pinocytosis*. The cell membrane invaginates while a ring of cytoplasm simultaneously projects outward as a pseudopod (Figure 2.25), forming a depression. Finally, the membrane encircles the material and pinches off to form vacuoles, which then move to the interior of the cell. Pinocytosis is an important mechanism for the uptake of large molecules, but quantitatively it is probably not an important transport mechanism for water and low molecular weight solutes.

Specific substances induce pinocytosis, usually proteins, polypeptides, or even amino acids. Inducers usually are cations; proteins are effective only on the acid side of their isoelectric point. Viruses sometimes can induce pinocytosis, but pure nucleic acids and carbohydrates generally are not effective. Once pinocytosis has been started, it continues for 15 min to 1 hr and then often ceases abruptly. Pinocytosis is a metabolically dependent process that increases at higher temperatures, is accompanied by increased oxygen consumption, and can be inhibited by cyanide, carbon monoxide, or other metabolic poisons.

Pinocytosis begins after binding of the inducer to the cell surface. This can be demonstrated by the use of proteins to which fluorescent dyes have been conjugated. It is thought that acidic mucopolysaccharides on

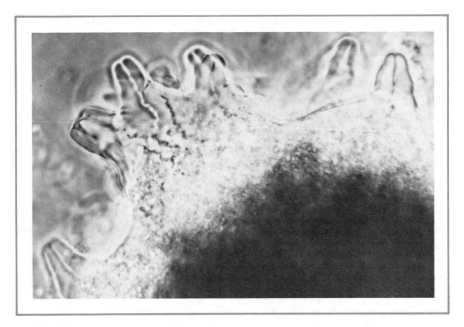

FIGURE 2.25 Light micrograph showing active pinocytosis in an amoeba. (Courtesy of David Prescott)

the cell surface act as specific binding sites. Invagination, pseudopod appearance, and vacuole formation follow. During active pinocytosis, diffusion through the cell membrane of other substances is also increased. For example, amoeba and white blood cells become much more permeable to glucose when pinocytosis is induced with bovine plasma albumin.

In addition to pinocytosis, which involves the formation of small vacuoles, many eukaryotic cells (such as polymorphonuclear white blood cells) can take up large pieces of particulate matter or even bacteria or whole cells by *phagocytosis*. *Chemotaxis* is a prominent feature of phagocytosis by the cells of higher organisms, where phagocytic cells quickly migrate to regions of injury or inflammation. Two serum factors, specific antibodies called *opsonins* and *complement system factors,* are required for very efficient phagocytosis in vertebrates. Phagocytosis also is dependent upon active metabolism and is accompanied by increased oxygen consumption and glucose utilization.

16 • INTERCELLULAR COMMUNICATIONS

Recently it has been found that the junctional complexes between adjacent cells, particularly epithelial cells, contain channels through which molecules as large as several hundred daltons, or even about 1000 daltons, can pass essentially unhindered. As a result of the free movement of water, ions, and low molecular weight solutes through the channels of the junctional complexes, electrical conductivity between adjacent epithelial cells is high (resistance is low). A row of epithelial cells behaves as though the cytoplasm were continuous electrically, forming a core conductor surrounded by an insulating sheath composed of the parts of the epithelial cell membranes exposed to the outside surfaces.

Intercellular channels are formed rapidly after appropriate cells make contact with each other. Using micromanipulation techniques, fully patent intercellular channels form between sponge cells in minutes after they are moved together, and in the case of newt embryo cells, they form in seconds. Formation of intercellular channels requires a significant concentration of Ca^{2+} ions in the extracellular fluid. The intracellular Ca^{2+} concentration must be low, a condition which normally is found owing to active outward transport of Ca^{2+}. When the intracellular Ca^{2+} concentration increases because of inhibition of metabolism and cessation of active Ca^{2+} transport or after microinjection of Ca^{2+}, the intercellular channels become disrupted. Thus intercellular coupling is labile.

Werner Lowenstein, who has carried out much of the experimental work on intercellular channels, found that they are absent or very unstable between cells of many epithelial cancers. Cell fusion of several types of cancer cells with normal cells produces hybrid cells that are not cancerous. Such hybrids, which show contact inhibition of growth, also have regained the capacity to form junctional channels. Thus the lack of intercellular channels may be associated with the uncontrolled growth that typifies cancer cells.

The cell membrane and intracellular membranes are complex cellular components, which use a variety of mechanisms to transport materials in a very selective fashion. Membranes in this way play the key role in maintaining the intracellular environment required for optimal functioning of the life processes and actually divide the cell into several domains with different environments for specialized functions. Specializations of the membrane provides means of adhesion and intercellular communication.

REFERENCES

Branton, D. Fracture Faces of Frozen Membranes. *Proc. Nat. Acad. Sci. U.S.* 55:1048, 1966.

Davson, H., and Danielli, J. F. *The Permeability of Natural Membranes.* New York: Cambridge University Press, 1952.

Dowben, R. M. *Cell Biology*. New York: Harper & Row, 1971.

Hoffman, J. F., ed. *The Cellular Functions of Membrane Transport*. Englewood Cliffs, N.J.: Prentice-Hall, 1964.

Nystrom, R. A. *Membrane Physiology*. Englewood Cliffs, N.J.: Prentice-Hall, 1973.

Rothfield, L. I., ed. *Structure and Function of Biological Membranes*. New York: Academic Press, 1971.

Singer, S. J., and Nicolson, G. L. The Fluid Mosaic Model of the Structure of Cell Membranes. *Science* 175:720, 1972.

Skou, J. C. Enzymatic Basis for Active Transport of Na and K across Cell Membranes. *Physiol. Rev.* 45:596, 1965.

Stein, W. D. *The Movement of Molecules across Cell Membranes*. New York: Academic Press, 1967.

Wallach, D. F. H., and Fisher, H. *The Dynamic Structure of Cell Membranes*. New York: Springer-Verlag, 1971.

NEUROPHYSIOLOGY

Bert Shapiro

Much of animal physiology is concerned with the control of the internal environment. For example, one of the major functions of circulation is to maintain the constancy of the internal environment. If a particularly heavy oxygen demand exists in one part of the body, then more oxygen must be delivered there. This and similar control mechanisms involve nerves and neural tissue. First, the organism must detect change in its internal state. Sense organs exist throughout the body to measure such parameters as temperature of skin and brain, acidity and salinity of the blood, and stretch of the gut. These organs then convert the information to a neural signal. Next, some processing must occur. Finally, a course of action must be dictated. Commands go to muscles or glands to move or secrete. In higher animals many systems are subject to a wide variety of neural control. Heartbeat, sweating, salivation, breathing, and gut movement are all under such control. It would be inconceivable to study any of these systems without also studying nerves.

On a wider scale, animals receive information about the world through sense organs. This information is also processed, decisions are made, and commands are given for action. Here again, each animal's behavior can be understood only through a knowledge of its neurophysiology. The wide variety of behavior, the enormous diversity of niches and life styles found throughout the animal kingdom, are possible only through a corresponding richness and variety of sense organs and central processing. As one example, some electric fish can maneuver, avoid predators, and catch prey in muddy waters. This has been achieved through evolution of an elaborate neural system for initiating electric pulses, and an even more complex system for detecting changes in these pulses and analyzing these changes rapidly to ascertain what sorts of objects exist. Since the nervous system is so important to the survival of every metazoan, we must understand the system to understand that survival.

It is equally important to understand the basic principles

of neural function. To do so we must also grasp those principles underlying membrane transport and function. The study of membrane transport is closely entwined with that of cellular neurophysiology. Both involve many of the same thermodynamic principles, and some of the same researchers study both fields. Similarly, many of the mechanisms studied most extensively in nerves are also found in muscles, glands, and neuroendocrine organs. Muscle cells share with nerve cells a special signal-conducting mechanism. This has been best studied in nerve cells. Muscles and glands are directly excited by neurons and respond to neural transmitters. Neurosecretory organs are modified neurons and probably release their hormones by the same mechanism neurons use for release of transmitter molecules. Neurons are just one of several classes of cells that actively pump ions across their membranes. Finally, like all other cells, neurons are metabolically active. They need glucose, make ATP, and synthesize molecules. In short, any study of the basic function of nerve cells requires a knowledge of the general principles of cell structure and function; in turn such a study contributes more than its share to our knowledge of these principles.

1 • FUNCTION AND EVOLUTION OF NERVOUS SYSTEMS

For most people it is hard to imagine an animal without a nervous system. Coordinated movement and behavior are generally associated with nervous systems. Endocrine systems provide slower responses to changing environment but would hardly be adequate to regulate and sequence even the simplest slow movements of jellyfish. The neuron doctrine of Cajal states that all nervous systems are built up of numerous cells, called *neurons*, which are specialized to transmit and integrate signals. Yet even protozoa, which could not possess a multicellular system, display coordinated behavior. In this case the rapid coordination involves transmission between organelles.

Such communication within cells could not possibly be the ancestor of systems of communication *between* cells. Parker and others, in the early part of this century, suggested that independent effectors are antecedent to true neurons. Nematocysts in coelenterates, oscular myocytes in sponges, many ciliated cells, and smooth muscles in the irises of some vertebrates can indeed function independently. With the exception of the oscular myocytes, however, these effectors exist alongside perfectly good and highly evolved nervous systems in the same organisms. Many muscle systems, such as the vertebrate heart, can function without any nervous input. The electrical characteristics of such muscle membrane greatly resemble those of neurons. Since nervous systems require effectors to act on but are not themselves necessary, Parker probably was correct.

2 • STRUCTURE OF NERVE CELLS

All metazoa possess a nervous system that extends throughout most of the body. The nerve cells, or neurons, contact each other to form a vast complicated network, which in higher organisms achieves staggering complexity. Estimates of the number of neurons in man varies, but 10^{11} is a reasonable figure. Most of these make contact with several cells. Much of neurophysiology is involved in the anatomy and behavior of single cells and small circuits primarily involving a few cells.

Some neurons are so large that it is possible to dissect the cell from the nervous system. This allows one to observe the shape of the cell but not its place in the nervous system. Development of histological techniques for selectively staining some neurons was accomplished in the nineteenth century. One, developed by Camillio Golgi, was refined to a high art by the great neuroanatomist S. Ramon y Cajal, whose classic work remains a source of anatomical detail. Figure 3.1 shows a neuron from the cerebellum stained with the Golgi method. Notice the incredible complexity of its shape. It contrasts sharply with a polyhedral liver cell or an epithelial cell. What is the functional significance of such extensive branchings? Neurons are the site of extreme convergence and divergence of pathways. Many neurons may feed information into a given cell. This is the case here. Also, one neuron may feed information to many other cells. One example is a mammalian motoneuron, which might excite a thousand muscle fibers.

Figure 3.2 shows a variety of vertebrate neurons. Much of the structure of individual cells in the central nervous system (CNS) was revealed using the Golgi method. In addition to the elaborate branching, most of the cells display one, or possibly two, long processes known as *axons,* which are specialized to carry impulses over great distances. Within a brain the axons may run in tracts to affect neurons in a region remote from that of the rest of the cell. Axons may reach astounding lengths, making neurons the longest cells in the animal kingdom. In the vagus nerve some axons run from the skull to part of the small intestine. In a giraffe or whale this is a trip of several meters.

The extensive branching and long axon are characteristic of neurons. Like all good synthetic cells, neurons possess extensive synthetic machinery in the form of ribosomes, often attached to a rough endo-

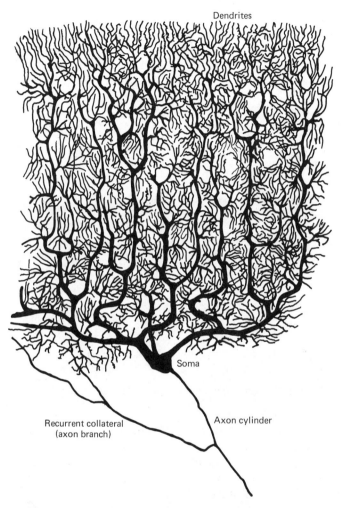

FIGURE 3.1 Purkinje cell from the human cerebellum. Method is Golgi silver stain. (Drawing by Cajal, 1901.)

plasmic reticulum. Although a mature neuron does not undergo mitosis and generally cannot be replaced, it contains a nucleus. This is presumed to be the source of messenger and ribosomal RNA. The region around the nucleus, containing most of the RNA, is called the cell body or *soma*. As can be seen from Figure 3.2, the soma is a clearly defined region in most types of neurons. From the soma arise one to several processes. Neurons are often classified by the number of such processes: uni- or monopolar, such as the cutaneous sensory neuron; bipolar, such as the optic or olfactory neurons; or multipolar, such as Purkinje cells or spinal motoneurons.

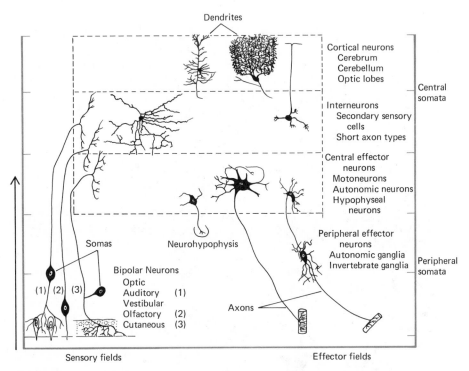

FIGURE 3.2 Variety of neurons in man. Note degree of branching of some axons as well as dendrites. (Redrawn from D. Bodian, *Cold Spring Harbor Symp. Quant. Biol.* 17:1–13, 1952.)

In the past 25 years techniques have developed for putting ultrafine glass pipettes into cells, most often to monitor the electrical potential of the cell (see Section 4 of this chapter). Such pipettes can also be used to inject marker substances into the cell. Cobalt can be injected into a cell, and, since it is electron-dense, the ramifications of that cell can then be followed with the electron microscope. A fluorescent dye, Procion Yellow, may also be injected. It spreads throughout the cell and becomes fixed to the cell proteins. This technique was originally developed using invertebrates and has spread to the mapping of vertebrate neurons. Figure 3.3 shows a typical Procion Yellow picture of a crayfish neuron and an artist's reconstruction of the entire portion of the cell in this ganglion. This illustrates a notable characteristic of *invertebrate* neurons. Typically, they are unipolar with a process connecting the soma to the highly branched portion of the cell. In contrast to multipolar or bipolar cells, the soma is not in the pathway of the signal. In fact in some invertebrates the cell body may be detached from the rest of the cell with little loss of function for weeks.

The branches of the cell that receive signals from other neurons are usually referred to as *dendrites* (from the Greek *dendron,* meaning "tree");

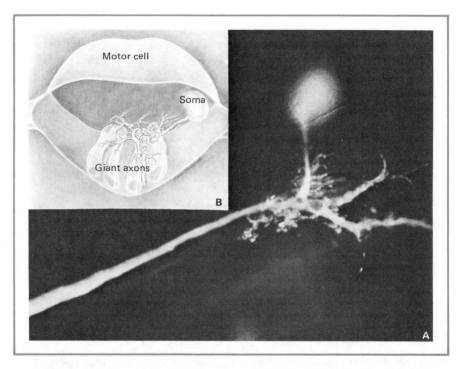

FIGURE 3.3 Motoneuron from crayfish nerve cord. (**A**) The cell has been injected with Procion yellow dye and caused to fluoresce. (**B**) Artist's drawing of part of this cell is based on reconstruction from pictures such as the one on the left. (After A. I. Selverston and D. Kennedy, *Endeavour* 28: 107–113, 1969.)

see Figure 3.2. On the output side of a multi- or bipolar cell the processes are called axons. For vertebrate sensory unipolar cells some people would call the long, thin process peripheral to the soma a dendrite. It is probably best to call this part of the axon and confine the term "dendrite" to the branching region. In invertebrates unipolar axons the terminology can become very confusing because the axon may extend over several segmental ganglia and branch in each one to receive inputs and deliver outputs. In this case the term "dendrite" should be confined to the branched area designed to receive input.

Coelenterates have many bipolar cells in their nerve nets. In higher invertebrates unipolar cells predominate. Although a few multipolar cells exist, such as the crayfish stretch receptors, they are scarce. The vertebrate CNS is filled with multipolar cells. Some of the trends within this form have been described by Cajal. Figure 3.4 shows the increasing complexity of pyramidal cells of the cortex paralleling the increased complexity of the cortex itself. One type of vertebrate multipolar neuron, often called *association neurons,* is distinguished by the presence of short axons and den-

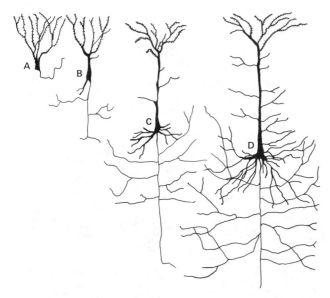

FIGURE 3.4   Complexity of comparable neurons from different vertebrates. Pyramidal cells from (**A**) frog, (**B**) lizard, (**C**) rat, and (**D**) man.

drites. Electron microscopy shows the presence in some of these processes of features that are both pre- and postsynaptic—that is, axonal and dendritic. Cajal has stated that the biggest difference between the cortex of monkey and man is the number of these association neurons.

There are many well-recognized classes of multipolar cells. Purkinje fibers are one. They are distinguished by the large number of dendritic branches and the flat nature of the branching. All Purkinje cells have a dendritic network oriented in a single plane, like an espaliered tree. In vertebrates the specific cell types are confined to specific regions and layers of the CNS, and they generally have similar functions. All the millions of Purkinje cells are located in the same layer of the cerebellum. They receive input from climbing fibers from cells outside the cerebellum and from basket and granule cells within the cerebellum. The axons end both within the cerebellum and in other specific regions of the brain but always inhibit their target neurons.

2.1   Giant Cells

There is a continuous spectrum of axon and soma diameter in the animal kingdom. The largest neurons are famous. Squid giant axons can attain a diameter of 1 mm. Many snails—marine, freshwater, and terrestrial —display huge somas up to 1 mm in diameter. The function of the large axons throughout the animal kingdom is to allow faster signal conduction. Calculations based on the mechanism of impulse conduction show that the velocity increases approximately as the square root of the diameter. Thus these axons are useful in escape responses where speed is essential. The reason for the large cell bodies is not known. In some cases the huge nucleus can be shown to be polyploid, but the function is uncertain.

In the vertebrates the rule that the thicker the axon the faster the signal also holds true. But for myelinated axons such large diameters are not needed to achieve fast conduction. In general few somas in the vertebrates even approach the size of medium-sized invertebrate cell bodies.

Whatever the primary purpose of large neurons, an additional use exists. It is easier to stick wires and micropipettes into them. It is even possible to dissect out single somas and do involved chemical analyses. And during an experiment individual cells are visible in a dissecting microscope or even to the naked eye. The pantheon of neurophysiological preparations is filled with the likes of squid and lobster giant axons and snail and leech giant somas.

2.2 Glia

The nervous systems of nearly all animals are not composed of neurons alone. Many other cells are in proximity to the neurons. These cells, called *neuroglia* or simply *glia,* may nearly completely surround all somas in a region of ganglion or spinal cord. Peripheral nerves also contain many glial cells. Myelinated and giant axons are individually surrounded by several glia. Vertebrate C fibers and the smaller invertebrate axons may run in bundles surrounded by a glial cell so that the inner axons are not in contact with any glia.

There is a clear series of increasing glial diversity and volume as we ascend the phylogenetic scale. In general, the more complex the nervous system the more varied and numerous the glia. Coelenterate nerve nets, planarian nerve fibers, and enteropneust giant fibers are naked. Coelenterate ganglia have few glia. In many lower invertebrates there is only one type of glia. In insects and cephalopods there are several distinct types which occupy a significant percentage of the space in the central nervous system. In mammals glial cells greatly outnumber neurons and occupy 40 to 50 percent of the brain volume.

The most remarkable glial cells are those that form myelin sheaths around some vertebrate axons. Figure 3.5 shows such a sheath. It is composed of several very fine membrane layers enclosing little cytoplasm. This may amount to several thousand stacked membranes in the sheath of a large axon. The myelin does not extend continuously over the entire length of the axon but, rather, is regularly interrupted by a region of naked

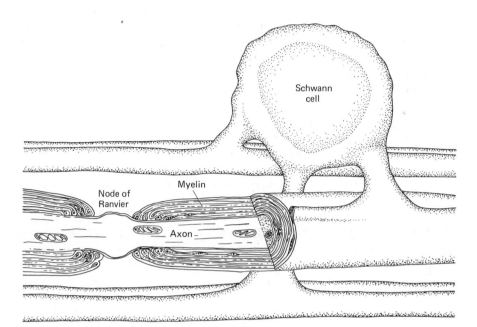

FIGURE 3.5 One reconstruction of myelination in the brain. Layering of Schwann cell membrane between nodes is illustrated in partial cutaway. (Redrawn from M. B. Bunge, P. B. Bunge, and H. Ris, *J. Biophys. Biochem. Cytol.* 10:67–94, 1961. With permission of The Rockefeller University.)

or nearly naked axon membrane. The spacing of these unmyelinated regions, called *nodes of Ranvier,* is regular. The thicker the axon the longer the length between nodes. Axon branches and in the CNS, synaptic bulbs, occur only at the nodes of Ranvier and the unmyelinated terminals.

Roles of Glia.   We know very little about glia. Since they make up so much of a nervous system, one might suppose they have a role in signaling or information processing. Such a role has never been shown. Studies of the electrical activities of glia in both leech and salamander show that they undergo slow electrical changes. These are essentially passive responses to the change in ionic composition of the extracellular fluid. Such changes are due to activity in neighboring neurons. Glia may be in extremely close contact with each other, but they are separated from neurons by 150 Å. There is a possibility that their electrical activity affects regions of the nervous system, but it has not been demonstrated. Recordings of gross potentials from large portions of the nervous system, as in electroretinograms (ERGs), may detect the passive electrical changes in glial cells. There is some evidence that glia play a role in neuronal nutrition in insects. Yet here too the evidence is scanty.

Less exciting, but more evident, is the obvious function of glia as mechanical supports for neurons. Electrical isolation and insulation is also provided. This is also a function of myelination. The stacking of membranes increases the resistance between the inside of the axon (axoplasm) and the extracellular fluid. At the same time it reduces the capacitance between the two (see Section 4). This, combined with the regular interruption of the myelin by nodes of Ranvier, serves to increase the axon's conduction velocity many times over. This advantage is so important that it is doubtful that behavior as complex as ours could have ever evolved without myelin.

3 • AXONAL TRANSPORT

Nerve cells have their synthetic machinery for proteins confined to a region around the nucleus, and this raises a problem. How does material synthesized in the soma get to the end of the axon? Diffusion is too slow over such distances. Diffusion time for a meter is measured in years. Therefore a means must exist to move material down the axon. By introducing a radioactively labeled substance into one end of the neuron, it is possible to show an active transport down the axon. For a great variety of species axonal transport, from soma to axon terminal, moves in two phases. A slow phase, typically about 1 to 2 mm/day, transports most types of substances in the axon. It may represent the gradual gross replacement of the cell. A faster phase, typically 5 to 50 cm/day, includes most of the protein, mitochondria, and vesicles. Also included in some cases are suspected neurotransmitters, perhaps bound to larger particles.

In some cases, particularly when an axon is cut, changes spread slowly in the opposite direction. This has led some people to suspect a transport in the opposite, retrograde, direction; in fact this has been shown in a number of cases. A typical rate for axonal transport in the visual system of the chick is about 7 cm/day.

4 • ORIGIN OF BIOELECTRICITY

There is a voltage across the membranes of all cells. Transmission in nerves and muscles as well as contraction in muscles involves voltage changes. The discovery of these facts entailed centuries of work and was closely entwined with the elucidation of the nature of electricity. Some of the giants in the study of bioelectricity were famous in the history of physics: Galvani, Volta, von Humbolt, and Helmholtz. Although the shock delivered by electric fish has been known since prehistory, its identity with the electricity produced by rubbing two objects together was not discovered until the nineteenth century. The ability of electric currents to stimulate a muscle directly or indirectly through its nerve was demonstrated by Galvani. But the interpretations of his experiments were debated fervently until von Humbolt untangled the wealth of phenomena described by Galvani.

One of the phenomena observed by Galvani was that of injury potentials. We know that muscle cells have a potential across them; the inside is negative relative to the outside. Cell injury produces a short circuit so that positive current flows through the extracellular fluid from the uninjured to the injured region and then into the cell. However, measurement of the potential between the injured and intact regions cannot give the full voltage between the inside and outside.

In the 1930s Osterhout and Blinks, with many co-workers, succeeded in measuring the voltages across the cell of several large algal cells: principally *Nitella, Halicystis,* and *Valonia*. The giant cells are up to 15 cm long! They can be impaled with a glass capillary and the puncture allowed to heal (this may take a day). Then the potential between the cell vacuole and the outside could be measured for *over a week*. The magnitude and sign of the potentials varied with the species, but the establishment of the existence of a stable potential was important. Many of the techniques and experiments that later became so important in neurophysiology were first performed on plant cells.

What causes these potentials? The principal cause of the voltage is a phenomenon by no means limited to living systems. Two things are required: (1) dissimilar solutions are separated by a membrane, and (2) the membrane is selectively permeable to only some of the ionic species existing in the solutions (see Chapter 2, Section 7). A simple case is shown in Figure 3.6. On the left side of the flask is a solution containing 100 mM KCl and 10 mM NaCl. On the right side is a solution of 10 mM KCl and 100 mM NaCl. If the membrane between the two is not permeable to either Na^+, K^+, or Cl^-, then no potential exists. Now let us open up channels in the membrane that are selective. K^+ can pass through them, but Na^+ and Cl^- cannot. K^+ ions will tend to diffuse down the concentration gradient—from a region of high density to one of lower density.

As the potassium cations, each positively charged, diffuse to the right, they create a charge imbalance on both sides; the left side becomes negative and the right positive. This voltage tends to pull the positive ions back to the left. The more potassium ions that flow from left to right the greater the voltage created. The greater the voltage the greater the force pulling back the potassium ions. Eventually an equilibrium is achieved.

There are several ways of deriving the equilibrium value. One way is from the work required to move a mole of particles across the membrane. At equilibrium no net work is required to move particles across the

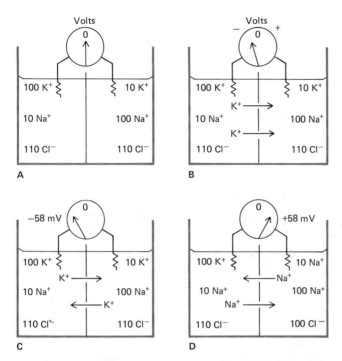

FIGURE 3.6 Potential (voltage) recorded between two solutions separated by selectively permeable membrane. (**A**) Membrane impermeable to K⁺, Na⁺, Cl⁻. (**B**) Membrane just made permeable to K⁺. (**C**) Membrane permeable to K⁺ and at equilibrium. (**D**) Membrane permeable to Na⁺ and at equilibrium.

membrane. If we imagine moving potassium ions from right to left, we must oppose the tendency to diffuse in the opposite direction. This requires work and is given, from thermodynamic considerations, by the expression

$$RT \ln \frac{[K^+]_{\text{left}}}{[K^+]_{\text{right}}}$$

where R is the universal gas constant and T is the absolute temperature. The work involved in moving a mole of particles across the electric gradient of the membrane is zFE, where z is the number of charges per particle, F Faraday's constant (the number of coulombs per mole, or 96,500), and E the electrical potential gradient.

At equilibrium the sum of the electrical and concentrating work should be 0.

$$zFE + RT \ln \frac{[K^+]_{\text{left}}}{[K^+]_{\text{right}}} = 0$$

or

$$E = -\frac{RT}{zF} \ln \frac{[K^+]_{\text{left}}}{[K^+]_{\text{right}}}$$

The latter is known as the Nernst equation (see Chapter 2, Section 8). Here E is defined as the potential of the left side minus that of the right:

$$E = E_{\text{left}} - E_{\text{right}}$$

For ease of handling this is often converted to $\log_{10}$, and the values of the constants are put into the equation as follows:

$$E = \frac{58}{z} \log_{10} \frac{[\text{ion}]_{\text{left}}}{[\text{ion}]_{\text{right}}}$$

for a temperature of 18°C.

For our original case, a gradient of K^+ ions with a membrane selectively permeable to K^+, the left-hand side will become 58 mV more negative than the right-hand side.

What would have happened had we made the membrane selectively permeable to sodium rather than potassium ions? Sodium ions are more concentrated on the right side than on the left. They would have diffused to the left and made the left-hand side more positive. E would be + 58 mV. Had we made the membrane selective for Cl^- alone, then there would be no voltage across it because the concentration of the ion is the same on both sides.

The first question asked by most students is, "If potassium diffuses from left to right, doesn't this change the concentration of the potassium ions on both sides?" The answer is yes, but not very much. What determines how many ions cross the membrane before an equilibrium is achieved? This is a property of the membrane separating the two solutions. One of the electrical properties of any object is its capacitance, which is defined as

$$C = \frac{Q}{V}$$

where Q is the charge on the capacitor and V is the voltage across it.

This means if we know the voltage and capacitance of the membrane, we know how much charge is required to produce that voltage. All natural, as well as most synthetic, lipid membranes (see Chapter 2, Section 3) have a capacitance of 10^{-6} F/cm². With this value, a potential of 58 mV, and assuming the ions are monovalent, we can calculate the charge separation. The answer is 6×10^{-13} mole/cm^2 of membrane. Thus if 1 cm^2 of membrane separated two solutions of 1 liter each, the concentration of potassium ions on the left after equilibrium would be $100 - (6 \times 10^{-10})$ mM. That on the right would be $100 + (6 \times 10^{-10})$ mM. If the volumes separated were only 1 ml each, the concentrations would be $100 - (6 \times 10^{-7})$ mM and $10 + (6 \times 10^{-7})$, respectively.

In real life no membrane is completely selective. A variety of different channels exists in the membrane, selective for different ions (see Chapter 2, Section 7). None of these channels is perfectly selective. Clearly the actual voltage measured across such a system is some sort of weighted average of the potentials due to each ion alone. The more permeable the membrane is to a specific ion, say, K$^+$, the closer the potential will approach E_K, the Nernst potential for potassium. A commonly used expression for the membrane potential of a system involving several ionic species is the Goldman, or Goldman-Hodgkin-Katz equation (see Chapter 2, Section 8):

$$E = -\frac{RT}{zF} \ln \frac{P_K[\text{K}^+]_{\text{left}} + P_{Na}[\text{Na}^+]_{\text{left}} \text{ and } P_{Cl}[\text{Cl}^-]_{\text{right}} \cdots}{P_K[\text{K}^+]_{\text{right}} + P_{Na}[\text{Na}^+]_{\text{right}} + P_{Cl}[\text{Cl}^-]_{\text{left}} \cdots}$$

Here P_K, P_{Na}, and P_{Cl} are the specific permeabilities of the membrane to the three ions involved. If other ions are significantly permeable and present on one or both sides of the membrane, then they must be included.

Let us take our previous example but in this case make the membrane equally permeable to K$^+$, Na$^+$, and Cl$^-$ ions. Then the solution for the potential is 0 mV. At this potential Cl$^-$ is at equilibrium and has no tendency to cross the membrane. Potassium diffuses from left to right and sodium from right to left. Since the gradients and permeabilities for the cations are equal, the flow rates are equal. Therefore, although there is a net flow of *each* cation, there is no change in voltage. The rates of *total* charge movements are equal and opposite. But the voltage does not attain an equilibrium! The system is running down, and eventually the concentration of all species on either side of the membrane will be equal. The potential given by the G-H-K equation is a steady-state one. It will change only slowly as the ionic concentrations change, but it is not an equilibrium.

Intracellular fluid is nearly always more negative than the extracellular medium. The development of techniques to measure the intracellular potential of nerves and muscles was central to the establishment of modern neurophysiology. The most commonly used technique is illustrated in Figure 3.7. This is merely a refinement of the capillary tubing method

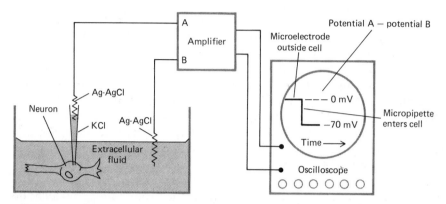

FIGURE 3.7 Typical method of recording electrical events from inside of cell. Conventionally the potential is given as the potential inside minus the potential outside, so that at rest neurons are negative. Oscilloscope trace shows record of potential as micropipette penetrates the cell. Ag-AgCl and KCl are usually used to reduce potentials between the electrode and the recording medium.

used to record from giant algal cells. For animal cells the tubing is heated and pulled down to a fine tip. The external diameter of such a tip is typically 0.1 to 0.5 μ. The bore in the tubing is filled with a conducting solution such as 3 M KCl. Ag-AgCl wire leads from the tubing, and they both extend to a special amplifier. The signal from the amplifier is displayed on an oscilloscope. This allows one to follow the entry of the micropipette into the cell and any subsequent changes in intracellular potential. Such a system often can resolve potential changes of 100 μV and can measure events occurring in less than 100 μsec.

Other important methods of intracellular recording are less widely used. For giant axons, such as those in squid and the marine worm *Myxicola,* a wire may be pushed down the axis of the axon cylinder. In fact a pair of such wires is in standard use for most experiments. Another technique involves measuring the potential between an injured and intact area of a muscle or nerve fiber. However, the problem of short-circuiting between the two regions must be solved. This is done by insulating the area between the electrodes with a nonconducting medium, such as isotonic sucrose, and by using electronic instruments that guarantee minimal leakage of current between the two electrodes.

As the micropipette enters a nerve or muscle cell, there is a sharp and pronounced drop in the potential. After a short period—from a few seconds to an hour—the cell membrane seals around the pipette, and a stable voltage can be measured. The constancy of the potential and the fact that it is associated with no neural signaling gives rise to its name, the *resting potential.* This is as much a misnomer as is the old "resting phase" for the period in a cell cycle between mitotic movements.

The origin of most of the resting potential is the existence of gradients in concentration of several species of ion. Table 3.1 shows for some cells the intra- and extracellular concentrations of some important ions, the Nernst potentials for these ions, and the resting potentials observed (see Chapter 2, Section 1, and Chapter 6, Section 1.3). Several generalizations can be made. If the cell membranes were selective for either K^+ or Cl^- alone, the resting potentials would be negative. Were they selective to Na^+ or Ca^{2+}, the potentials would be positive. (The estimates for the effective concentration or, more precisely, activity of intracellular calcium is difficult. A large part of the total intracellular calcium is bound and not free in solution.) The resting potentials are negative. But in these and a great number of other neurons and muscle fibers, the potential is less negative than E_K, the Nernst potential for potassium. It is rarely identical with E_{Cl}. It would appear from this that the membrane is permeable to several ions but more so to K^+ and/or Cl^- than Na^+ and Ca^{2+}. Other experiments indeed show that cells are usually most permeable to K^+, although Cl^- permeability can also be significant. At rest there is a small but measurable sodium permeability. Calcium permeability varies a lot. Paramecium, for example, is moderately permeable to calcium.

The failure of the resting potential to always follow the Nernst relationship for potassium requires comment. Figure 3.8 shows curves of resting membrane potential, E_m, for one preparation as a function of potassium concentration. The curved lines are drawn using the Goldman equation and the straight lines using the Nernst equation for potassium. For high external potassium the Nernst relationship fits fairly well. For a wider range of $[K^+]$ the G-H-K equation gives a better fit. Even this fails to fit the data for the entire range measured. We now know this is because the ratio $P_K:P_{Na}:P_{Cl}$ is not constant. P_K changes, increasing when the resting potential decreases (that is, becomes more positive).

The resting potentials are primarily due to diffusion of a few ionic species. Yet this is not an equilibrium. A cell permeable to sodium and potassium ions will run down, accumulating sodium and losing potassium. To maintain a constant intracellular composition, the cell must pump ions across the membrane. This movement of sodium ions up a steep electrochemical gradient requires work and thus ATP. (This is discussed in greater detail in Chapter 2, Section 12.) Evidence from red blood cells indicates that the sodium pump also pumps potassium into the cell. A common figure is that three sodium ions are extruded and two potassium ions taken up per ATP split. We might think of the cell as a battery with the resting potential as its voltage. However, it is a slightly leaky battery and requires continuous charging. In the case of the cell the charging process, the pump, removes more sodium than it replaces with potassium. Thus the cell loses positive charge and becomes more negative. The pump is called *electrogenic* because it generates a potential that contributes

TABLE 3.1 Ion Concentrations, Nernst Potentials, and Resting Potentials of Some Important Cells

Preparation	[K+] Potassium Concentration, mM	E_K Potassium Potential, mV	[Na+] Sodium Concentration, mM	E_{Na} Sodium Potential, mV	[Ca2+] Calcium Concentration, mM	E_{Ca^+} Calcium Potential, mV	[Cl−] Chloride Concentration, mM	E_{Cl} Chloride Potential, mV	E_m Resting membrane potential, mV
Crab muscle	344	−84.3	65	+49.7	(7)	>23.1	140	−33.2	−70
"Blood"	12.1		468		17.5		524		
Squid axon	335	−88.4	52	53.6	0.4	>81.1	135	−36.7	−60
"Blood"	10		445		10		580		
Frog muscle	124	−101	3.6	85.4	(4.9)	>+10	1.5	−99.4	−90
"Blood"	2.2		109		2.1		77.5		
Human erythrocyte	136	−81.5	19	52.9	—	>0	78	−9.11	−8
Plasma	5.35		155		3.2		112		

Notes:

[ion]: Concentration of ions inside or outside several animal cells.

E_{ion}: The Nernst potential for that ion across that membrane. This is the potential difference across the membrane that would prevail if the membrane were permeable only to that ion species. The potential difference is defined as the potential inside the cell minus that outside.

E_m: The actual potential difference observed across the cell membrane.

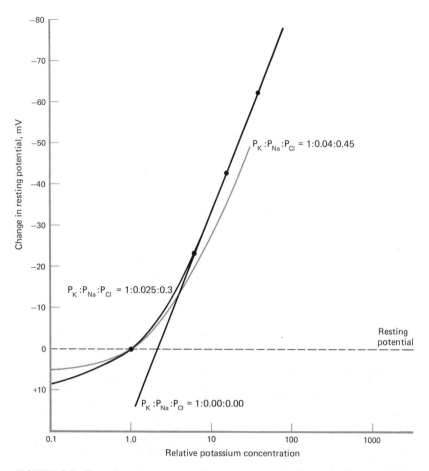

FIGURE 3.8 The relationship between the resting potential of a squid giant axon and the concentration of K$^+$ in the bathing fluid. The straight line is what we would expect if the cell was permeable to K$^+$ alone (Nernst equation). The two curved lines are what we would expect if the cell was permeable to K$^+$, Cl$^-$, and Na$^+$ in given ratios (Goldman equation). The potential follows neither perfectly because the permeability ratio changes as the potential across the membrane changes.

to the total resting potential. The size of this electrogenic component varies with the preparation. Some typical values are: squid giant axon, 1.8 mV; leech soma, 5 mV; nudibranch snail giant soma, 15 mV; and some vertebrate smooth muscle, 10 to 20 mV. In the latter case the electrogenic pump may produce half the resting potential. Action potentials, the conducted impulses passing along axons, are produced by the opening of special channels in the membrane and are thus *not* due directly to the electrogenic pump. However, during an action potential the rate of sodium entry is much greater than at rest, so the cell demands faster sodium extrusion to maintain constancy. Even cells in which the pump makes only a small contribution to the resting potential may show a pronounced hyperpolarization after a volley of action potentials, and this has been shown in some cases to reflect increased activity of the electrogenic pump. For active neurons many more pump sites than channels are required since each pump extrudes sodium ions more slowly (10^2/sec) than they enter through each channel during action potentials (10^8/sec).

5 • PASSIVE SPREAD OF ELECTRICITY

Signals in axons involve changes in potential across the cell membranes. Suppose that we changed the potential across the cell body of a large unmyelinated neuron, reducing it to 0 mV. Such a change would require passing a current across the membrane sufficient to neutralize the charge on the membrane capacity. How would such a change spread down the axon? Qualitatively you might judge that the amount of axon membrane is important, since that determines the charge which must be changed. Also, the electrical resistance of the membrane must be important, because without it current would flow immediately through the membrane instead of spreading along the axon. If we apply current at a particular point on the membrane and measure the induced voltages, we find that the voltage change decreases as we go down the axon and away from the site of current application. If we plot change in voltage versus distance, we find that voltage decreases exponentially. The distance at which the voltage is decreased to 37 percent (that is, $1/e$) of its original value is known as the space constant of the membrane, or λ. As we progress along the axon, not only is the voltage change smaller but the time it takes to reach its peak level is greater. This type of voltage spread is called *passive* or *electrotonic*.

The situation is analogous to the spread of heat along a metal. If we hold one end of a knitting needle in a flame, the other end may stay cool. Partway down the needle is warmed somewhat, and it takes a while for it to reach even that temperature. It is a poor way to send a signal over any distance of needle, and passive electric spread is equally poor for axons. It is useful for distances smaller than λ. Most of the signal spread is passive in dendrites and cell bodies as well as in certain sense and muscle cells. Therefore much work has been devoted to determining λ and its dependence on other factors. To a reasonable approximation,

$$\lambda = \sqrt{\frac{r_m}{r_i}}$$

where r_m and r_i are the resistances per unit length of membrane and axoplasm, respectively.

6 • ACTION POTENTIAL MECHANISMS

Signal transmission over a distance in a nervous system requires action potentials. Figure 3.9A shows the membrane potential changes that constitute the action potential. The cell depolarizes (becomes less internally negative) and generally reverses polarity, becoming internally positive. It then returns to the resting potential, although it may exhibit a hyper-polarized (more negative) phase before it finally returns. This event, the action potential or spike, is explosive and stereotyped. It also exhibits a clear threshold. Depolarization below a critical level fails to initiate spikes; above this level a standard spike is initiated whose form and size is not dependent on the shape or size of the stimulus.[1] Most important, the spike propagates without decrement.

Since the spike is large (about 100 mV) and brief, it entails a pronounced local current into one area of the axon. Later, after conduction, the spike is located farther along the axon, and the current enters the membranes there. Figure 3.9D shows the record obtained from the two "extracellular" electrodes hooked around an axon in the air. When the action potential wave is under the first electrode, it is negative, since positive charges (sodium ions) are leaving the bath near the electrode and entering the cell. When the wave is under the second electrode it is more negative than under the first electrode. When the wave has passed completely, the electrodes are at the same potential. It is common practice to display the voltage as the difference between the potentials at the two electrodes. Hence a biphasic wave would be observed. Extracellular recording still is quite useful. It is only possible to impale a small fraction of the types of neurons that exist. Often interesting events occur in regions of the neuron significantly removed from the impaled portion.

At this point we may be able to make an intelligent guess about the cause of the action potential. The cell initially is mainly permeable to potassium ions. Since the potential approaches the Nernst potential for sodium during the spike, it must indicate a high sodium permeability at that time. When it returns to resting condition, the sodium permeability must also return. Since the Nernst potential for calcium is also positive, a high calcium permeability is also possible. Both Ca^{2+} and Na^+ are suffi-ciently abundant in the extracellular fluid that there is no problem in their providing enough current to depolarize the membrane so rapidly. Let us go even further to determine what opens up sodium-selective channels. To fire the cell we must depolarize it. Somehow a decrease in the potential difference across the membrane must open the sodium channels. For a small change in voltage not many sodium channels are opened. When the

---

[1] This is not strictly true for a nonpropagating action potential. But any differences at the point of initiation disappear after the spike has traveled several length constants.

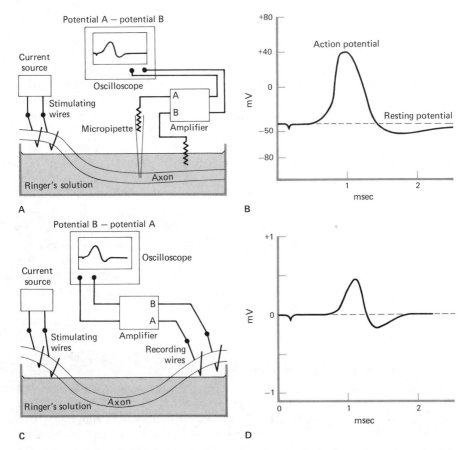

FIGURE 3.9 (**A**) and (**C**) Schematic representation of two methods of recording action potentials from a squid giant axon. (**A**) Intracellular recording as in Figure 3.7. Axon is best stimulated if it is removed from the bathing solution in the region of stimulation. (**C**) As in (**A**), but recording is with two extracellular wire electrodes. The signal is best if the axon is lifted out of the bath in the recording region. (**B**) and (**D**) Action potentials recorded from axons as shown in (**A**) and (**C**). (**B**) Recorded intracellularly. Slight downward deflection indicates time of stimulus. Delay to action potential caused by wave taking time to reach the recording electrode. Cell depolarizes to the point of becoming positive and then returns to resting level. Hyperpolarization to −65 mV, seen here, is not present in all types of axons and muscle fibers. (**D**) Potential recorded extracellularly from same axon. The two phases are due to the fact that the wave passes over first electrode A and then electrode B. Absolute magnitude depends on wire electrode contact with the axon and resistance between two recording electrodes.

stimulus stops, the cell obeys the Goldman equation and hyperpolarizes. This closes the sodium channels back down, and we are back where we started. For a voltage change above threshold many sodium channels are opened. Now if the cell obeys the Goldman equation, it will depolarize more. This in turn opens up more sodium channels, and the explosion is under way. It is much like any other explosion in this respect (see Chapter 1, Section 4). Imagine some scattered gunpowder. If we heat up some of it, we may cause a bit of it to burn, but that will not generate enough heat to ignite more gunpowder. If we ignite a sufficient amount, it heats up faster than the heat can be dissipated. This ignites more gunpowder, in turn heating yet more, and we have an explosion.

The problem with an explosion is that it is hard to study. How do we find out how voltage opens Na channels if this event in turn changes the voltage? The trick is to use electronic feedback circuits to maintain the voltage of a patch of membrane at the desired level. Then the parameter measured is the current crossing that patch of membrane at that voltage. This technique, now common, is known as *voltage clamping*. Since current is carried in solution by ions, we are measuring the flow of ions through the membrane. If sodium channels open up, we can watch the sodium current move down the sodium electrochemical gradient. Potassium current is the passage of potassium ions across the membrane down their electrochemical gradient. If the voltage is constant, then so is each electrochemical gradient. Thus the currents tell us relatively how many sodium and other channels are open at various times. Such analyses and techniques are very sophisticated indeed. Nevertheless, this and much more was achieved in the elegant experiments of Hodgkin and Huxley on squid giant axons. The analyses were carried to the point of writing and solving a system of simultaneous differential equations (see Chapter 1, Section 6). When the equations are solved for the response to a large stimulus, the result is an action potential. The threshold and a host of other distinctive features of action potentials and axonal membrane in general are derivable from these equations. The analyses are as elegant as the experiments!

The most important features of the analyses are the following.

1. During an action potential, ions move down their electrochemical gradients. This means, for example, that sodium moves into the cell since the inside of the cell is negative and sodium is more concentrated extracellularly. Naturally, this cannot go on forever unless the ionic gradients are maintained. This is the job of the sodium-potassium pump.

2. The permeability of the membrane to some ionic species can change. The change is due to a change in voltage across the membrane. Depolarization—making the cell more positive inside—initially increases the permeability to sodium. This was described previously and is what

makes the action potential explosive. Depolarization also increases the permeability to potassium. This should make the cell stay near E_K, which is near the resting potential. The reason an action potential can still take place is that the increase in potassium permeability takes longer to occur. A sudden change in voltage first increases sodium permeability and then later potassium permeability.

3. If only the phenomena just described were true, an action potential would be a once-in-a-lifetime event. The cell would never return to its resting potential. But a change in voltage changes the sodium permeability in another way as well. It causes the permeability to shut down. Thus depolarization has *two* effects on sodium permeability: it increases it rapidly and decreases it slowly. These two processes occur in parallel. They are often likened to two gates in series in a sodium channel. Both must be open for sodium to enter or leave the cell. Depolarization opens one in less than a millisecond while closing the other in more than a millisecond. Figure 3.10A shows the sodium and potassium conductances in a squid axon

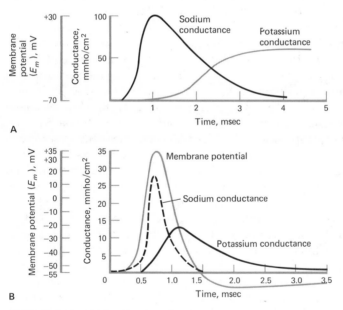

A

B

FIGURE 3.10   (A) Conductance changes in a squid giant axon following changes in the voltage across a membrane. Notice that the sodium conductance declines even though voltage is constant. Potassium conductance remains constant after its initial increase. **(B)** Conductance changes in a patch of squid membrane during an action potential. The increase in sodium conductance initiates the action potential. The reduction in sodium conductance and increase in potassium conductance are responsible for the hyperpolarization at the end of the action potential.

maintained at a voltage sufficient to produce a large sodium current. Here the electrical term "conductance," or g, is used. This is merely the reciprocal of ohms so that Ohm's law would be $I = gE$. This is a measure of how readily the ions flow with a given voltage pushing them. Ionic conductances are not identical to permeabilities but are nearly proportional to them under ordinary conditions. Since the experiments deal in volts and amperes, conductance in mhos is what we can measure directly.

Figure 3.10B shows the conductance for sodium and potassium ions during an action potential. Sodium conductance has decreased to near zero before the spike is even over. Meanwhile, potassium conductance is high, and the membrane hyperpolarizes. We pointed out that the sodium conductance declines because of the second, slower inactivating process. Just as this inactivation is relatively slow to block sodium permeability, it is slow to unblock after the action potential is over. There is a period in which another action potential cannot be generated. This is called the *absolute refractory period*. A longer phase, during which the threshold is high but returning to normal, is called the *relative refractory period*. These phases are a consequence of the slow inactivation process (preventing sodium permeability from increasing again) and the slow decrease in potassium conductance (tending to keep the cell from being depolarized). The refractoriness limits the frequency of action potentials. Although a few cells may be able to fire at 1000/sec, most cannot exceed a few hundred per second. Some slow cells cannot even fire at 100/sec. Action potentials are the means of coding information in the nervous system. This limits the rate at which information can be transmitted.

It is now possible to extrude the axoplasm from squid axons and perfuse the cell both internally and externally. Electron microscopy shows that such preparations may be mere membrane-bound bags. These isolated membranes give excellent action potentials and currents. The nature of these currents, due to passive flows of ions down their gradients, is confirmed by such experiments. One advantage of the internally perfused axon is that a great variety of agents can be applied on either side of the membrane. Local anesthetics, for example procaine (novocaine), have been shown to act by decreasing sodium permeability and often also potassium permeability. Present belief is that such anesthetics, often weak bases or acids, must actually enter the cell before acting (see Chapter 2, Section 10). Many experiments have tended to confirm the separate identity of the paths for sodium and potassium ions. Most dramatic are those with a substance from the puffer fish known as *tetrodotoxin* (TTX). TTX, applied to the axon, blocks all sodium currents. The tetraethylammonium ion (TEA) can specifically block potassium currents.

How do the ions cross the membrane? We have reiterated the fact that the forces driving them across are electrochemical. We might wonder whether they crossattached to some special enzyme or carrier. Several lines of evidence indicate that this is not the case. Instead, the ions cross through channels in the membrane (see Chapter 2, Section 3). These channels are small—of ionic dimensions. There are separate channels for sodium and potassium ions. These two types of channel are selective, but the selectivity is imperfect (see Chapter 2, Section 7). For example, lithium traverses sodium channels about as well as does sodium. The selectivity of these and other channels is essential for the generation of action and synaptic potentials. The number of channels has been estimated by experiments with radioactively labeled tetrodotoxin, which binds moderately specifically to sodium channels. The best estimate for unmyelinated rabbit axons is 27 sodium channels/μ^2, crab and lobster axons are somewhat less, and garfish olfactory nerve is 3 sites/μ^2.

How does an action potential propagate? Imagine a patch of membrane depolarizing during an action potential. This occurs because sodium ions enter through open sodium channels. The current generated by this action potential spreads down the axon. This is passive spread. It results in depolarization farther down the axon. If we imagine the axon having a resting potential of −60 mV, threshold of −40 mV, and action potential to +40 mV, we can make a simple estimate. The 100-mV spike decays passively to 37 mV after one space constant. Thus one λ away it will be sufficiently depolarized to generate an action potential. An action potential in this patch in turn can serve as a source of depolarization. Actually the process is continuous. The wave of depolarization travels at a constant velocity down the axon. We can extend the gunpowder analogy by thinking of a fuse. A long line of gunpowder is made and one end ignited. The heat generated from explosions in one region heats up the neighboring region above its threshold. The wave of ignition then travels along the fuse.

The rate of conduction is a function of the space constant, since this describes how well depolarizations spread along the axon. $\lambda = \sqrt{r_m/r_i}$, which in turn is proportional to the diameter. Thus conduction velocity is roughly proportional to the square root of fiber diameter. For invertebrates this means axons must be 100 times as wide to conduct 10 times as fast. This is the reason why such huge axons exist in the invertebrates. In squid the giant axon conducts an impulse to distal parts of the mantle musculature. This then causes a contraction of the mantle cavity. The axons to the more proximal mantle muscles do not need to conduct quite as fast. The set of axons forms a consistent series with the square root of the diameter of each axon proportional to its length. This makes the mantle contraction synchronous and causes such a powerful jet of water that some squids have been known to accelerate in the air—the only jet-propelled airborne animal.

Conduction in myelinated axons is different from that in unmyelinated axons. The myelin is an excellent insulator and severely restricts current flow out of the axon in the region between nodes. The current generated during a spike at one node depolarizes the next one above the threshold. In this case the impulse jumps from node to node, and the wave is discontinuous. The same effect can be produced with our fuse model (Figure 3.11). Instead of a continuous layer of gunpowder, we could have local concentrations. Connecting these "nodes" of gunpowder are fine, copper wires surrounded by a vacuum bottle. The wire and vacuum serves as a well-insulated conductor of heat with a low capacity. The heat from one exploding patch would be transmitted along the wire to the next patch. Such a system is faster and more economical of gunpowder. This is analogous to myelinated neuron conduction. It is faster and requires fewer sodium channels. This means less energy for the Na-K pump to restore the ionic concentrations. The consequences are profound. A 20-μ (including myelin) frog myelinated fiber is faster than a 1-mm squid giant axon and has a far lower metabolism per unit length. Imagine, then, the size and energy consumption of the human brain if all the myelinated tracts (white matter) were replaced by comparably fast unmyelinated axons.

The conduction velocity in myelinated axons is roughly proportional to their diameter, as is the distance between nodes. Myelination is really only efficient when the axon is above a minimum size. Table 3.2 shows a classical grouping of axons in the frog sciatic and cat saphenous nerves. A clear gap in conduction velocity is seen between B and C fibers. This is because the numerous fine C fibers are unmyelinated. Our information about extreme hot or cold skin temperatures is carried by C fibers and may take half a second to reach spinal cord synapses.

What relevance does the mechanism of action potential generation in squid axons have to human physiology? The mechanism elucidated by Hodgkin, Huxley, and others exists in all axons that display action potentials. Many cells have been voltage-clamped, and the same basic mechanism has been observed in squid, lobster, cockroach, and annelid worm unmyelinated axons and in frog, bird, and mammalian nodes of Ranvier. Similar mechanisms are also present in vertebrate striated muscle (see Chapter 4) and invertebrate neural somas. In each of these cases there are two voltage-dependent sodium gates, which first open and then close the sodium channels upon depolarization. Even many details of the kinetics are alike. The basic mechanism is all that is necessary for a spike. The voltage-dependent potassium channels are not necessary but do change the spike characteristics; as would be expected, the number and behavior of these channels is more variable.

Lately more and more cells (or parts of cells) with calcium action potentials have been discovered. In these cases voltage-dependent calcium channels replace or supplement sodium channels. It seems that calcium channels are inherently slower and stay open longer than the sodium channels. Therefore sodium channels are preferable for axonal conduc-

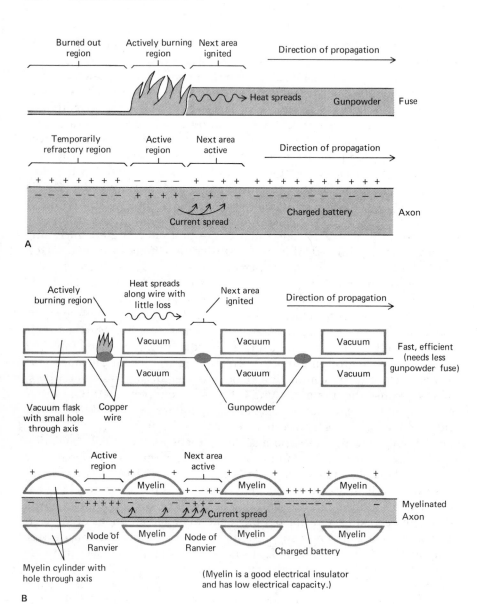

FIGURE 3.11 Spread of excitation in an axon and a fuse. Spread of heat in fuse is analogous to spread of current in axon. Increase in temperature is analogous to change in potential. (**A**) Unmyelinated axon compared to ordinary fuse. (**B**) Myelinated axon compared to specially modified fuse. Wire conducts heat between patches of fuse. Internode conducts current between nodes of Ranvier. Vacuum insulation and low heat capacity correspond to high resistance and low electrical capacity of the myelin.

TABLE 3.2 Conduction Velocity of Axons of Different Diameters

Animal	Axon	Diameter, μm	Myelation	Velocity of Conduction, m/sec
Crayfish	Axon in leg nerve	36	No	8 at 20°C
Squid	Giant axon	500	No	30 at 15°C
Frog	A fibers	11–18 (includes myelin)	Yes	17–42 at 20°C
	B fibers	2 (includes myelin)	Yes	4
	C fibers	2.5	No	0.3
Cat	A (α)	12–22 (includes myelin)	Yes	60–120
	(β)	8–12 (includes myelin)	Yes	40–60
	(γ)	2–8 (includes myelin)	Yes	10–40
	B	3 (includes myelin)	Yes	3–15
	C	1	No	2

tion. However, the effective intracellular concentration of calcium is very low, and many processes are very sensitive to any increase. Much interest has arisen in the role of intracellular calcium as a modulator of some cellular activities. Increases in intracellular calcium are the normal means of initiating contraction in all muscles (see Chapter 4, Sections 4 and 6). Calcium increases are strongly implicated in synaptic secretion, glandular secretion, ciliary beat modulation, and some luminescent flashing. In all these cases, except for vertebrate striated muscle, calcium action potentials have been observed. These muscles are too thick and too fast to produce a synchronous twitch if the calcium source were extracellular. However, slower muscles, such as vertebrate smooth muscles, and even some fast, striated crustacean muscles do have calcium action potentials.

The squid giant axon need fire only once to produce a large muscle twitch, which propels the animal. Most axons, however, fire repetitively in normal circumstances. When a constant current is passed across the membrane, the neuron fires repetitively often at a constant frequency. The frequency is linearly proportional to the current density over a wide range. Modifying this almost invariably is some accommodation—the frequency declines a bit during a constant stimulus (Figure 3.12). In some cells this is only during the initial part of the burst; in others the cells may eventually fail to fire altogether. Both the Hodgkin-Huxley model and the squid axon itself fail to fire repetitively in this way. There are other, longer term changes in membrane permeability that have been explored in only a few

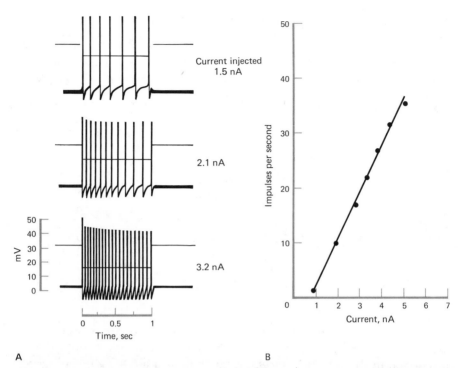

FIGURE 3.12   Repetitive firing induced in a neuron when current is injected into the soma. (**A**) Current (top trace) and potential (bottom trace). For a constant current the cell fires repetitively. It starts at a slightly higher frequency and then stabilizes at a constant rate. (**B**) Plot of the steady firing frequency against injected current for this cell. (From M. G. F. Fuortes and F. Mantegazzini, *J. Gen. Physiol.* 45:1163–1179, 1962. With permission of The Rockefeller University.)

cases, most notably some snail neurons. These may account for some of the features of repetitive firing.

The spike is a stereotyped response that travels down the axon without loss. It is clear that the only information it can communicate is its presence or absence—not its shape. The simplest code one can imagine is a simple mean-frequency code. The higher the frequency the brighter the light, stronger the pull, louder the sound, and stronger the resulting muscle contraction. This frequency code is approximated in many cases. Other information has been known to be communicated by other characteristics of spike patterns: burst duration, peak frequency, rate of change of frequency, phase (with respect to some standard), probability of no spikes (during a fixed interval). Unless we know the code(s) in each case, we cannot interpret the spike activity we monitor.

Spikes have drawbacks for information transfer; that is, they are stereotyped and possess a refractory period. Over short distances there are corresponding advantages to passive electrotonic spread. Many dendrites, some somas, some muscle fibers, and even some axons do not spike. In addition, there are some cells, most notably many invertebrate muscle fibers, that give only weak spikes.

7 • SYNAPTIC TRANSMISSION

Even the shortest reflex paths in mammals contain sensory, motor, and muscle cells. The signals must be transmitted between nerve cells and from motor nerve to the muscle fibers. There is no cytoplasmic continuity at the junctions, but, rather, distinct regions called *synapses*. It is at the synapses that signals are transmitted, and the nature of the synapses and their location and abundance on the cell determine the nature of transmission. Every spike in a sensory cell does not necessarily generate one in a motoneuron. The synapse is the site of much of the signal integration and processing.

Neurons can synapse with other neurons in a variety of ways. Most commonly in vertebrates the signal travels along the axon of a presynaptic neuron and into fine axonal branches. These make contact with the dendrites of another, postsynaptic, cell (Figure 3.13A and B). This is called an *axodendritic synapse*. Synaptic morphology is diverse; many types of contact are observed. One of the most common is between expanded bulbous (bouton) endings of the fine axons branches and similarly expanded dendrite terminals.

The presynaptic terminal contains mitochondria and numerous vesicles—rounded or elliptical in cross section (Figure 3.13C). Over part of the area of contact is a region with a uniform gap, the *synaptic cleft*. In vertebrate axodendritic synapses the gap is about 300 Å. The postsynaptic membrane is much more electron-dense, and often there is additional electron-dense material extending into the cytoplasm. The region between the cells is also more electron-dense. There are other characteristic features that allow electron microscopists to identify synapses. There are several similarities between these synapses and some commonly observed contact regions between epithelial cells, the *zonula adherens*. As the name implies, the *zonula adherens* serves to attach cells in epithelial sheets. Attachment at synapses is also strong, and when nervous tissue is broken up by ultrasonic vibrations, bits of pre- and postsynaptic membrane often are still attached.

At synapses the mitochondria and, most important, the vesicles are concentrated on the presynaptic side. Many of the vesicles are concentrated directly opposite the gap, heightening the appearance of morphological asymmetry. Synapses are also functionally polarized. Transmission is in one direction alone—from the vesicled presynaptic side toward the postsynaptic element. The same sorts of polarized synapses are seen in all

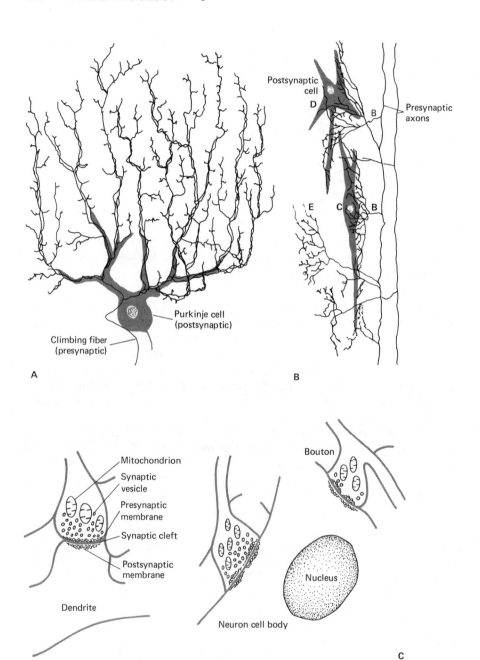

FIGURE 3.13 (**A**) Synapse between multiple terminals of the axon of a climbing fiber and the dendrites of Purkinje cells in a human cerebellum. (**B**) Synapses in the cat's medulla. Each axon branches to synapse with many cells, and each cell receives synaptic input from several cells (not all shown here). (**C**) Semischematic drawing of two types of synapse in vertebrate brain. Axodendritic synapses (left) may be excitatory and axosomatic synapses (right) may be inhibitory.

nervous systems. Coelenterates possess symmetrical and presumably unpolarized chemical synapses.

Synaptic anatomy is correlated with physiology in other cases. We now know, for instance, that certain vesicle types are peculiar to certain transmitters or classes of transmitters. The common clear, spherical vesicle of about 300 to 400 Å is found in all cholinergic synapses—synapses in which acetylcholine is released presynaptically. Larger vesicles, 500 Å in diameter, with a dense core are characteristic of catecholamines, such as norepinephrine (noradrenaline). Very large vesicles over 1000 Å and filled with an intensely staining core are diagnostic of neurosecretory axons.

The 5 percent of vertebrate synapses that are axosomatic have a distinct synaptic morphology. The synaptic gap is only 150 to 200 Å, and the vesicles are flattened or elliptical. In several cases these have been shown to be inhibitory synapses.

Not all synaptic transmission is chemical. In some cases the currents from the presynaptic cell affect the membrane potential of the postsynaptic cell directly. The most common example is where the currents flowing into the presynaptic cell flow across the synapse and out of the postsynaptic cell. Passive electric spread will depolarize the postsynaptic cell and typically cause it to generate a spike. In chemical synapses electrical isolation of the two cells is important, and the extracellular space is continuous with the synaptic cleft or gap. Until recently the sites of electric transmission were thought to be tight junctions. These are junctions which consist of an actual partial membrane fusion. Now more people believe that tight junctions are designed to prevent current or ionic species from passing along the intercellular space between cells. Instead it seems that electric transmission is associated with *gap junctions*. These are true electrical synapses. They consist of patches of closely apposed membrane separated by a roughly hexagonal lattice of unique protein. Within the lattice are channels through which can pass ionic current and large molecules (see Chapter 2, Section 4).

Most *neural* synapses are chemical; electrical synapses are utilized for a few distinct functions. However, many nonnervous cells, such as epithelia and smooth muscle, are electrically connected by gap junctions. In early embryos all cells are electrically connected. In many ways what makes the nervous system special is that the cells are electrically *disconnected*, and those connections are replaced by integrative chemical synapses.

7.1   Principles of Synaptic Transmission

Most of our knowledge of synaptic transmission does not come from vertebrate CNS. Instead, vertebrate and invertebrate neuromuscular junctions have been studied. Neuromuscular junctions have a large postsynaptic cell, the muscle fiber, which can be seen easily and impaled with microelectrodes. It is the only type of synapse in the region. Moreover, the presynaptic cell, the motoneuron, can be stimulated conveniently and unambiguously with extracellular electrodes on the conspicuous nerve trunk. The best investigated synapses between neurons are primarily in the invertebrates. There individual cells can be identified from animal to animal. Large axons and somas allow insertion of several microelectrodes.

The frog neuromuscular junction has been investigated in great detail and has provided a large proportion of our knowledge of synaptic function. Bernard Katz and his collaborators, particularly Ricardo Miledi, stand out as giants in the field. First, what is striking about these neuromuscular junctions is that each neuron spike produces a spike in the muscle fiber. Each synaptic potential is suprathreshold and thus produces a spike and twitch. The 1:1 relationship is different from the integration and convergence found in most synapses. Why then have a chemical synapse at all? Why not have an electrical continuity between the two cells and allow the action potential to spread from axon to muscle fiber? Too much current is required, and the axon cannot possibly provide it. An analogous problem would be to try to heat up a cannonball with a hot knitting needle. Here the primary purpose of the chemical synapse is current amplification.

The main events in the sequence have been well worked out. (For the rest of the sequence see Chapter 4, Section 4.) An action potential travels along the axon and produces a large depolarization of the fine nerve terminals. The depolarization causes a compound, acetylcholine,

$$(CH_3)_3N^+CH_2 - CH_2 - O - \overset{\overset{\displaystyle O}{\|}}{C} - CH_3$$

to be released into the synaptic cleft. The acetylcholine (ACh) diffuses to the postsynaptic membrane where it combines with a specific receptor molecule. This combination opens up channels in the postsynaptic membrane. These channels are permeable to the cations sodium and potassium. When they are open, they tend to make the membrane potential approach about -10 mV (about halfway between E_K and E_{Na}). Enough of these channels are opened to depolarize the membrane above the spike threshold. The spike then propagates along the muscle fiber in both directions. The acetylcholine does not remain attached to the receptor. It diffuses away and is hydrolyzed by the enzyme acetylcholinesterase (see Chapter 2, Section 3). This splits the molecule into acetate and choline. Much of the choline is reabsorbed by the presynaptic endings, where it later is recombined with acetate.

One characteristic of all chemical synapses is the transmission delay. The delay from presynaptic to postsynaptic depolarization in the frog neuromuscular junction is about 0.6 msec. This delay is almost entirely presynaptic. For circuits in the central nervous systems, involving short axons and many synapses, synaptic delay is a large part of total signal delay. Mounting evidence points to the following presynaptic sequence: depolarization of the axon terminal opens up some voltage-dependent gates to calcium channels. Calcium diffuses into the cell. This somehow causes the release of ACh from the vesicles. The ACh is found only in the vesicles and is released by exocytosis—a process approximately the reverse of pinocytosis (see Figure 3.14 and Chapter 2, Section 15). Less certain, but with some evidence, is the recycling of the vesicles. In this model the vesicle membranes are only temporarily incorporated into terminal membrane. The membrane then recycles and forms new vesicles, which then become loaded with ACh.

The ACh receptor is very specific, and, although many analogs are known that can mimic ACh, they are not as potent. The reaction is very much like that of an enzyme with its substrate. The receptor is the enzyme to which the ACh binds by enzymelike kinetics. Some compounds, such as d-tubocurarine, compete with ACh for the active site. The channels opened by binding of receptors with transmitter are distinct from those opened by changes in the membrane voltage. The postsynaptic ionic channels are about the same size as the voltage-dependent sodium channels in axons. The best estimate for the electrical conductance of a single endplate channel is 2×10^{-11} mho; for sodium channels in frog muscle fibers it is 11×10^{-12} mho. The two most striking things about axonal Na^+ channels is their voltage dependence and selectivity. The synaptic channels are less selective; they let in Na^+ and K^+ equally. Yet they do effectively exclude chloride. The number of opened channels is not affected by voltage to any appreciable extent but is very sensitive to the presence of ACh. TTX, which blocks voltage-sensitive sodium channels, has no effect on synaptic channels. If we know the size of the conductance or permeability for each channel, we can calculate how many channels are opened following the release of ACh after a presynaptic impulse. In the frog neuromuscular junction the number is about 1 million. This represents a large short circuit of the postsynaptic membrane.

In order to study synaptic events we must look at synaptic responses alone and eliminate the muscle spike. One way this has been done is to add d-tubocurarine to block some receptor molecules. The synaptic potential is then reduced below the spike threshold. We can also study the currents passing through the synaptic region by voltage-clamping the muscle membrane. Figure 3.15 shows the synaptic potentials and the currents that cause them. The excitatory postsynaptic po-

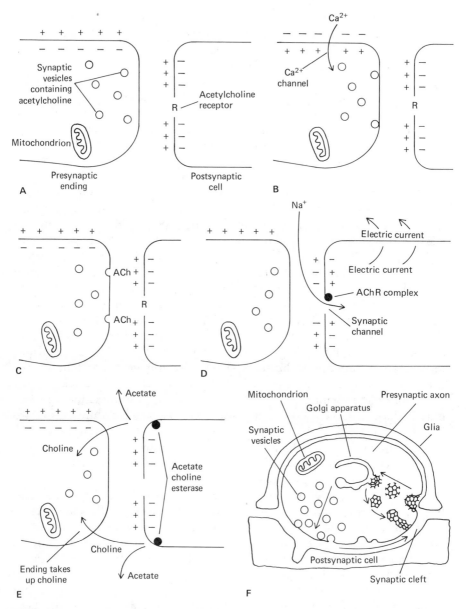

FIGURE 3.14 (**A–E**) Stages of transmitter release. (**A**) Before release. (**B**) After presynaptic action potential. (**C**) Transmitter released into cleft. (**D**) Transmitter combines with receptor and opens channels which let sodium into the postsynaptic cell. This produces synaptic current and excitatory postsynaptic potential (EPSP). (**E**) Acetylcholine is split by cholinesterase. Some of the choline is reabsorbed by the presynaptic neuron. (**F**) Single drawing depicting scheme for entire cycling of vesicles. After transmitter release the vesicle membrane (now incorporated into the cell membrane) is coated and reconstituted in the Golgi apparatus.

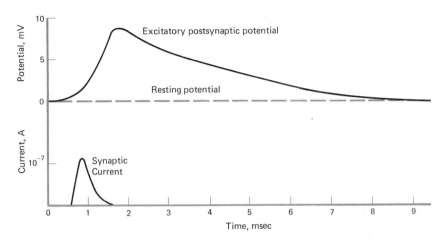

FIGURE 3.15 Current and potential changes in postsynaptic cell following single presynaptic action potential. The synaptic current reflects how long the postsynaptic sodium channels are opened. The EPSP lasts somewhat longer because of the electrical properties of the postsynaptic cell. The sizes of EPSPs vary greatly from synapse to synapse, even on a single cell. The amplitude also depends on many other conditions. This example is a frog neuromuscular synapse partially blocked by curare.

tential (EPSP)[2] outlasts the current because of the passive electrical properties of the muscle fiber. The currents and EPSPs are depressed by curare and prolonged by eserine, a chemical that blocks acetylcholinesterase.

Similar recording from junctions, even in the absence of axon spikes, reveals the presence of very small potential changes that have the same shape as EPSPs. These too are blocked by curare and prolonged by eserine. There is no question but that they are the result of release of ACh from the nerve terminals. These miniature EPSPs (or "minepps," as they are called) do not come in all sizes. Rather they exist in multiples of a basic unit, the *quantum*. The quantal amplitude is what would be produced by the release of several thousand ACh molecules reacting with the receptor to open several hundred channels. Most physiologists believe each minepp represents the release of the contents of one vesicle.

When the presynaptic membrane is slightly depolarized, the frequency of minepps is increased. The spontaneous quantal release is nearly random. The elevated release during axonal depolarization is also random, but the frequency is higher. A decrease in presynaptic membrane potential merely increases the probability of release of each quantum. Ingenious means have been devised to depolarize the presynaptic membrane and estimate its membrane potential. However, it is very

---

[2] At the neuromuscular junction this can be called an end-plate potential (EPP).

difficult to penetrate the presynaptic endings with a micropipette. One of the few synapses in which both pre- and postsynaptic cells may be impaled is the synapse in the squid stellate ganglion (Figure 3.16). At this synapse the relationship between the pre- and postsynaptic depolarizations has been described in detail. Chemical synaptic transmission de-

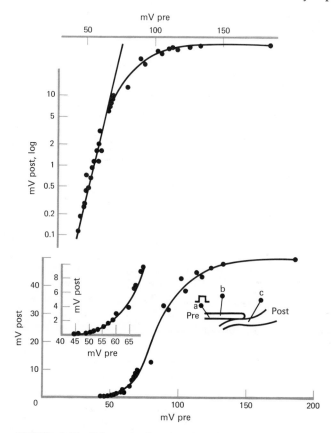

FIGURE 3.16 Relationship between the presynaptic and postsynaptic potentials in a synapse in the stellate ganglion of the squid. Inset in graph shows that two electrodes (a and b) in the presynaptic axon change and measure the presynaptic potential. One electrode (c) monitors the postsynaptic potential. This is one of the very few preparations in which both pre- and postsynaptic cell processes can be impaled. The graph shows that after sufficient presynaptic depolarization the relationship is close to exponential. An action potential of 100 mV produces a 40-mV EPSP. If the action potential is reduced to 75 mV, the EPSP is only 15 mV. At an increase to 200 mV it is only increased to about 48 mV because the curve levels off and then begins to decrease. (From B. Katz and R. Miledi, *J. Physiol.* (*London*) 192:407–436, 1967. By permission of Cambridge University Press.)

pends on the presence of external calcium. If appropriate tricks are played (the sodium spike is blocked and the potassium channels obstructed), then the voltage-dependent calcium channels found at the presynaptic terminal will produce a calcium spike. This spike is confined to the presynaptic terminals and is produced by the entry of calcium into the region. This calcium entry is thought to be a necessary link in coupling the change in potential in the terminal with the release of vesicle contents.

The role of vesicles in transmission is confirmed by two types of experiments. First, packets of vesicles can be isolated from whole brains. These are called synaptosomes and contain most of the ACh in the brain. The synaptosomes can be broken up and the vesicles isolated. These indeed contain much of the brain ACh. Recently techniques and preparations have been developed that permit exhaustion of the vesicles. That is, if the presynaptic axons are stimulated at a high frequency, the number of vesicles seen in the electron microscope decreases dramatically. When this happens, the synaptic transmission also fails.

No phenomenon is more basic to neurophysiology than that of selective ionic channels. Great interest centers around the nature of the postsynaptic receptor, which either gates or is part of the postsynaptic channels. Recently there has been considerable progress in purifying and studying the receptor. Several snake toxins bind tightly and specifically to the ACh receptors. Radioactive labels, such as ^{131}I, can be attached to the toxin, and the toxin-receptor complex can then be purified. Another use of labeled toxin is in locating and counting the ACh receptors during development, in tissue culture, during regeneration, or in a normal synapse.

We can see that there are several steps involved in synaptic transmission. It is far from an all-or-none process. ACh must be synthesized, packaged, and released. Calcium must enter the cell and cause the release. The ACh must then combine with specific receptors, which open membrane channels. If the presynaptic cell fires repetitively, then ACh synthesis and packaging must keep up with the rate of release. Indeed, all these processes do not continue unchanged during repetitive firing. Often the second EPSP in a train is actually larger than the first. At higher frequencies there may be a progressive decrease in the size of EPSPs—a sort of exhaustion or fatigue. The relationships between frequency and size of the EPSP are rarely simple and vary with the synapse. Yet such relationships have important consequences for normal function. This synaptic complexity is expressed under normal functioning and is part of what makes the passage of information across a synapse so special. It is apparent that synaptic behavior could be altered in many ways—for example, by synthesizing more receptor or more presynaptic calcium channels. This is the reason why much speculation on the long-term changes occurring in the brains of higher animals revolves around synaptic function.

7.2   Postsynaptic Inhibition

All that is needed to convert the excitatory synapse to an inhibitory one is to change the selectivity of the channels opened when the transmitter combines with the receptor. If, instead of allowing both Na^+ and K^+ through, it allowed only K^+, then the equilibrium potential for this process would be about -90 mV, not -15 mV. The first is well below spike threshold, the latter well above it. ACh is released by axons of the vagus nerve and inhibits the heart. The fibers become more permeable to potassium. This opposes the spontaneous depolarizations that initiate heartbeats.

Among invertebrates there are many easily studied inhibitory synapses. Inhibitory neuromuscular synapses are also abundant and have been used for experiments comparable to those on excitatory frog neuromuscular synapses (Figure 3.17). These fibers are dually innervated—they receive excitatory and inhibitory innervation. If the excitatory axon is stimulated, then a depolarizing EPSP is observed. If we stimulate the inhibitory axon alone, we may observe an I (inhibitory) PSP. This may be hyperpolarizing if the resting potential is low or even depolarizing if the potential is high. The sign of the IPSP reverses not at E_K but rather at E_{Cl}. When the excitatory and inhibitory axons are excited simultaneously, there is a reduction in the amplitude of the EPSP. This is an important point. Even if the IPSP is slightly depolarizing, it tends to push the membrane potential towards E_{Cl}. Therefore it counteracts the EPSP, which tends to push the potential toward -15 mV. In general, postsynaptic inhibition is effective because it tends to hold the membrane potential below spike threshold and/or reduce the effects of other depolarizations. At most

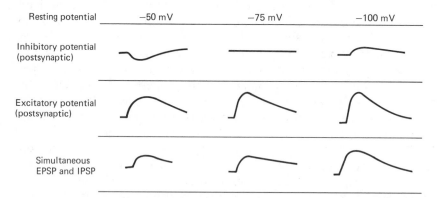

FIGURE 3.17   IPSPs and their efficacy in reducing EPSPs. The size and sign of the IPSP vary with the membrane potential, but stimulation of the inhibitory axon tends to reduce the EPSP at all potentials shown. This is because the postsynaptic inhibition makes the membrane potential of the postsynaptic cell approach -75 mV. The EPSP makes it approach -10 mV.

inhibitory synapses the inhibitory channels are Cl^- selective or Cl^- and K^+ selective. Purely K^+ selective channels are rarer.

7.3 Presynaptic Inhibition

In some synapses inhibition is greater and has a different timing than would be calculated from its postsynaptic effect. This has turned out to be due to presynaptic inhibition. The causes of presynaptic inhibition have been elucidated through some beautiful experiments by Dudel, Kuffler, and the Takeuchis. The inhibitory axon has been shown by others to terminate on both the excitatory axon and the postsynaptic muscle fiber as well. Stimulation of the inhibitory axon releases transmitter onto receptors on the excitatory axon. This makes the excitatory terminals more permeable to chloride and thus reduces the spike depolarization when the excitatory axon is stimulated. In squid stellate ganglion (Figure 3.16) and other synapses it has been shown that the amount of transmitter released from the terminal depends on the depolarization of the terminal. The bigger the depolarization the more quanta released. Presynaptic inhibition reduces the size of the presynaptic spike and thus reduces the amount of transmitter released. In this preparation the neurotransmitter is the same for both pre- and postsynaptic inhibition, gamma aminobutyric acid.

Presynaptic inhibition exists; why not presynaptic excitation or facilitation? There is reasonable evidence for such a phenomenon in the snail *Aplysia*. Stimulation of one excitatory axon increases the synaptic release of transmitter from another axon. This phenomenon is called *heterosynaptic facilitation*.

7.4 Other Transmitters

ACh is only one of several transmitters. It was the first one for which strong evidence was obtained about its role. Since then the biochemical pathways for its synthesis and destruction have been well worked out. Another peripheral transmitter in vertebrates, norepinephrine, has also been investigated extensively. No one doubts that these are transmitters in the vertebrate periphery. Both are also found in vertebrate central nervous system (CNS), where they are presumed to function as transmitters. There evidence is weak except in a very few cases. ACh and, to a lesser extent, epinephrine have been implicated in neurotransmission in invertebrates. Before there was even scanty evidence for ACh transmission in arthropods, anticholinesterases such as parathion were found to be excellent insecticides. Now we know that invertebrates utilize some of the same neurotransmitters as man. It is in invertebrates that γ-aminobutyric acid (GABA) has been established as a transmitter. Other substances, such

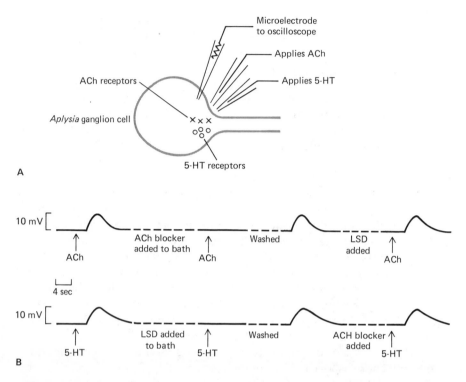

FIGURE 3.18 A cell in *Aplysia* (**A**) with two types of excitatory synapses (**B**) each with its own transmitter, presumed to be ACh and 5-HT. Here the two substances are applied to the cell separately by pipettes. Although the responses are similar, each is blocked independently by a distinct blocking compound: hexamethonium (blocks ACh receptor) and LSD (blocks 5-HT receptor).

as 5-hydroxytryptamine (5-HT), glutamate, and dopamine, are strongly suspected of being transmitters in lower animals. All of these also occur in some neurons in mammals. Glycine and aspartate are putative transmitters in vertebrates but so far have not been implicated in other animals. For most neurotransmitters research in diverse organisms is complementary. GABA pathways were first elucidated in crustacea; LSD was implicated as a postsynaptic blocker in mollusks. Despite the host of neural differences, reserpine tranquilizes not only people but also clams.

We noted how synapses are particularly sensitive points in signal transmission. We should expect many substances that act on nervous systems to act on synapses. Some snake, spider, bacterial, and mushroom toxins, arrow poisons, nerve gases, and insecticides profoundly disturb synaptic function. Also hallucinogens, tranquilizers, amphetamines, and some barbiturates act largely on synapses.

Many cells receive multiple inputs from several axons, releasing a total of two or more transmitters. An individual cell may have receptors for several transmitters on it. In contrast, it is generally assumed that an individual neuron releases only one type of transmitter from all its branches. Figure 3.18 shows such a cell from *Aplysia*. In this case two receptors are stimulated by the application of transmitters from a semimicropipette positioned near the synapse. 5-HT and ACh application both open up channels in the membrane. There are different channels near two discrete receptor types. The selective block of the receptors by two compounds is one strong piece of evidence.

7.5 Spike Generation by Synaptic Activity

If we stick an electrode into a neuron soma in an animal's CNS, we can monitor the synaptic activity and the cell's response. Figure 3.19 shows a hypothetical recording. The membrane potential will fluctuate with the inputs. Some excitatory inputs will open many synaptic channels and some only a few. Some will facilitate greatly and some fatigue. Although an individual bouton may produce a tiny synaptic potential, one axon might make hundreds or perhaps thousands of synapses on the postsynaptic cell. Inhibition also will vary, the IPSP exhibiting a large range in amplitude. When all these effects combine, one other factor looms large: the location of the synapse on the cell. A synapse at a distal dendritic branch may produce a large depolarization of that branch. But this potential will decay as it spreads passively toward the soma. A more proximal EPSP of equal size will decay less and produces a greater somatic depolarization. Another way of looking at this is that at excitatory synapses there is an inward ionic current. As the current spreads from a distal dendrite, some of it leaks across the membrane. Only a fraction remains to depolarize the soma.

In order for this cell to spike it must reach threshold. Numerous clever experiments by J. C. Eccles and his colleagues demonstrated that the area of some cells that fires first is the initial segment or axon hillock region (Figure 3.19A). This is for two reasons. Since the cell narrows down at that point, the ionic current pathway is compressed and the current density increases. Also, the voltage threshold of this area is probably lower. When the potential of the initial segment reaches threshold, the cell spikes, and that spike travels along the axon. The soma itself spikes after the initial segment. It is the initial segment's distinctive membrane parameters that determine the relationship between current and spike frequency. This segment integrates the currents originating from synaptic

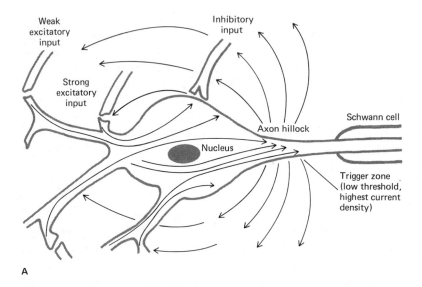

A

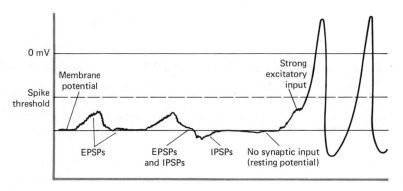

B

FIGURE 3.19 (**A**) Semischematic representation of a vertebrate motoneuron. Two of numerous excitatory inputs and one inhibitory input are shown. Synaptic current is represented by arrows. Note inhibition short-circuits some of the excitatory currents. Cell fires first at "trigger zone" where current density is high and voltage threshold low. Synapses vary in their potency, and, all other things being equal, the closer to the trigger zone the more potent the synapse. (**B**) Recording from motoneuron. Actual spike firing occurs only when synaptic excitation is sufficiently strong. Yet potential in period before spikes varies considerably, and even a small additional synaptic input, if properly timed, could initiate another spike.

activity all over the cell and determines spiking. Monopolar invertebrate neurons do not have a comparable dendrite-soma-axon order. Rather, in some the integration occurs at a region of the axon known as the "integrating segment." This receives both axonal and dendritic branches. In some long interneurons, as in crayfish, there is one integrating segment per ganglion. This means the cell can start an impulse at any of several points along it. The signals along such an axon may represent the sum of all spikes generated at several dendritic branch points.

We mentioned before that chemical synapses, although the rule in the nervous system, are the exception in the body. However, we do find gap junctions in some nervous tissue and much muscle tissue. The gap junctions provide a low-resistance path for current. An action potential in cell A can provide a source of current to cell B and depolarize it. Perhaps enough current will pass between the cells to produce an action potential in cell B as well. We may also imagine the reverse happening and the action potential propagating from B to A. In some giant axons this is what happens. For example, the giant axons in the earthworm nerve cord result from the fusion of several neurons. Although they consist of electrically connected separate neurons, for all intents and purposes the system of axons behaves as if it originated from a single cell.

A good number of electrical synapses have been observed and explored. Figure 3.20A shows recordings from two such electrically coupled cells. Depolarization of one cell stimulates the other and vice versa. The two cells fire together normally. This is a way of assuring identical spike

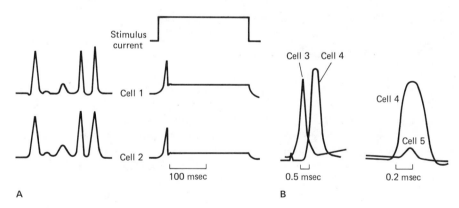

FIGURE 3.20 Some properties of electrical synapses. (**A**) From brain of puffer fish. (Left) When spinal cord is stimulated, both cells 1 and 2 are excited simultaneously with identical patterns. (Right) When cell 2 is stimulated by current from an intracellular electrode, cell 1 is also stimulated after a very slight delay. (From M. V. L. Bennett, *Ann. N.Y. Acad. Sci.* 137:509–539, 1966.) (**B**) Contrast in synaptic delays at chemical and electrical synapses in the hatchet fish. Cell 3 chemically synapses with cell 4, and the transmission delay is about 0.5 msec. Cell 4 electrically synapses with cell 5 to produce a subthreshold depolarization with a delay of about 0.05 msec. (From A. A. Auerbach and M. V. L. Bennett, *J. Gen. Physiol.* 53:183–210, 211–237, 1969. With permission of The Rockefeller University.)

patterns. Electrotonic synapses often serve this function. Moreover, since the synapse only requires electrotonic spread between cells, it may be very fast (Figure 3.20B). Chemical synapses are slowed by the intricate presynaptic processes. When chains of synapses are required, the cumulative difference can be appreciable. Thus where synchronous output is critical, as in the discharge of electric fish or the vibration of some sound-producing organs, we often observe electrical junctions. Electrical synapses are characteristically bidirectional in contrast to chemical synapses, which are usually polarized. However, there are some polarized electrical synapses. Current passes across the synapses more easily in one direction than another. One of the most famous electrical synapses is between the giant interneurons and large motor axons of the crayfish. The now classical experiments by Furshpan and Potter showed this to be highly polarized. Excitation spreads only from interneuron to motor axon. Many other properties that we associate with chemical synapses, such as inhibition, can exist at electrical synapses.

To summarize, the advantages of electrical synapses are for speed and perhaps reciprocal action. Chemical synapses allow temporal changes in efficacy, long summation, extreme unidirection, and amplification. Electrical synapses are not effective if the postsynaptic cell is much larger than the presynaptic. Chemical synapses avoid this by releasing a chemical messenger that provokes the postsynaptic cell to provide its own stimulating current.

8 • NEUROSECRETION

Neurosecretory cells are, as their name suggests, neural and secretory. The terminals end on blood sinuses but otherwise greatly resemble axonal terminals. The neurosecretory vesicles are large and densely staining. In invertebrates, particularly crabs, and in the urophysis of fish it has been possible to record propagating action potentials. Intracellular recording from crab neurosecretory terminals shows that the spike propagates to the end. Calcium is required for neurosecretory release. Repetitive stimulation of a neurosecretory axon causes the depletion of neurosecretory vesicles. In some invertebrates the contents of the granules have been found in the extracellular space, and there is good evidence for exocytosis. All these observations suggest that neurosecretion is like the release of synaptic transmitter. One difference is that the hormone is released into the circulation rather than into the extracellular space, and it may act at distant targets as well as at adjacent postsynaptic cells. (See Chapter 9 for further details.)

9 • TROPHIC EFFECTS

It is a common observation that nerves do more than stimulate muscles and other nerves. Vertebrate striated muscle fibers atrophy if the nerves to them are cut. Protein synthesis decreases markedly, and a characteristic atrophy occurs. After reinnervation the fibers can regain their original size and metabolism (see Chapter 4, Section 10). Some of these effects are due to the simple lack of muscle activity since they are reduced by direct electrical stimulation of the muscle fibers. The degree of atrophy due to simple lack of activity varies with the muscle.

One of the most striking effects of denervation is the spread of ACh sensitivity along the muscle fiber. The fiber free of synapses distributes its synaptic receptors over the entire cell surface. In some cells this change can be reversed or prevented by direct muscle stimulation. In frog muscles this is not so. Perfusion of the whole muscle with ACh for several days fails to prevent spread. If the nerves to a rat muscle are blocked rather than cut, the ACh sensitivity spreads. Stimulation of the axons beyond the block prevents the spread. But the synaptic transmission itself cannot be the sole cause of the restriction of ACh receptors. When the cut nerve regenerates enough to remake contact with denervated muscle fibers, these fibers begin to lose their dispersed ACh receptors. This happens before the synapse is electrically functional.

This and other experiments and observations, especially on changes in fetal muscle, indicate that there are one or more distinct trophic factors that affect the morphology and metabolism of the muscle fiber. Nearly all workers would agree on the existence of such substances, which could carry information not only on the presence of a synapse but also on the type of axon that is making contact. Such trophic substances are a long way from being isolated. Their properties and roles are central questions in development, but they lie outside the scope of this book.

10 • SENSE ORGANS

10.1 General Principles

Animals must obtain information about their environments. Sense organs provide their windows on the world. The job of the sense organs is to convert this information into electrical signals and pass it on to the rest of the animal. The sense organs process and relay only some of the information in the environment. Some animals lack organs for certain types of information; for example, few animals can "hear" airborne sounds. Even when specialized organs, such as eyes and ears, are present, only some of the information is received and relayed to the rest of the nervous system. We can see many colors, but not ultraviolet; we confuse pairs of superimposed colored lights with shades of gray; when we look at an airplane propeller, we cannot tell the direction of rotation. Thus even vertebrate eyes, the most sophisticated sense organs in the animal kingdom, relay only a fraction of the information present in the visual environment.

The central problems of sensory reception are apparent. First, how is the stimulus converted into electrical information? Second, what information is obtained? Third, how does the nervous system encode and process this information? We have learned much already but are still just beginning to solve these all-important problems of sensory function.

Receptors may be surprisingly sensitive. In some cases receptors are more sensitive than the best instruments. The sensitivity may attain the theoretical maximum. The thermoreceptors of boa constrictors are more sensitive to microwaves than the finest man-made detectors. Mammalian ears can resolve movements of the basilar membrane as small as 10^{-3} Å. Shark electroreceptors can detect fields of 10^{-8} V/cm (for comparison the electric field in the air near a flashlight battery is of the order of 0.1 V/cm). Photoreceptors can detect single quanta of light. Chemoreceptors on the antennae of some moths can detect single molecules of the pheromone bombykol.

Some of the same receptors or receptor organs that display great sensitivity also can respond to a large range of stimulus intensities. Vertebrate ears respond to stimuli over an intensity range of 10^{12} (120 dB). Eyes are roughly comparable. These are both complex sense organs, so they can exceed the range of any one receptor cell. Nevertheless, even single receptor cells may vary their response to intensities spanning a 10^{10}-fold range. How is this done? If the intensity is coded by spike frequency, then there is a problem. Axons cannot fire at a rate above about 1000/sec. Suppose that 10^{10} light quanta/sec are required to produce this rate. If the firing rate is proportional to light intensity, then the response to 1 quantum would be at a rate of 10^{-7}/sec (one action potential every four months). This would not work. Receptors with a large dynamic range fire at a rate roughly proportional to the *log* of intensity. With such a system the photoreceptor would increase its firing rate by 100/sec for every tenfold increase in light intensity. This system sacrifices resolution to obtain increased range. We can distinguish 1000 quanta from 2000; we cannot distinguish 1×10^{10} from 1.0000001×10^{10}, which also differs by 1000 quanta. This nonlinearity of response is a crucial characteristic of many receptors.

What information do we want from our receptors? Do we need continuous information about the stimulus amplitude? Some receptors do indeed provide such information. There are numerous position receptors in our joints. Each receptor fires maximally and continuously at a specific joint angle. On the other hand, most sense cells signal changes in stimulus intensity. Cells that respond to a continuous stimulus with a constant output are known as tonic receptors. Those that respond only to changes are known as phasic receptors. The process whereby the cells' response decreases during a constant stimulus is called *adaptation*. Many receptors display some adaptation but continue to respond to a constant stimulus for a long time. Such cells are both phasic and tonic. Some organs have tonic and phasic receptors in parallel; the crayfish muscle receptor organ is a good example.

Animals have many sense cells for several categories of stimulus. The information about which type of stimulus is occurring—sound, light, chemical—is not contained in the pattern of action potentials. Instead, different sensory modalities are carried in different axons. Action potentials in the optic nerve are interpreted as visual patterns, those in the auditory nerve as tonal patterns. Even if the optic nerve is stimulated electrically or by pinching, the organism "sees" a pattern. Many finer features of the sensory stimulus are signaled by specific classes of axons in a sensory nerve rather than by the type of neural activity sent along the collection of axons. For example, different sound frequencies produce action potentials in different axons of the auditory nerves.

10.2 Generator Potentials

By now we should expect changes in ionic selectivity, and hence membrane potential, to be involved in all neurophysiological processes. It comes as no surprise that what stimuli do to sense cells is to open or close ionic channels. This produces a change in the membrane potential. Membrane potential changes produce changes in spike frequency or, in some cases, may directly modulate synaptic transmitter release.

We can study generator potentials in isolation by blocking the action potential mechanisms in a sense cell. Tetrodotoxin (TTX) blocks the channels that let sodium into the cell during the action potential. It is specific for the action-potential sodium channels and does not appreciably affect the generator potential. Suppose we impale a crayfish stretch receptor cell (Figure 3.21), put on TTX, and stretch it. What do we see? During the stretch the cell depolarizes. The greater the stretch the greater the depolarization (Figure 3.22). When we release the stretch, the cell returns to its original membrane potential. This depolarization, caused by the sensory stimulus, is the generator potential. What causes this depolarization? If we pass current into the cell and measure the voltage change it produces, we can calculate the cell's conductance using Ohm's law. We find that the depolarization is caused by an increased membrane conductance. The stretch opens up ionic channels in the membrane. By varying the

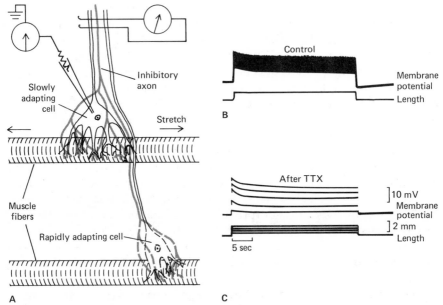

A

B

Control

Membrane potential

Length

After TTX

]10 mV Membrane potential

]2 mm Length

5 sec

C

FIGURE 3.21 (**A**) Crayfish stretch receptor. Dendrites of slowly and rapidly adapting receptor cells are each embedded in a single muscle fiber. Two muscle fibers are parallel. An inhibitory efferent runs to both cells. It is easy to record extracellularly from the nerve trunk and to impale the two sensory cells with micropipettes. (**B**) Intracellular record from the slowly adapting receptor during a constant stretch. The cell fires at a rate not resolvable on this time scale. (**C**) Same record, except that TTX has been added to block the action potentials. The depolarizations, called generator potentials, are greater the stronger the stretch. (**B** and **C** after S. Nakajima and K. Onodera, *J. Physiol. (London)* 200:187–204, 1969. By permission of Cambridge University Press.)

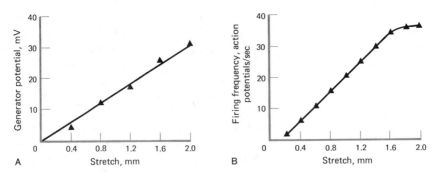

A

B

FIGURE 3.22 (**A**) The size of the generator potential in the slowly adapting receptor as a function of stretch (TTX added): Data are from an experiment similar to that in Figure 3.21C. (**B**) A plot of the firing frequency versus stretch in the same cell as in A. In this case the TTX had not yet been added. Experiment is similar to that in (**B**) of Figure 3.21.

composition of the bathing solution we can determine what ions can go through these channels. The channels seem to be permeable to Na^+ and some other cations. The reversal potential is close to -10 mV in the stretch receptor. This could occur if the channels were about equally permeable to sodium and potassium ions. The sequence of events then is: stretch, open ionic channels, let in Na^+, depolarize.

The unique property of a sense cell is its ability to convert a sensory stimulus (here it is stretch) into a change in membrane voltage. This generator potential may not increase linearly with the strength of the stimulus, or it may not remain constant during a constant stimulus. Such adaptation is readily seen in many receptors. In them the generator potential does not bear a constant linear relationship to the stimulus intensity. On the other hand, the generator potential in the crayfish receptor is nearly proportional to the degree of stretch and is almost constant during a prolonged stretch.

There are two basic types of information processing after the production of a generator potential. In some cases, such as vertebrate hair cells and fish electroreceptors, the generator potential produces a change in the release of transmitter from a synapse (see Section 11.3). The process sequence is: stimulus, generator potential, altered presynaptic secretion, altered postsynaptic permeability, postsynaptic potential. In other cases, such as the stretch receptor, the generator potential produces a change in the firing frequency of the sense cell. We discussed earlier (Section 6) this general property of neurons to raise their firing frequency when depolarized. Some sense cells, such as the crayfish slow-adapting stretch receptor and the *Limulus* lateral eye, fire at a frequency that is nearly proportional to the amplitude of the generator potential. Since the generator potential is proportional to the tension of the receptor organ, the firing frequency is also proportional to stretch (Figure 3.22). It is easy to decode the signal in the axon since the relationship between action potential rate and tension is relatively simple.

Most sense cells do not send a constant signal to the CNS during a constant stimulus. Instead they are phasic to some degree and tend to adapt to a constant stimulus. In some cases this adaptation can be shown to be due to a decrease in the size of the generator potential. In some other cases the sense cell may stop (or decrease) firing even during a prolonged generator potential. We have emphasized that in most cells the signal is proportional to the *log* of stimulus intensity. In many cases approximately the same relationship can be seen between log stimulus intensity and generator potential or generator current.

Vertebrate photoreceptors display a different type of response to stimuli. When a rod or cone is illuminated, its membrane potential increases; the cell hyperpolarizes (Figure 3.23). Measurement of the resis-

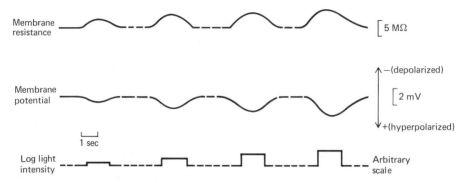

FIGURE 3.23   One of the first measurements of membrane resistance and potential changes of a vertebrate rod during illumination. This is from the lizard Gekko. Illumination increases the membrane resistance (closes ionic channels) and hyperpolarizes the cell. The greater the light the greater the effect, except that there is some saturation at high light intensities. (After J. Toyoda, H. Nasaki, and T. Tomita, *Vis. Res.* 9: 453–463, 1969.)

tance of the cell shows an *increase* during illumination. The sequence of events is: light, close ionic channels (permeable to Na⁺), hyperpolarize. Invertebrate photoreceptors *open* ionic channels during stimulation. The advantages and evolution of the difference are still items of speculation.

11 • RECEPTOR TYPES

There are many classes of receptors and a countless variety of examples of each type. We shall discuss only a few interesting or commonly studied receptors.

11.1   Thermoreceptors

Thermoreceptors are found in many groups but have been little studied. Homeothermic vertebrates have numerous skin and deep temperature receptors. They use this information to maintain a constant body temperature (see Chapter 10). Mammals possess both cold and warm receptors, which respond maximally to cold or warm temperatures, respectively. We are most sensitive to temperature *changes,* and there are reports that the human threshold for sensation is a change of 0.003°C in 3 sec.

Some snakes are phenomenally sensitive to temperature changes. Boas use special thermoreceptors in their heads to detect small warm or cold spots in the landscape. In this way they can locate a warm-blooded animal, like a mouse, in the dark. The infrared radiation from a mouse is strongest at a wavelength of about 10 μ, and boas and pit vipers are sensitive to it. Boas respond equally well to a broad range of infrared frequencies, including microwaves. They are significantly superior to any man-made instrument as detectors of microwave radiation.

11.2 Electroreceptors

The world is filled with electric fields. Animals produce them continuously as a result of their nervous and muscular activity. Now we know that a variety of fish can detect electric fields; indeed, behavioral tests show that sharks can detect fields as low as 0.01 μV/cm (1 volt per 1000 km). Such sensitivity allows these fish to detect animals buried in the sand and may be sufficient to allow the fish to resolve its orientation in the earth's magnetic field and determine its own water speed. The actual electroreceptors have been found and studied extensively. They are of several forms, but all either are in pits just under the skin or are adjacent to canals that extend to such pits.

The electric eel and *Torpedo* are famous because of the strong, even lethal, shocks they can produce. However, many electric fish can generate mild electric pulses, and their electroreceptors, which sense these signals, are particularly tuned to detect alterations in them. This allows them to localize an object in the water even in the dark. A great deal of central processing is required for this, and all electric fish have an enlarged cerebellum. One species has a brain/body weight ratio nearly that of humans.

How do electric organs generate these fields? The organ is analogous to many batteries in series. The voltage of batteries in series equals the sum of their individual voltages. The generating cells, or electrocytes, are evolved from muscle cells. An electric organ consists of stacks of these flattened cells. One flat face of each electrocyte depolarizes while the other does not. Instead of generating an inward current as in muscle cells, the ionic current enters one side and exits passively from the other. The next cell also depolarizes on only one face, the comparable face, and so on. The way this works in the skate *Torpedo* is shown schematically in Figure 3.24. Only one side of each electrocyte is innervated, but its innervation is extremely dense. This makes the electric organs the best source of ACh receptors and membrane cholinesterase for neurophysiologists.

11.3 Mechanoreceptors

The best-studied vertebrate mechanoreceptor is the muscle spindle. This resembles the crayfish stretch receptor in that it consists of one or two sensory cells whose dendrites are embedded in a muscle fiber. Stretch of the fiber causes increased firing in the sense cell. Both crayfish and vertebrate stretch receptor muscle fibers receive efferent (from the CNS) innervation.

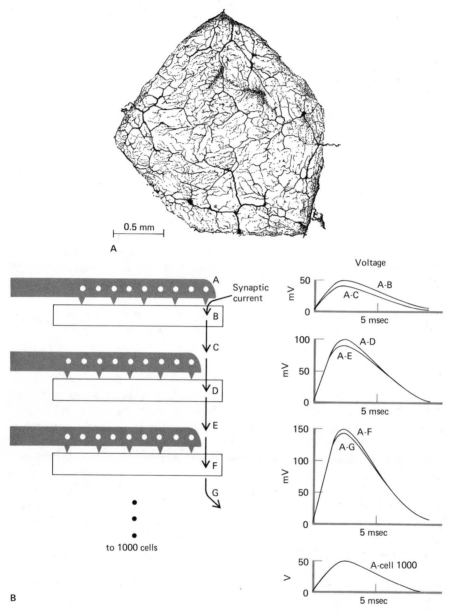

FIGURE 3.24 (**A**) Electrocyte (electric current generating cell) in a marine ray, stained to reveal fine nerve branches. The cell can be seen to be covered with branches and synaptic terminals. (After M. V. L. Bennett, *Fish Physiology,* W. S. Hoar and D. S. Randall, eds. New York: Academic Press, 1971.) (**B**) Schematic diagram of a stack of electrocytes and their innervation on one side. Each cell depolarizes primarily on one face. The graphs on the right show the synaptic potentials recorded between pairs of points: A–B, A–C, A–D, and so on. Most of the voltage across the innervated face (A–B) is seen across the entire cell (A–C). The electrocytes are in series, so the voltage increases as we include more cells in the measurement. The 1000 cells in the stack can generate about 1000 × 50 mV, or 50 V. The current is appreciable because there are many such stacks of cells, all discharging synchronously.

A typical spindle in the frog is shown in Figure 3.25. Mammalian receptors are more complex, reptilian ones slightly simpler. What happens if the spindle is stretched rapidly, with the stretch maintained for many milliseconds? The generator potential falls, even during constant stretch (Figure 3.25). The firing frequency also decreases; the frequency more closely follows the *rate* of stretch rather than *amount* of stretch.

The muscle fibers of the spindle are attached "in parallel" with the rest of the skeletal muscle, so when the other fibers contract, shortening the muscle, the spindle fibers slacken and the sensory axon is *less* excited.

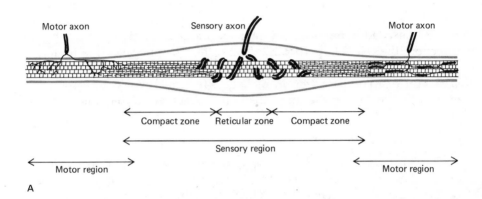

A

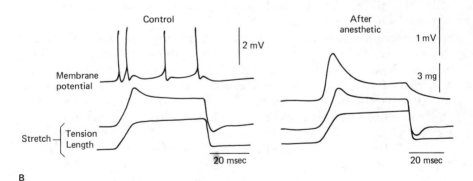

B

FIGURE 3.25 (**A**) Schematic diagram of the structure of a frog muscle spindle showing the Ia sensory neuron and the γ motor axons. The latter innervate the muscle fiber and cause a region *outside* of the sensory area to contract. Thus they cause a contraction which stretches the sensory dendrites. Other working fibers in the muscle are parallel to the spindle fiber. (After M. Ottoson and G. M. Shepherd, *Handbook of Sensory Physiology*, vol. I, W. R. Loewenstein, ed. New York: Springer-Verlag, 1971.) (**B**) Left: The spindle fiber was stretched to a given length (bottom trace), and the tension (middle trace) and intracellular potential of sensory cell (upper trace) were recorded. Right: A local anesthetic was added so that the intracellular recording revealed the generator potential rather than the spike activity. (After I. Ausmark and D. Ottoson, *Acta Physiol. Scand.* 79:321–334, 1970.)

If, instead, the tendon is stretched, spindle and other muscle fibers are all stretched together, and the neurons are excited. There are two efferents to the spindle fibers, both of which cause some contraction of the ends of the spindle fibers.[3] This in turn stretches the sensory region of the fibers and thus excites the sensory cell. Therefore if we excite the motor axons (efferents), we tend to stretch the dendrites and stimulate the receptor. In other words, commands from the central nervous system (along the γ-motor axons) can vary the sensitivity of the spindle organ to a given stretch.

One kind of mechanoreceptor found throughout the subphylum Vertebrata is the hair cell (Figure 3.26). When the hairs are bent in the direction of the kinocilium, the hair cell is depolarized, and the afferent nerve is excited synaptically. When the hairs are bent in the opposite direction, the cell is hyperpolarized, and the synaptic secretion (onto the nerve) is reduced. These cells are unbelievably sensitive. A hair cell depolarization or hyperpolarization of only a few microvolts can produce a measurable change in the firing frequency observed in the afferent axon.

11.4   Chemoreceptors

Nearly all neurons are chemoreceptors. They are sensitive to particular chemicals at low concentration and depolarize or hyperpolarize upon exposure. This is the mechanism of chemical neurotransmission, and it may serve as a model of chemoreception in general, although we know very little about the mechanism of chemoreception in sense organs subserving taste and smell in vertebrates.

Insects possess simpler chemoreceptive organs. Study of some of these classes of chemoreceptive hair has contributed much to our knowledge of chemoreception in general. Within each chitinous hair are the dendrites of several chemoreceptive cells. Each cell responds to a distinctive class of molecules. For example, in the blowfly one cell is excited by sugars, another by acids, alcohols, and salts. When the sugar-sensitive neuron is excited, feeding behavior is initiated. Excitation of the other axon initiates avoidance responses. In the carrion beetle, *Necrophorus*, some chemoreceptive neurons are depolarized (and hence excited) by carrion odor but hyperpolarized (and hence inhibited) by propionic acid and some other chemicals. The specificity of the receptors and the relationship between molecules that resemble each other in sensory excitation are unsolved problems. It is not hard to understand how one can design highly specific receptor molecules because such specificity exists in enzymes, toxins, and synaptic receptors. Superb examples of such specificity exist

---

[3] These are called gamma γ-efferents or γ-motoneurons to distinguish them from the large α-efferents, which innervate the regular power-producing muscle fibers.

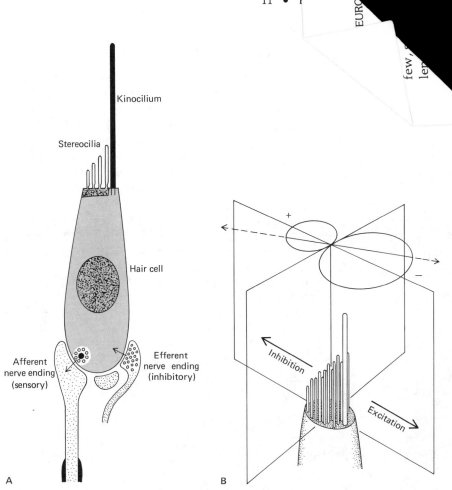

FIGURE 3.26 (**A**) Structure of a typical vertebrate sensory hair cell from the lateral line of a fish. (**B**) The hair cell is directionally sensitive. Bending in one direction results in depolarization (excitation); in the other, hyperpolarization (inhibition). Displacement in an intermediate direction depends on the components in the depolarizing or hyperpolarizing direction. (After A. Flock, *Handbook of Sensory Physiology*, vol. I, W. R. Loewenstein, ed. New York: Springer-Verlag, 1971.)

in insects, where appropriate pheromones initiate unique behavior patterns at very low concentrations. Less specific receptors are sensitive to classes or families of substance. What defines these classes? We do not have the answer yet.

11.5 Photoreceptors

Photoreception in the animal kingdom is confined to wavelengths between 300 and 1000 nm. Individual receptor cells do not span this entire spectrum. Many animals in fact are insensitive over part of this range; only a such as bees and some moths, display a behavioral response to wavelengths below 400 nm.

The spectrum of sensitivity in any receptor is determined by two things: the absorbance in the light path before it reaches the receptor and the absorbance of the visual pigment, rhodopsin. Wherever photoreceptors are found, there is rhodopsin or a closely related molecule. Rhodopsin consists of a carotenoid pigment, retinaldehyde, combined with a protein, or opsin, fragment. Retinaldehyde is the aldehyde of vitamin A_1 (an alcohol). Vitamin A deficiency can produce some blindness because of the loss of visual pigment. In general, the absorbance spectrum of rhodopsins depends on the protein half, or opsin. If the opsin is changed, the absorbance spectrum is also changed, and the visual cell is most responsive to different wavelengths. The absorbance spectrum of human rhodopsin is shown in Figure 3.27. Also plotted is the relative sensitivity of humans

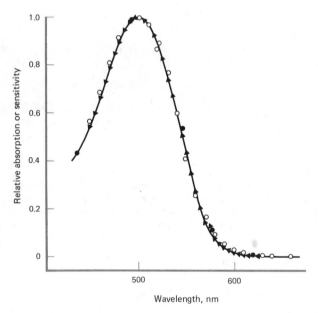

FIGURE 3.27 Absorbance of human rhodopsin in solution as a function of wavelength. Also plotted are the sensitivity of the normal human vision (corrected for lens and other pigmentation) and the sensitivity of vision in a person whose lens has been removed (because of cataracts). The three sets of points agree remarkably well. (From G. Wald and P. K. Brown, *Science* 127:222–226, 1958.)

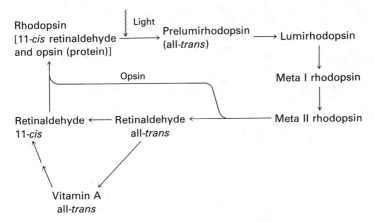

All-*trans* retinaldehyde₁

Vitamin A₂ or retinaldehyde₂

All-*trans* vitamin A

11-*cis* retinaldehyde₁

FIGURE 3.28 Retinaldehyde (retinal) and some related molecules. The double bonds in the carbon chain are in the 7, 9, 11, and 13 positions. The cycle at the bottom is a sketch of the course of rhodopsin bleaching in the vertebrate eye and its subsequent regeneration.

(after certain corrections) to flashes at different wavelengths. The agreement provides very strong evidence that the absorption of light by rhodopsin is an important early step in the conversion of light to the neural signal. Most rhodopsins are most sensitive to green light, with a wavelength around 500 nm. Some freshwater fish, amphibians, and reptiles have a rhodopsin that is most sensitive to light of 520 nm wavelength. This pigment has vitamin A_2 (see Figure 3.28), rather than A_1, combined with the opsin. Animals with color vision possess several visual pigments, in different cells, with distinctly different spectral sensitivities. These differences in spectral sensitivity are due to the differences in the opsin halves of the pigments.

The changes produced in rhodopsin are of great interest since they are necessary for vision. If a vertebrate has been kept in the dark, its rho-

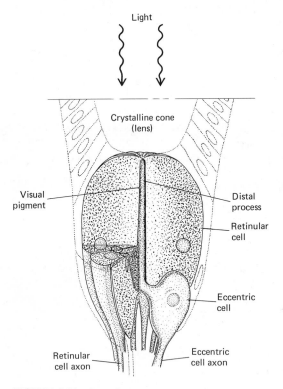

Light

Crystalline cone
(lens)

Visual
pigment

Distal
process

Retinular
cell

Eccentric
cell

Retinular
cell axon

Eccentric
cell axon

FIGURE 3.29 A single ommatidium in the eye of the horseshoe crab *Limulus*. There are about 15 retinular cells and 1 eccentric cell per ommatidium. Only the eccentric cell exhibits action potentials, and signals are sent down its axon to the CNS. The retinular cells produce generator potentials but do not send information along their own axons into the CNS.

dopsin exists primarily as one isomer. The retinaldehyde is in the 11-*cis* configuration (Figure 3.28). Light merely converts the retinal from 11-*cis* to the all-*trans* isomer. This is known as prelumirhodopsin. Prelumirhodopsin spontaneously goes through a series of transitions. In vertebrates it finally splits, with the protein separating from the retinal. In invertebrates such as squid and insects the rhodopsin does not split; rather the chain of reactions stops with the production of Metarhodopsin I. It is not known at what stage the changes in rhodopsin initiate a photoresponse in the receptor cells. There are some good reasons for thinking that it may be the transition from Metarhodopsin I to Metarhodopsin II in vertebrates.

Many photoreceptors have been studied. Of all of them the eyes of *Limulus,* a primitive arthropod, have a unique place. *Limulus,* in the hands of H. K. Hartline and his collaborators, has told us more about visual reception and processing than any other animal—and it told us first. The lateral eyes of *Limulus* are compound eyes. Compound eyes consist of many facets, each of which belongs to an *ommatidium* (Figure 3.29). Each ommatidium has its own lens, the crystalline cone, and a bunch of cells with enlarged somas and long axons leading, in *Limulus,* into the central nervous system. The rhodopsin is found in a region called the *rhabdom.* Two types of neurons are found, retinular and eccentric cells. The 10 to 15 retinular cells convert the light into a current and voltage signal. Their axons are fine and appear to be vestigial. The eccentric cell fires and sends information into the CNS. The retinular cells and the eccentric cell in each ommatidium seem to be electrically coupled—that is, connected by electrical synapses.

The response of a retinular cell to light is shown in Figure 3.30B. At very low levels of illumination (and if the eye is thoroughly dark-adapted), one can see small, randomly occurring voltage bumps. Each is due to the absorption of a single light quantum by a single rhodopsin molecule. As the light intensity is increased, the quantal responses sum to produce a larger generator potential. As we increase the light further, we see that the size of the quantal responses decreases. This means that at higher light levels each photon leads to a smaller potential change in the receptor. Because the quantal responses decrease with increasing illumination, the *total* generator potential is not proportional to light intensity. Instead it is proportional to the *log* of the light intensity. This is one form of adaptation.

We know that each quantal generator potential (these little bumps at very low light levels) is due to the absorption of a single photon by a single rhodopsin molecule. This absorption results in the isomerization of the retinal half of the molecule. Then the protein half spontaneously goes through a series of changes. The size of the generator current produced by the change in a single rhodopsin molecule varies. However, it involves the entry of more than a million sodium ions into the cell. This represents considerable amplification: one photon, one rhodopsin molecule, millions of sodium ions entering. The energy in this generator current is much larger than that in the photon. How is this possible? The rhodopsin must somehow be controlling a gate to an ionic channel. Only a little energy is needed to open the gate, but a lot of current can pass through the channel once it is opened.

The response of the eccentric cell soma to illumination is also shown in Figure 3.30B. The generator potentials in the retinular cells spread into the eccentric cell. The eccentric cell is capable of giving action potentials and fires at a frequency roughly proportional to the size of the generator potential. If we record from the axon of the eccentric cell (Figure 3.30A), we can see the action potentials because they are actively propagated. The generator potential is not.

Each receptor sends its signals centrally toward the CNS. Some receptors and sense organs also receive signals from the CNS. Spikes in these afferent axons may increase (as in the spindle) or decrease (as in hair cells) the firing frequency of the sense cell. There also are interactions *between* sense cells. Suppose that we record from the axons from two ommatidia in a *Limulus* eye and illuminate the ommatidia independently. If we illuminate the two ommatidia simultaneously, they inhibit each other (Figure 3.31). This is due to the existence of inhibitory synapses from the eccentric cell of each ommatidium onto the trigger zone (where action potentials originate) of the other eccentric cell. If the two ommatidia are farther apart, mutual inhibition still occurs, but the effect is weaker. In general, the farther apart the two ommatidia the less the inhibition.

Light intensity   Eccentric cell axon (in optic nerve)

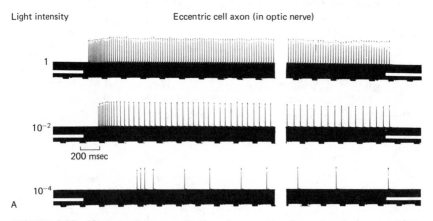

FIGURE 3.30 (**A**) Extracellular recording from the eccentric cell axon in a single ommatidium at different levels of illumination. (After H. K. Hartline, *The Harvey Lectures*, series XXXVII. Lancaster, Pa.: Science Press, 1941–42.

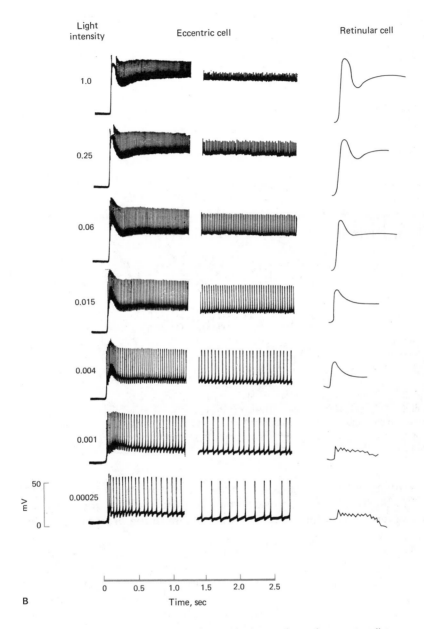

Light intensity   Eccentric cell   Retinular cell

1.0

0.25

0.06

0.015

0.004

0.001

0.00025

50

mV

0

0   0.5   1.0   1.5   2.0   2.5

Time, sec

B

FIGURE 3.30 (**B**) Intracellular records from both retinular and eccentric cell somas during illumination. Retinular cells do not spike but they are the source of the generator potential. They are electrically coupled to eccentric cells, which fire repetitively and send their signal into the central nervous system. (From M. G. F. Fuortes and G. T. Poggio, *J. Gen. Physiol.* 46:435–452, 1963. With permission of The Rockefeller University.)

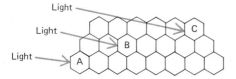

Action potentials seen in axon from eccentric cell in ommatidium A, B, or C

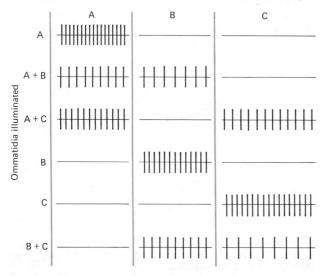

FIGURE 3.31   Illumination of nearby ommatidia inhibits the response of an eccentric cell. This is called inhibition. The closer the two ommatidia the greater the mutual inhibition. Thus B inhibits A or C more than A and C inhibit each other. The inhibition of A by B is roughly proportional to the firing frequency of B.

Why have such a system? Is there any advantage from this mutual lateral inhibition? Yes, it is so useful that it is universal; if it does not occur directly between receptors, it occurs somewhere in the processing of sensory data.

The advantages of lateral inhibition are as impressive as those of phasic, as opposed to tonic, receptors. We want to be maximally sensitive to changes. We are relatively uninterested in uniform and constant stimuli. If you look at a world that is white on the left side and grey on the right side, it is wasteful of information for every receptor to describe how bright its narrow segment of the world is. It is better for the ones at the edge merely to tell you how much of a change there is. Figure 3.32 shows an actual recording from ommatidia when half the *Limulus* eye was illuminated with bright light and the other half with dim light. The firing rate is greatest near the edge of the bright region, lowest near the edge of the dim

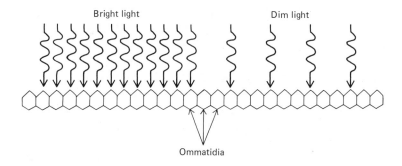

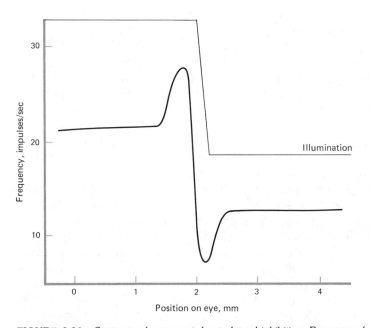

FIGURE 3.32 Contrast enhancement due to lateral inhibition. Response of eccentric cells from ommatidia at different points along a region half illuminated with bright and half with dim light. The cells near the edge are inhibited more (dim area) or less (bright area) than their neighbors. The edge is more pronounced when encoded in the level of cell activity than is the actual variation in light intensity.

one. Contrast is enhanced. This is reminiscent of the Xerox process. Edges are clearest and uniform shades reproduce less well. The same phenomenon can be demonstrated in dozens of ways in our own visual perception.

Some eyes can code for far more than the intensity of light. Some can transmit information on the plane of polarization and/or the spectral composition of light. For the past quarter century there has been strong behavioral evidence that some arthropods can perceive the plane of polarization. Much more recently we have been able to record from insect retinular cells and prove that their response depends on the orientation of the plane of polarization. It appears that in each retinular cell most of the rhodopsin molecules have a specific orientation. Each molecule absorbs light best if it has a certain plane polarization. Since the molecules tend to be aligned, the entire cell also displays this polarization sensitivity. In our eyes the rhodopsin molecules are not as well aligned, and we have very little polarization sensitivity.

We all know that the spectral composition of light is important information. We have color vision. Color vision occurs sporadically among crustacea, insects, cephalopods, and vertebrates. Most mammals have little or no color vision, but higher primates have excellent color perception. How is this achieved? The correct explanation was provided by Young and Helmholtz in the nineteenth century. Suppose that an animal had only one pigment, with a typical spectral sensitivity curve as shown in Figure 3.33A. A strong stimulus of 420 nm bleaches as much pigment as a weak one of 535 nm. The cell responds identically to the two stimuli. Suppose that we had two types of receptor cells, each containing its characteristic pigment. One absorbed maximally at 535 nm, the other at 565 nm. If we illuminate with light of a single wavelength (monochromatic), we obtain a unique ratio of responses in the two cells. The ratio depends on the wavelength of the stimulus. We can distinguish hue and intensity of a monochromatic stimulus. Suppose we illuminate with a white light; this will stimulate the two cells in some ratio—let us say, 1:1. This would be indistinguishable from a monochromatic light producing that same ratio. With three types of receptors and with three pigments we cannot duplicate the effect of white light with any single monochromatic stimulus. It should be possible to duplicate the effect with certain *pairs* of monochromatic stimuli. This is indeed true, and these are the pairs of complementary wavelengths (red-green, yellow-blue). Another way of looking at it is that with one pigment we can only perceive intensity and hue; with three pigments we can perceive not only intensity and hue but saturation as well.

We indeed have three different kinds of cones (one type of vertebrate visual receptor cell), each with its own pigment. The absorption maxima are around 440, 530, and 565 nm. Our spectral sensitivity is limited by the shapes of the absorption spectra. At 700 nm even the "red" cone is relatively insensitive, and the "blue" completely insensitive. At 1000 nm our threshold is so high that we can feel the heat on our face before we can see the stimulus. Our color sensation is still a weak red. Our sensitivity to short wavelengths—ultraviolet light—is limited by the absorption by the lens. Persons whose lenses have been removed are more sensitive than the rest of us to light below 400 nm. However, the color sensation is still of violet.

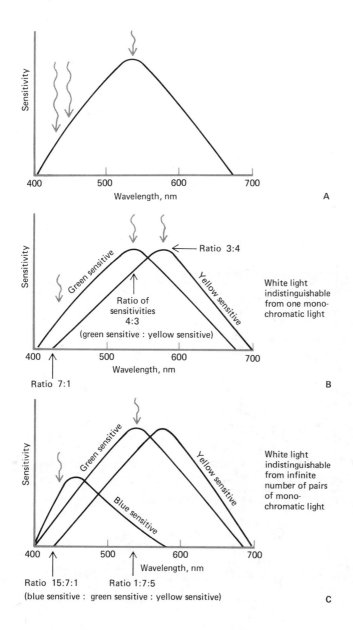

FIGURE 3.33 Color vision in animals with one, two, and three visual pigments. (**A**) With one visual pigment a bright light at 420 nm is fully equivalent to a weaker light at 535 nm. The animal is color blind. (**B**) With two pigments each wavelength excites the two pigments in a unique ratio. However, white light (all wavelengths) also excites in some ratio that corresponds to some monochromatic light. Animal has a poor color vision. (**C**) With three pigments it takes a pair of monochromatic lights (complementary colors) to mimic white light. Human color vision uses three visual pigments. Complex frequency analysis of light is not possible with only three pigments. Our hearing makes a far more sophisticated analysis of the sonic frequency composition of sounds than our vision makes of the electromagnetic frequency composition of lights.

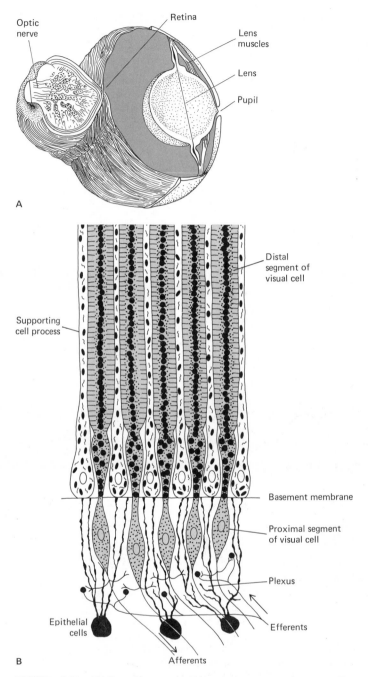

FIGURE 3.34 (A) Eye of the octopus. Note similarity to vertebrate eye. One notable difference is that the retina is not inverted, so the optic nerve does not leave through a hole in the retina (contrast Figure 3.35). (B) Cell structure of octopus eye. (From J. Z. Young, *The Anatomy of the Nervous System of* Octopus vulgaris. New York: Oxford University Press, 1971.)

Eyes of great complexity have evolved independently several times. The lateral eye of *Limulus*, with its many ommatidia, each with its own lens, is called a compound eye. Other arthropods have eyes of far greater complexity. They display better spatial resolution and may possess color vision and detect the plane of polarization. Other animals, including vertebrates, possess camera eyes. Such an eye has a lens and other light-focusing structures. Moreover, it can change the plane of focus. The image is focused on a layer of photoreceptors in part of the retina. The amount of light reaching the retina can be varied by several means, such as variable pupil and the migration of pigment cells. The most advanced camera eyes are found in cephalopods, such as the octopus (Figure 3.34), and in vertebrates (Figure 3.35).

An octopus can direct its eye toward a target. It can move the lens back and forth to change the focus. A pupil adjusts to the light level; the two eyes can vary pupillary size independently. The density of receptor cells in octopus is about 70,000/mm² compared to the 60,000/mm² peak density in man. (The receptor density in one region of the eyes of eagles may reach 1,000,000/mm².) In humans about one-third of the light hitting the retina, even at 500 nm, is absorbed by rhodopsin. In squid the figure is between 50 and 90 percent. This high absorption is also achieved in some fish. In summary, a squid or octopus eye is comparable optically to typical vertebrate eyes such as our own.

Focusing by a lens depends on the differences in refractive indices of the lens and surrounding materials. The cornea has a refractive index of 1.38; air, 1.00; the aqueous humor, 1.33; the lens, 1.45. Thus in terrestrial vertebrates the cornea does most of the focusing; the lens performs fine focusing. Fish and amphibians move the lens back and forth to adjust the

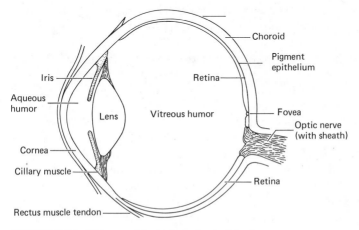

FIGURE 3.35 Diagram of the basic features of the human eye.

focus. Higher vertebrates alter the lens curvature. The eye is focused at infinity unless the ciliary muscles contract. Their contraction actually results in a release of the tension on the lens. The lens then rounds into a more convex shape, and the eye is focused on nearer objects. The lens stiffens with age and no longer springs back completely to its convex shape upon ciliary contraction. This is the cause of farsightedness in older people.

The refractive index of water is 1.33, so the corneas of aquatic animals do little focusing. Instead, fish lenses perform most of the focusing. Their indices of refraction may be 1.67. This is obtained by precise layering of materials with refractive indices ranging from 1.33 to 1.53. Remarkably, fish lenses are very nearly spherical and yet are nearly free of both spherical and chromatic aberration.

We have duplex vision. By this is meant that we have two parallel systems in each eye: rod and cone systems. Rods contain a rhodopsin that is maximally sensitive to light of about 500 nm. The majority of receptors over most of the retina are rods. Cones come in three types, depending on the wavelengths maximally absorbed by their visual pigment. There are blue, green, and red (really yellow) cones. They are most numerous in a region near the center of the retina, the fovea. The fovea has only cones, no rods. As we go from the fovea toward the periphery, cones constitute an ever-decreasing fraction of the receptors. Cone vision is color vision. Rod vision is color-blind. We are essentially color-blind over most of our eye. However, when we look at something, we focus the image on our fovea, and therefore objects appear colored.

There is another reason we focus on the fovea. A single cone may provide the sole input to a ganglion cell, whose axon runs in the optic nerve into the brain. As many as 1000 rods may feed into one ganglion cell. Because of overlap the ratio of rods to ganglion cells is about 100:1 in man. This means that foveal vision has far greater acuity. We can resolve two points separated by an angle of 26 seconds. Acuity in the periphery is much, much poorer.

We use rods in dim light and cones in brighter light. The rod system is more sensitive because of the convergence of the information as it is carried centrally. Also, individual rods themselves appear to be more sensitive than cones. On the other hand, rods adapt more to moderately bright light, as does the neural network leading from rods to the CNS. The result is that in dim light we see mainly with peripheral rods, in bright light with foveal cones. In light intensities between bright moonlight and dim twilight we use both systems. The switch from rod to cone vision accounts for some of our adaptation to bright lights. By changing the pupil area, the light striking the retina can be varied perhaps tenfold. Still there are other mechanisms involved, since we can see over a light intensity range greater than 10^{10}. Some of this adaptation occurs in the photoreceptors themselves.

Some of it is due to adaptation of other cells in the pathway through which information reaches the CNS.

11.6 Hearing

Mechanoreceptors are used for things other than simple stretch and pressure detection. Vertebrates and some invertebrates possess statocysts. These consist of a fluid-filled compartment with a row or more of hair cells. One or several pieces of heavier material is in the fluid and presses on some of the hairs. The animal can determine which way is down by which hairs are stimulated. In 1893, Kreidl replaced some of these otoliths ("ear stones") in a shrimp statocyst with iron filings. When the shrimp was exposed to a strong magnetic field, it oriented accordingly. "Down" was in the direction of the magnet.

Some hearing organs bear many resemblances to equilibrium organs. Mammalian ears have evolved from the equilibrium organs and are part of the same inner ear structure. Sound detection is a problem of vibration detection. We can detect low-frequency sounds by placing our fingertips on a vibrating surface. Good ears are far more sensitive to airborne sound and may detect higher frequencies. All ears do this by converting vibrations of air to those of a solid or liquid. Sound in air consists of both a pressure wave and a displacement wave. Pressure in a gas is exerted in all directions, in contrast to the highly directional gas displacement. All mammalian ears are pressure detectors. Pressure reception requires a massive chamber that is opaque to sound. This should be closed by a very stiff diaphragm whose displacement is very small. We can hear sounds that move our ear drums only 10^{-11} cm! In contrast, insect hearing uses displacement receptors. For this the diaphragm should (and does) have a low mass and move on hinges of low resistance.

In either, the movement of air is converted to the motion in a liquid. The power contained in a large volume of moving air can move only a small volume of water. Therefore there is always mechanical reduction. For example, air moves mammalian eardrums or insect tympani. The surface area of the tympanum is much larger than the tissue or fluid which is moved.

Figure 3.36 shows the structure of a simple insect sound receptor organ. Airborne sound moves the tympanic membrane. This is only a few microns thick and "floats" back and forth in response to the stimuli. It serves as a piston that moves a smaller but more rigid object, the tip of the capsule that contains the dendrites of the two sensory neurons. If we record from the sensory neurons, we can see an increase in firing frequency during sonic stimulation. The receptor in this moth responds well to frequencies between 30 and 60 kHz—beyond our range. This is the

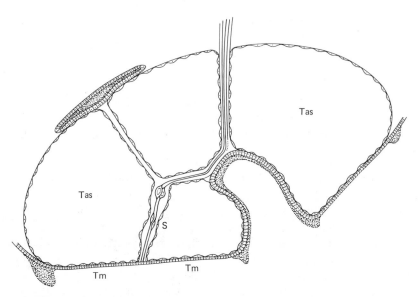

FIGURE 3.36 Tympanic organ of a noctuid moth. The displacement wave moves the tympanic membrane (Tm) and thus the dendrites of the two sensory cells (S) which send their axons down nerve III N1b. The tympanic membrane is light, compliant, and backed by tympanic air sacs (Tas). (From K. D. Roeder and A. E. Treat, *J. Exp. Zool.* 134:127–157, 1957.)

frequency range of the cries of some bats that feed on moths. Bats can detect and intercept moths when they are about 10 feet away. Fortunately for the moths, they can hear bats 100 feet away and take evasive action.

It is one thing to detect sounds and another to discriminate frequencies. Male mosquitoes have hearing organs that are sharply tuned to the frequency of the wing beat of the females. However, they respond identically to other tones if these are loud enough. Crickets and their relatives emit complex calls. However, they seem to listen to chirp duration, repetition rate, and perhaps some amplitude changes of the tone within each chirp, but not to the tone itself. How might tones be discriminated? If the neurons responded with an action potential for each pressure or displacement wave, then the firing frequency would represent the sonic frequency. Another mechanism is analogous to that for color vision. If several neurons or organs responded optimally to different frequencies, the animal could deduce the tone by comparing the responses. Longhorn grasshoppers have good hearing organs. Each organ has many sensory neurons with different optimal tones. The central processing of the tones is what one would expect if the animal were discriminating frequencies.

However, there is scant evidence that even these insects can indeed discriminate tones.

Mammals have the best hearing. We can hear from 20 to 20,000 Hz (if we are still young) and distinguish tones 0.3 percent apart. We can discriminate between two pure tones differing by 10 percent even if we hear only two cycles of each. We also are very sensitive to modulation of tones (FM). This ability seems to be central to our understanding of speech. We can hear over a large intensity range: 10^{12}, or 120 dB. No other sense organ in the entire animal kingdom matches this range.

Sound enters the external ear and moves the eardrum (Figure 3.37A). Since this is a pressure transducer, it moves very little. The vibrations are then transmitted to the cochlea via a series of middle ear bones. Over 50 percent of the energy in the moving eardrum is transmitted to the cochlea. The ratio of areas of drum to cochlear cross section is the same as the relative compliance of air and cochlear fluid. This means that the stapes, at its attachment to the cochlea, vibrates about the same distance as the drum. The intact eardrum and ossicles also will vibrate if the sound is conducted through the skull rather than through the air. This is particularly true for low frequencies.

The ossicles insulate the cochlea from excessive vibration since the joints buckle during very loud sounds. We also have muscles in the middle ear region. Contraction reduces sensitivity, and we reflexly contract them while talking. Bats emit piercing cries to echolocate objects and then listen to the weak echos. Their inner ear muscles contract just before each cry and relax just after, so their hearing is not adapted to high intensities.

Vibrations of the oval window, adjacent to the stapes, initiate compressional waves in the cochlear fluid known as perilymph (Figure 3.37A and B). At any but very low frequencies this causes vibrations in the basilar and Reissner's membranes. Vibration of the basilar membrane causes a shear between the tectorial and basilar membranes. The hair receptors are located at the juncture of these two membranes. These hairs are very similar to the hair cells of lateral line organs (Figure 3.26) and chemically synapse onto neurons whose axons run in the auditory nerve and enter part of the brain, the medulla. The hair cells are also innervated by efferents presumed to be inhibitory. The entire system—tectorial and basilar membranes, hair cells, primary sensory neurons—is known as the organ of Corti.

Recently some workers have recorded intracellularly from the nerve terminals in the organ of Corti. From this and related work the following picture emerges. The hair cells are much like lateral line hair cells. They depolarize in response to a displacement in one direction and hyperpolarize to the opposite movement. They are tonic and respond to displace-

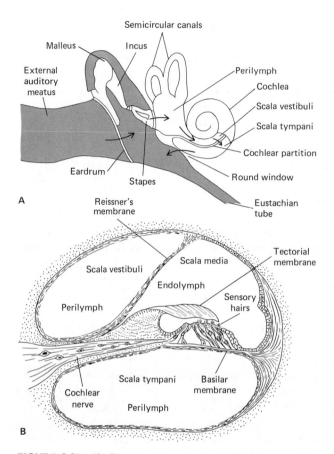

FIGURE 3.37 (**A**) Diagram of the structure of the inner ear. The eardrum moves the malleus, incus, and stapes, and ultimately generates waves in the perilymph in the cochlea. (After Helmholtz, 1863.) (**B**) Cross section through organ of Corti. The compressional waves in the perilymph cause movement of the basilar membrane relative to the tectorial membrane. This excites the hair cells mechanically. (After Politzer, 1873.)

ment, not velocity. The change in hair cell potential modulates excitatory synaptic secretion onto the nerve terminal.

If we monitor the firing pattern in the sensory axons, we will see that they fire spontaneously, typically from 10 to 200/sec. At very low vibration frequencies the entire basilar membrane may move in synchrony with the compressions of the perilymph. The axons tend to fire in synchrony. Above approximately 50 Hz the basilar membrane ceases to vibrate in a simple fashion. Rather, traveling waves of vibration pass down the membrane. However, at 500 Hz an individual hair cell is bent in each direction 500 times each second. This can cause the sensory axon to fire at a rate of 500/sec. Thus individual axons follow the sonic vibrations, but all the axons are not in synchrony since the basilar membrane does not move up and down uniformly. Recordings from the acoustic nerve show some axons following on a 1:1 basis up to perhaps 1500 Hz. These cells fire at the highest frequency in the mammalian nervous system and must have a very short refractory period. Unfortunately, a cell firing 1000 times per second to a 1000-Hz stimulus cannot code for intensity at the same time. Also, even humans can hear frequencies up to 20,000 Hz; dogs, 30 to 50 kHz; and echolocating bats and porpoises, 150 kHz.

How then do we code for high frequencies? The great physicist and physician Helmholtz proposed that the basilar membrane resonated to the different frequencies. Each tone would cause a different part of this membrane to resonate and stimulate different hair cells. Amplitude would be coded by the firing frequency or action potential number. The crucial concept is that the "what" of frequency is coded in terms of "where" on the basilar membrane and "which" hair cells and axons depolarize. This would explain why there are so many hair cells along the basilar membrane. The information they carry into the CNS would not be redundant according to this scheme.

Georg von Bekesy, another physicist and physiologist, carried the model farther and provided what is the generally accepted interpretation of frequency coding. The individual segments of the basilar membrane do not resonate. The compressional waves in the perilymph initiate a traveling wave in the basilar membrane (Figure 3.38A). Traveling waves are familiar to anyone who has played with a jump rope. If we shake the rope up and down a few times at one end, a wave of movement travels down the cord. The basilar membrane is not like a rope. It is not uniform in diameter, stiffness, and elasticity. A wave traveling down it varies in amplitude. The position of maximum amplitude depends on the sound frequency. The end near the stapes is displaced more than other points at high frequencies; the displacement of the other end exceeds other regions at low frequencies. Plots of peak movement versus basilar membrane position exhibit broad maxima (Figure 3.38B). The peak is broad enough so that for us to resolve frequencies as well as we do, we would have to have mechanisms for sharpening the maximum.

Much of the sharpening of frequency resolution is accomplished centrally. The principle, however, is one we are familiar with. It involves lateral inhibition. Recall the three rules of such inhibition. Sensory cells inhibit their neighbors. The closer the cells the greater the inhibition; the greater the firing rate of a cell the more it inhibits its neighbors. This will sharpen a broad, blunt peak of neural excitation in a mass of neighboring cells to a more pronounced and distinct maximum.

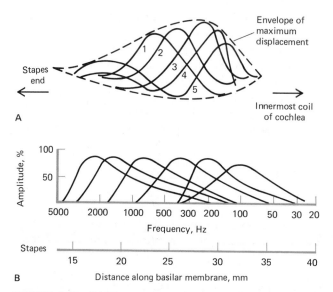

A

B

FIGURE 3.38 (**A**) Vibration pattern in a mammalian basilar membrane. During a pure tone a wave travels along the basilar membrane in the sequence shown—1 through 5. The dotted line shows the envelope of maximum displacement for this tone. (**B**) Envelopes (scaled to equal peak amplitude) of vibration for six pure tones. At high frequencies the maximum is at the stapes end; at lower frequencies it moves down the basilar membrane.

12 • REFLEXES

Stimulation of receptors can result in excitation of effectors. The simplest circuit involves no neuron-neuron synapses. In nematodes a strong touch stimulates receptors in the head. Axons from these synapse directly on the muscle. These may be the only cases of "no-synaptic" reflexes. In other organisms the simplest reflexes are monosynaptic; that is, they involve only one central, neuron-neuron synapse.

A few monosynaptic pathways have been studied in invertebrates such as the "medicinal" leech. Mechanical stimuli to the leech's skin cause contraction of some muscles in the same segment. There are mechanoreceptors in the skin responsive to touch (T), pressure (P), and noxious (N) stimulation. The T cells electrically synapse with the motoneurons. The P cells synapse both electrically and chemically. Why some of the reflexes are via electrical and others chemical synapses is not known.

Vertebrates have countless reflexes. Most of these reflexes are mediated in the spinal cord and can occur in a decapitated animal—the proverbial "chicken with its head cut off." There are also reflexes involving circuits in the brain: sneezing, blinking, pupil adjustment, to name a few. Of all vertebrate reflexes the best studied is the muscle stretch reflex. We have all observed this. The doctor hits our knee tendon with a hammer. This stretches the attached muscle and reflexly initiates contraction of the same muscle.

Much of the classic work on reflexes in general was done by C. S. Sherrington and was well summarized in his book published in 1908. In the 1920s Sherrington observed a new type of phenomenon. When he stretched the leg muscle of a cat whose brain had been removed, the muscle pulled back. If he cut the nerve to the muscle, the response disappeared. These reactions required a reflex arc: sensory neuron to spinal cord, spinal events, motoneuron to muscle. The muscle only pulled back when stretched. It *actively* resisted deformation. The reflex tended to keep the muscle length (and hence joint angle) constant. This has subsequently been shown in the posture of both cat and man. If, for example, your legs start to buckle, the muscle is stretched, and reflex contraction follows. This is also an example of a negative feedback mechanism (see Chapter 1, Sections 3 and 4). A sensor (the spindle fiber) detects a departure from the norm and sends an error signal to the central data processing unit (CNS), which can send out a command (action potentials in the motoneuron) to correct the error (contract the muscle).

When a muscle is stretched, all the fibers in it are stretched. These include the spindle fibers as well as those responsible for the power exerted by the muscle. Stretch of the spindle fibers sends signals along the Ia sensory fibers (Figures 3.25 and 3.39). These synapse only on motoneurons that excite the same muscle or a close synergist (another muscle cooperating in the same movement). Each Ia afferent signal produces only a small EPSP (0.5 mV) in any one motoneuron. However, a stimulus as strong as a tendon tap stimulates enough Ia afferents to depolarize a motor axon above spike threshold. Note that spindle receptors follow the *rate* of stretch (Section 11.3). This is an example of derivative control feedback (see Chapter 1, Section 4).

This is fine for a strong stretch, as with a reflex hammer. For most situations—for example, to maintain our posture—we require finer control. How does weak stimulation of the spindle initiate weak muscular contractions? This is possible because some motoneurons are always very near threshold while others are more hyperpolarized. The cells near threshold are part of the "subliminal fringe"—cells just below firing threshold. As we stimulate more and more Ia afferents more and more strongly, we increase the number of motoneurons firing. Simultaneously we bring others nearer threshold. In this manner the number of motoneurons reflexly excited is finely graded with the stimulus intensity.

The firing rate of the Ia neurons is not only controlled by the stretch but also by the activity in the γ-efferent (Section 11.3). When we give a

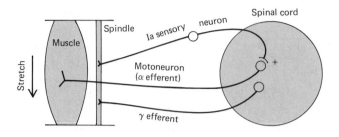

A

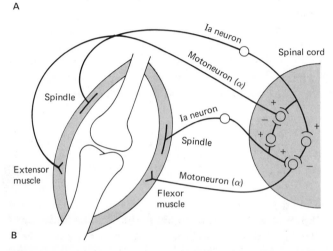

B

FIGURE 3.39 Mammalian tendon-stretch reflex. (**A**) Stretch initiates spiking in the Ia neuron, which reflexly excites the α-motoneuron. (**B**) Stretch of the flexor muscle excites the Ia neurons from its spindles. They reflexly excite the α-motoneurons to the flexor. The Ia fiber also excites interneurons that inhibit the α-motoneurons leading to the extensor muscle. This is a disynaptic reflex. The inhibition is called "reciprocal inhibition," since extensor stretching inhibits flexor contraction as well.

command to the α-motoneurons, we simultaneously send a signal down the γ-efferents to the spindle fibers. Although the muscle is contracting, the stretch receptors continue to fire because of this γ-efferent activity. If the muscle contraction is too slow, the signal from the receptors will reflexly speed up contraction. If the contraction is too fast, the Ia fibers become silent and thus reduce motoneuron activity.

The Ia afferents also send branches to a group of interneurons, that is, neurons that lie between the sensory and motor neurons in a neural pathway. These interneurons are inhibitory to the motoneurons that innervate the antagonistic muscles. As an example, when we stretch the knee tendon we reflexly (monosynaptically) excite a muscle that straightens the leg. We also reflexly inhibit (disynaptically) muscles that bend the knee. This is a general property of spinal organization. A signal that excites one class of motoneurons inhibits those to the antagonistic muscle. However, we can override this with sufficiently strong commands from the brain.

Control from the motor centers of the brain descends the spinal cord and acts on motoneurons directly and indirectly through interneurons. Descending signals can switch from one spinal pattern to another. They could also directly dictate motoneuron and γ-efferent activity or merely alter the excitation level of the motoneuron pool.

In the tendon stretch reflex we must first provide a sensory stimulus to initiate muscle contraction. One could build up a system of enormous complexity based on the reflex principle. In such a system a stimulus initiates a reflex that straightens the leg. This straightening in turn initiates a reflex that bends the leg and so on to produce a walking movement. On the other hand, complex sequences of movement can be preprogrammed in the central nervous system. If we cut the sensory nerves from the wings in a locust, we can still obtain the proper sequence of signals in the motor axons to produce flight. Some marine snails exhibit prolonged stereotyped escape behavior. The ganglia of the snail can produce the proper motor sequence and intensity even when they are isolated in a dish of seawater. In these and other cases the central nervous system of the animal possesses a network of specific connections and neuronal interactions that can produce a highly stereotyped but fairly complex behavior. What is the role of sensory input in such a system? First, sensory input can initiate this pattern or, in some cases, terminate it. It also can initiate a switch from one pattern to another. Second, sensory information can slightly modify or adjust the pattern in some cases. For example, signals from the mechanoreceptors in the locust wing can accelerate the wing beat. They also can initiate in-flight adjustments to compensate for any roll or yaw.

13 • BRAINS AND CENTRAL NERVOUS SYSTEMS

A "higher" organism is one belonging to a group with a well-developed nervous system. Certain trends are observed as we go from simple to advanced nervous systems. First is the concentration of cell bodies into plexuses and ganglia. Second is the tendency of ganglia to coalesce into a central nervous system. The more peripheral nervous circuits lose their autonomy and become more subject to central control. Third, there is a movement of nervous tissue forward and near the complex sense organs in the head. Among the invertebrates this cephalization reaches its peak in cephalopods, such as the octopus (Figure 3.40). In higher vertebrates new

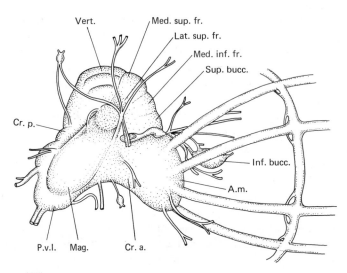

FIGURE 3.40   The octopus brain. Major lobes shown are: p.v.l.—pallioviseral lobe; cr.p.—posterior chromatophore lobe; mag.—magnocellular lobe; vert.—vertical lobe; med. sup. fr., lat. sup. fr., med. inf. fr.—median superior, lateral superior, and median inferior frontal lobes; sup. bucc.—superior buccal lobe; inf. bucc.—inferior buccal lobe; a.m.—anterior suboesophageal mass; cr.a.—anterior chromatophore lobe. (From J. Z. Young, *The Anatomy of the Nervous System of Octopus vulgaris.* New York: Oxford University Press, 1971.)

nervous tissue has evolved in certain areas (Figures 3.41 and 3.42). Control likewise shifts to the brain, and it is there that the highest processes—learning and memory—occur. In locusts, however, it has been shown possible to classically condition the ganglia in an isolated thorax. In invertebrates the higher processes may be less confined to the brain.

A distinguishing feature of most invertebrate nervous systems is the paucity of cells. Whereas vertebrate brains perform with millions of neurons, invertebrate ganglia may contain hundreds or thousands. Typically these are arranged with the cell bodies in an outer ring or rind. The synaptic region is separated from the somas in these monopolar cells. The contacts are in a central core or neuropile. Vertebrates have mainly multipolar cells; somas, dendrites, and axons exist side by side. There are large tracts of axons, often myelinated. These tracts glisten because of the fatty myelin—hence the term white matter. The gray matter may have a layered appearance because of the layering of cell bodies.

FIGURE 3.41   (Opposite page) (**A**) Brains of a cod, goose, and horse. Roman numerals are cranial nerve stumps. (**B**) Some of the known and assumed connections (or wiring) within the mammalian brain. (From A. S. Romer, *The Vertebrate Body.* Philadelphia: Saunders, 1955.)

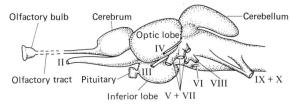

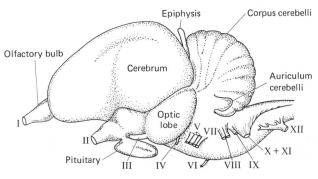

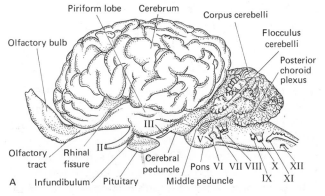

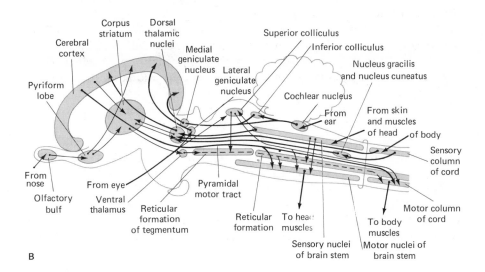

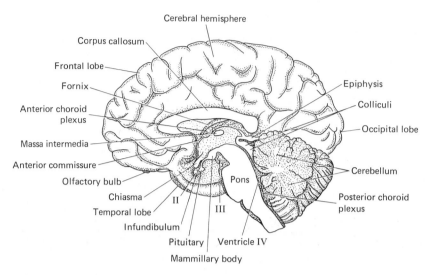

FIGURE 3.42 Right half of human brain as seen from midline. Most of temporal lobe is located on the lateral surface and hidden in this view. (From A. S. Romer, *The Vertebrate Body.* Philadelphia: Saunders, 1955.)

Many neurons in invertebrate ganglia are individually identifiable. The neurons in the ganglia of the marine snail *Aplysia* serve as fine examples. Somas are characterized by position, size, color and firing pattern, afferent input and efferent output, and neurotransmitter sensitivity. A map of one ganglion is shown in Figure 3.43A and B. As an example, we know that cell R_{15} displays bursts of action potentials and responds to local application of ACh with very long inhibitory potentials. Because of the identification of individual cells in each animal, we can deduce some of the circuitry in the ganglion (Figure 3.43C).

We can contrast this situation with that in vertebrates. Figure 3.44 shows some of the cell types in the mammalian cerebellum. There are no individually identifiable cells. Rather, there are identifiable cell types or classes. The cells of one class are morphologically indistinguishable and perform similar functions over a substantial region of the brain. For example, countless similar Purkinje cells receive excitatory inputs from countless similar granule cells and countless similar climbing fibers. They also all receive inhibitory input. Why are there so many similar cells? One reason is that different Purkinje cells may process information about different parts of the body. Indeed, the anatomy of the entire vertebrate brain is highly patterned. If we know a cell's type and location, we know approximately to what neurons and in what regions it will make contact. We also know that its neighbor, if of the same cell type, will make similar connections, but the neighbor's input and output will predictably differ

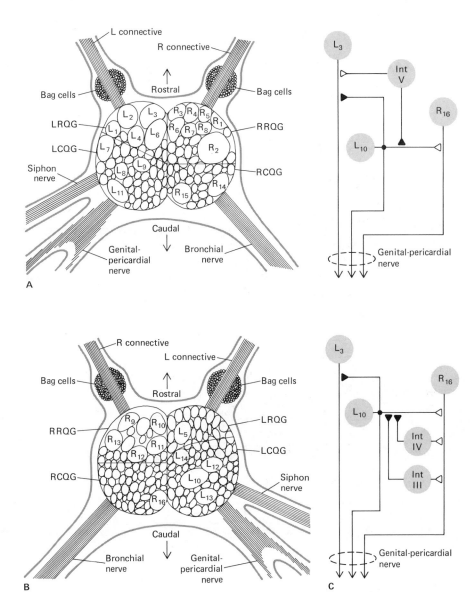

FIGURE 3.43 (**A**) and (**B**) Schematic representations of the anatomy of the abdominal ganglion of *Aplysia*. (**A**) Dorsal view, (**B**) ventral view. The location of some of the identified cells is shown. $L_1—L_{13}$ are cells whose somas are located in left half ganglion; $R_1—R_{16}$, in right half ganglion. RRQG—right rostral quarter ganglion; LRQG—left rostral quarter ganglion; RCQG—right caudal quarter ganglion; and LCQG—left caudal quarter ganglion. (After W. T. Frazier et al., *J. Neurophysiol.* 30:1288–1350, 1967.) (**C**) Schematic representation of two of many known circuits in this ganglion. Open triangles denote excitatory connection; filled triangles are inhibitory synaptic connections. Int refers to visually unidentified interneurons. (After E. R. Kandel et al., *J. Neurophysiol.* 30:1351–1376, 1967.)

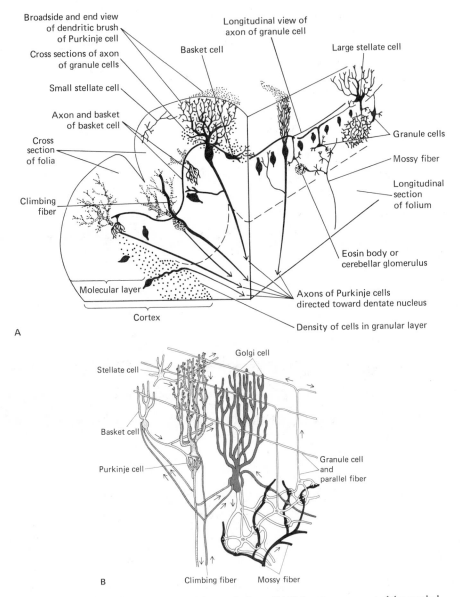

FIGURE 3.44 Synaptic organization of the cerebellum. (**A**) Only a tiny segment of the cerebellum is shown. Structure is repeated countless times as we progress along the longitudinal axis of the folium (perpendicular to the plane of Purkinje cell dendrites). Cell body layers comprise gray matter; axon tracts are white matter. (From J. Minckler, *Introduction to Neuroscience,* J. Minckler, ed. St. Louis: Mosby, 1972; after Ramussen.) (**B**) Interconnections of the neurons in the cerebellar cortex. The pattern is elaborate but stereotyped. Arrows indicate direction of nerve conduction. See Figures 3.2 and 3.15 for Purkinje cell branching and innervation by climbing fibers. (From The Cortex of the Cerebellum by R. R. Llinás. Copyright © 1975 by *Scientific American, Inc.* All rights reserved.)

slightly in a systematic way. Suppose that the cells we are discussing process sensory information, such as touch. Then, as we go from cell to cell in this brain layer, we may find a systematic variation in the parts of the body that each cell deals with. One cell may process touch information from the tip of a toe, and, as we progress along this axis, we successively

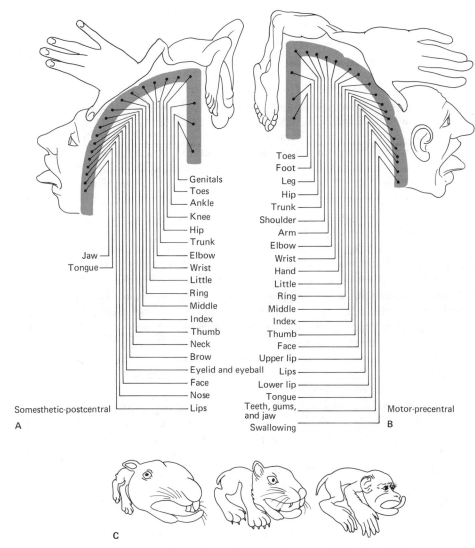

FIGURE 3.45 Map of sensory (**A**) and motor (**B**) regions in the cerebral cortex. Representation is ordered and heavily weighted—differently for sensory and motor regions. (**C**) Weighted drawings of rabbit, cat, and monkey based on the relative areas devoted to the body parts in the ventrobasal thalamic complex. (After J. E. Rose and V. B. Montcastle, *Handbook of Physiology,* Neurophysiology Section, vol. I, H. W. Magoun, ed. Baltimore: Williams and Wilkins, 1959.)

encounter cells processing information from the middle of the toe, then the base of the toe, then the ball of the foot, and so on (Figure 3.45A).

Studies of brain function have been concerned with several levels of localization. On the grossest level, scientists have tried to determine the role of the entire brain itself or of gross subdivisions, such as the cerebellum (Table 3.3). Much of this information comes from actual removal or destruction of large neural areas. On a finer level, neurophysiologists have long recognized that major brain areas are functionally subdivided. For example, the cerebral cortex includes regions devoted to visual processing (visual cortex) and others devoted to auditory and motor processing (auditory and motor cortices). We have obtained a great deal of information on this finer regionalization by selective destruction, stimulation, or gross electrical recording. Specific information about localization in the human brain has been obtained from the wealth of neurological data from persons with brain tumors, clots, aneurysms, and wounds. There also has been some direct stimulation and recording from human brains during surgery. Since the neuron is the unit from which nervous systems are constructed, there has been much work devoted to ascertaining what information individual neurons carry. Clearly this depends on the neurons' location and class. Extracellular recordings from single cells in well-defined brain regions have been very informative.

There is regional specialization in the central nervous systems of invertebrates. However, although the role of entire ganglia is often known, we have little information on regionalization within a ganglion. Such grouping may indeed be minor. For example, in *Aplysia*, where we know the properties of many individual cells, there is little apparent functional grouping within ganglia. *Octopus*, however, has a large brain with perhaps 150 million neurons. J. Z. Young and his associates have elucidated some of the pathways for particular functions. There are some specific regions, such as the optic lobes, devoted to equally specific functions, such as memory storage.

Vertebrate brains are divided into three main divisions: forebrain, midbrain, and hindbrain (Table 3.3). These sections are present throughout vertebrate evolution, but some regions are more elaborate in higher vertebrates (Figures 3.41 and 3.42). During development the brain starts as a fluid-filled tube. In higher animals large outgrowths of this tube develop. These include the cerebrum (forebrain, tectum), midbrain, and cerebellum (hindbrain). They are associated with three major sense organs, nose (forebrain), eye (midbrain), and ear and lateral line (hindbrain). In mammals most visual processing has moved to the forebrain.

Table 3.3 lists some of the regions and associated functions. Some functions, such as skeletal movement, involve many regions. We discussed the simple monosynaptic spinal reflexes. Other, more complicated

TABLE 3.3 Roles of Brain Subdivisions

Brain Region	Some Functions
Hindbrain:	
Medulla	Part of reticular system: sleep and attention; vital functions such as respiration, cardiovascular function; cranial nerves IX, X, XI, XII; VIII (see Chapter 5)
Pons	Cranial nerves IV, V, VI, VII; some reticular system; motor and sensory nerves to eye muscles, jaw, and face
Cerebellum	Control of equilibrium and movement, particularly in smoothing movements
Midbrain:	
Part of reticular system	Sleep, attention
Motor nuclei	Connect with cerebellum and other centers, oculomotor (III)
Superior colliculus	Visual (and auditory) information to control eye movements (?)
Inferior colliculus	Auditory processing
Forebrain:	
Diencephalon	
Thalamus, including:	Dialogue with cortex
Lateral geniculate	To visual cortex
Medial geniculate	To auditory cortex
Other centers	To motor cortex
Hypothalamus	Temperature regulation; involved in appetite, thirst, sleep, sexual drive, emotions; tied into pituitary system (see Chapters 5, 9, and 10)
Pituitary	See Chapter 9
Telecephalon	
Corpus striatum and related nuclei	Motor regulation, muscle tone
Hippocampus and related nuclei	Some memory, emotions, olfaction
Neocortex, including:	
Primary sensory	Touch, position
Secondary sensory	Further processing
Primary motor	Movement control
Supplementary motor	Movement control
Auditory cortex	Auditory processing
Visual cortex	Visual processing
Wernicke's area, Broca's area	Speech
Other regions	Other functions

spinal reflexes also occur. The major role of the cerebellum is also control of movements. When animals have cerebellar lesions, we see a distinctive, telltale syndrome. The animals make errors in rate, range, force, and direction of movements. The errors appear when the muscles are moved; there is no tremor at rest. The more joints involved and the more precision required the greater the tremor. Patients with such lesions report no alteration in sensations. The cerebellum appears to be responsible for taking orders from the motor cortex and modifying the orders to achieve smooth, coordinated execution.

There also is a region of the cortex clearly devoted to control of movement. This is the motor cortex. If we stimulate a region, including a clump of neurons, we initiate localized movement. This might be a finger twitch, knee flexion, or toe wiggle. The left cortex controls the right half of the body, the right cortex the left half. If we plot which musculature is moved by which critical region, we obtain a picture of a person (Figure 3.45B). Some regions, such as the hands, involve a large cortical area. Others, such as the back of the torso, are controlled by a small region. If we stimulate touch, heat, and other somatic sense receptors and ask which areas of the somatosensory cortex we excite, we obtain a similar picture (Figure 3.45A). In fact such maps, both sensory and motor, occur in several places in the brain. There are two complete body maps in the mammalian cerebellum. The weighting for different regions of the body differs with the species (Figure 3.45C).

Why are there such maps? Why have such a systematic projection of axons from region to region? If adjacent brain regions control adjacent regions of the body or receive information from adjacent regions of the environment, then coordinated movements of adjacent regions might be easier. Lateral inhibition and other data processing might also be simpler to organize.

Motor control by interneurons is different in invertebrates. Sensory and motor maps have not been found. Single interneurons can mediate complex postural or behavioral responses. For example, stimulation of a single cell in the crayfish cephalothorax evokes a defensive posture. Claws are raised, the body reared up, and the antennae brought together. Each of these motions requires the proper contraction of several, even many, muscles. The interneurons that mediate these complex, coordinated movements have been called "command" fibers. Command fibers have been found that initiate or dictate a wide range of postures in crustacea. In some mollusks one set of electrically coupled command neurons triggers an entire 30-second sequence of escape behavior. With a few notable exceptions, such cells do not exist in vertebrates. Rather, control depends on the integrated and coordinated output of large populations of cells.

What sort of information is carried by individual cells in the vertebrate brain? We have recently learned a great deal about the pathway for visual information. Let us first consider the retina itself. This originates as an outgrowth of the brain and is really part of the central nervous system. Figure 3.46 schematically depicts a vertebrate retina and displays the types of cells and some of their known synaptic contacts. When light hits a receptor cell, it causes a hyperpolarization. This modulates (probably reduces) synaptic release of transmitter. Information can travel toward the brain along the pathway: receptor, bipolar, ganglion cell. The ganglion cell axons run in the optic nerve into the brain. Some convergence is apparent from the diagram; many receptors may synapse with each bipolar. We can also see that information can travel laterally. Horizontal and amacrine cells extend laterally over a distance underlying many receptor cells. We discussed previously how we should expect lateral interactions between receptors or between cells carrying information from different receptors. This can provide lateral inhibition or a related means of contrast enhancement. Other points to note are that in the retina only the ganglion cells give repetitive action potentials. The other cells generally display slow hyperpolarizations and depolarizations. Also, the synaptic interactions can be highly complex and have not been completely elucidated. At some points three cells may come into close apposition, and more than one may be postsynaptic or presynaptic.

What is the output of the entire system? What is required to change the firing level in a ganglion cell? Let us take one type of cell as an example. There is a small region in the cat's entire visual field that, when illuminated, causes an increase in firing. This "on" center is typically 0.25 to 1° in diameter. Immediately outside this region, in the *surround,* comparable illumination may produce a burst when the light is turned off. This "off"

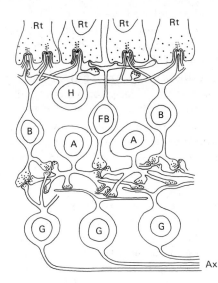

FIGURE 3.46 Diagram of the synaptic arrangements in a typical vertebrate retina. Rt—receptor (rod or cone) terminal; H—horizontal cell; B—bipolar cell; FB—flat bipolar; A—amacrine cell; G—ganglion cell; Ax—axons running through optic nerve and into brain. (From J. E. Dowling, *Invest. Opthalmol.* 9: 655–680, 1970.)

response is weaker the farther the light is from the "on" center. When we are finally at a point where illumination has no effect, we say we are outside of the *receptive field* of this ganglion cell. If we illuminate both "on" center and "off" surround simultaneously, we get a weak "on" response. Illumination of the surround inhibits the response from the center. There is an antagonism between center and surround that is analogous to the lateral inhibition seen in *Limulus*. It enhances contrast. All ganglion cells are not "on" center and "off" surround; some are "off" center and "on" surround.

There are other types of ganglion cells. For example, some encode color information. One type found in primates possesses an "on" center that is excited maximally by red light and an antagonistic surround most responsive to green light. All possible center-surround pairs of the three cone types exist. We discussed previously how color vision must involve comparison of responses from all three cone types.

The details of vertebrate retinal anatomy vary with the class. Retinal processing, exclusive of color coding, is more extensive in some lower vertebrates. This is consistent with the rule that the more complex the nervous system the more centralized the information processing. To cite one case, frogs have several distinct classes of ganglion cell. The cells of one class, called *net convexity detectors,* respond best when a convex edge is moved through the cells' receptive field. The greater the convexity of the leading edge the greater the response. Rapid movements are most effective. Cells of another class, *moving edge detectors,* respond best to a straightedge moving through the receptive field. In the frog all classes of ganglion cells send their axons into the tectum. In mammals this pathway is used primarily for controlling eye and head movements—for visual tracking. Mammals possess a parallel visual pathway: retina, lateral geniculate body, visual cortex. In lower mammals the retina-to-tectum pathway is still very important, but in cats, monkeys, and man the retinal-geniculate-cortical pathway carries most of the visual information.

Each ganglion cell receives information from receptors in its region of the retina. A map of the receptive fields of ganglion cells reflects the image of the environment projected on the retina. This map of the environment on the neurons is maintained along the visual pathway. Thus in the visual cortex there is a topographic projection of the visual environment. You may recall that in the somatosensory and motor cortex the amount of cortex devoted to each body region is not equal. In our visual cortex a much larger region, with far more cells, handles data from the center of the visual field than a comparable segment of the periphery. Proportionally more cortex is devoted to information from the cones in the fovea than the rods in the periphery of the retina.

What is required to stimulate a geniculate cell in a cat? Geniculate cells resemble ganglion cells in possessing a receptive field with a center and antagonistic surround. However, the inhibition of the center response by the surround is more complete. Illumination of the entire receptive field—"on" center and "off" surround together—results in a very weak response.

Three major classes of cells in the visual cortex have been described by D. Hubel and T. Weisel, the men responsible for much of our knowledge of the physiology of the visual cortex. The simplest, called "simple" cortical cells, resemble geniculate cells in several respects. They are barely excited by diffuse light. Each receptive field is divided into excitatory and inhibitory regions. One can predict the response to complicated stimuli by plotting out the excitatory and inhibitory regions with a small spot (Figure 3.47A). Unlike geniculate cells, the borders separating excitatory from inhibitory regions are straight lines. The orientation of the boundary varies with the cell: vertical, horizontal, and everything between. Some simple cells respond best when a slit or edge is moving in a given direction.

Complex cells do not respond to diffuse light or small spots. The best stimulus is a slit, bar, or dark edge, depending on the cell. The orientation of the slit or edge is critical. A complex cell can be fussy about the width of a slit or bar but not the location in the receptive field (Figure 3.47B).

Hypercomplex cells resemble complex cells in that the best stimulus is a line, bar, or edge of specific orientation. Often a moving slit is most effective. The line must be of a critical length. If it is longer, the orientation of the extension is crucial (Figure 3.47C).

The topographic representation of the environment on the cortex means that the position of the receptive field of neighboring cells is likely to be very close. However, as we move horizontally (tangential to the surface of the cortex), the preferred orientation of cells—simple, complex, or hypercomplex—varies. As we go into the cortex vertically, at right angles to its surface, we find that all cells have the same preferred stimulus orientation.

Hubel and Wiesel have postulated that many geniculate cells synapse on each simple cell, many simple on each complex, and many complex on each hypercomplex. A typical simple cell is excited only by a vertical dark line in a specific part of its receptive field. A complex cell is excited by the proper vertical line anywhere in its field. A hypercomplex cell is excited by a proper vertical line anywhere in its field provided that the line is not too long. We can imagine how we could construct immensely sophisticated cells after an entire chain with further additional synaptic steps. Whether or not this is how visual perception is actually constructed is not known.

The receptive field for all the cells we have discussed is some fixed portion of the animal's total visual field. For example, the "on" center of a geniculate cell might be in the upper right quadrant of the visual field, 45°

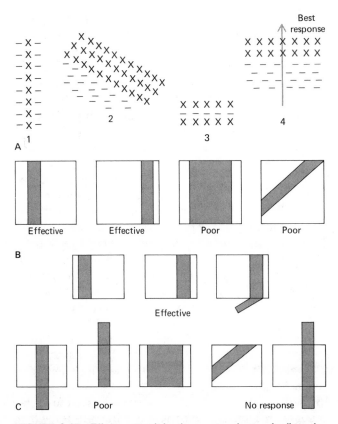

FIGURE 3.47 Effective stimuli for three major classes of cells in the visual cortex of the cat. (**A**) Receptive fields of four types of simple cells (1–4). In all cases very small spots of light produce either excitation and "on" responses (crosses) or inhibition and "off" responses (dashes). Borders between crosses and dashes are always straight. The arrow indicates the preferred direction (if there is one) of the movement of the stimulus through the field. In simple cells the best large stimulus may be light bars (1), edges (2), dark bars (3), or moving light bars (4) or edges. (**B**) Receptive field of a typical complex cell. Lines of a given width and orientation are best. Location of the line in the field has little importance. (**C**) Receptive field of a typical hypercomplex cell. The cell is fussy about the orientation, width, and length of the line. The position of the line in the field is not important.

out from the center. What happens if the cat turns its head? Stimuli that had been effective no longer are because they are in another region of the visual field. There are some cells in the visual pathway of crayfish that behave differently. As the animal tilts its head, the cells' receptive field changes accordingly. Thus if a cell is stimulated by a "vertical" stimulus, this means "vertical" in the environment, not the vertical axis of the head. These cells receive information from both eyes and statocyst (gravity receptor organ).

We mentioned that in mammals there is a parallel pathway for visual information. Some retinal fibers run to the superior colliculus (tectum), which also gets input from the cerebral cortex. The colliculus seems to contribute to the control of eye movements. One common type of collicular cell has a large receptive field and requires a moving stimulus. The cell fires when an object in the center of the visual field moves toward the periphery. The colliculus also receives input from auditory and tactile pathways. There are some cells that are excited by *both* auditory and visual stimuli. The receptive fields of the two modalities are related. The stimuli are most effective if they arise from the same region of the environment. In some cases it has been possible to determine the receptive field of a collicular cell and then stimulate the cells in its region by using very small currents. Sometimes this causes the animal to direct its gaze toward the point in space where the cell's receptive field had been. This may be part of a true visual tracking circuit. If an object is in the cell's receptive field and begins to move out, the animal moves its eyes and/or head to keep the object in the same part of the visual field.

13.1   Higher Processing

The sort of neural circuitry and processing we have been describing has not yet told us how the highest processes occur in the brain. How do we remember, learn, handle language and other abstractions? We don't know There have been some interesting findings, however, about the localization of some higher processes in man. Penfield and others have stimulated the brains of patients during surgery. A patient can experience unusual sensations when points on the superior or lateral surface of the temporal lobe are stimulated. He or she may report vivid hallucinations. For example, one woman said, "Yes, I hear voices. It is late at night, around the carnival somewhere—some sort of traveling circus. I just saw lots of big wagons that they use to haul animals in." This is typical of these hallucinations. They are made up of elements from the individual's past experience. The subject relives a period of the past although still aware of the present. Movement occurs in time and is accompanied by sights, sounds, interpretations, and emotions. In short they have many elements of multisensory, even emotional, recollections.

Language processing in humans appears to occur mainly in certain regions of the left cerebral cortex. According to one model, when a word is heard the output of the auditory cortex is received by a specialized speech region, Wernicke's area. If a word is to be spoken, the pattern is transmitted to Broca's area, which in turn excites the neighboring facial region of the motor cortex. Lesions in specific regions cause correspondingly specific types of speech defects. Persons with damage in Wernicke's area show normal speech rhythm, grammar, and articulation. Unfortunately, the actual words make no sense in context. Often the person must use highly circumlocutory phrases to express himself. Persons with damage in Broca's area display another syndrome. In short it does appear that speech elaboration is highly localized and involves a clear sequence of steps, each involving distinct cerebral regions.

Learning processes may be the most elusive. Considerable effort has been expended to find the physiological or anatomical correlates of learning, but success has been very limited. One approach has been to look at some "learninglike" processes in invertebrates, whose simpler nervous systems can be studied more easily. Habituation and dishabituation are found throughout the animal kingdom. When a stimulus is repeated many times, the response decreases. This is habituation. A strong, different stimulus can restore the response to the initial type of stimulus. This is dishabituation. Lately this has been studied in *Aplysia*. A monosynaptic gill withdrawal reflex has been found that clearly habituates. A sensory neuron makes excitatory synaptic contact with motor cells L_7, LDG, $L_9(1)$ and $L_9(2)$. Habituation has been localized to the synapses between the sensory cell and identified neurons. Dishabituation has been identified as due to heterosynaptic facilitation at the same synapse. It also has been possible to classically condition certain mollusks. The conditioned response seems to be retained in the isolated ganglion. This opens up many possibilities. These snails have identifiable neurons and synaptic interactions. We may be able to find the physiological correlates of conditioning even to the point of identifying all the cells involved and the changes they undergo during training. Although we might be able to identify subcellular changes accompanying certain kinds of training, we are a very, very long way indeed from understanding the workings of the mammalian brain.

REFERENCES

Aidley, D. J. *The Physiology of Excitable Cells*. New York: Cambridge University Press, 1971.

Bullock, T. H., and Horridge, G. A. *Structure and Function in the Nervous Systems of Invertebrates*. Vols. I and II. San Francisco: Freeman, 1965.

Cooke, I., and Lipkin, M., Jr., eds. *Cellular Neurophysiology. A Source Book*. New York: Holt, Rinehart and Winston, 1972.

Eccles, J. C. *The Understanding of the Brain*. New York: McGraw-Hill, 1973.

Fuortes, M. G. F., ed. *Handbook of Sensory Physiology* VII-2. New York: Springer-Verlag, 1971.

Hall, Z. W., Hildebrand, J. G., and Kravitz, E. A., eds. *Chemistry of Synaptic Transmission. Essays and Sources.* Portland: Chiron Press, 1974.

Hille, B. Ionic Channels in Nerve Membranes. *Progr. Biophys. Mol. Biol.* 21:1–32, 1970.

Hodgkin, A. L. *The Conduction of the Nervous Impulse.* Springfield, Ill.: Thomas, 1964.

Jacobson, M. *Developmental Neurobiology.* New York: Holt, Rinehart and Winston, 1970.

Katz, B. *Nerve, Muscle, and Synapse.* New York: McGraw-Hill, 1966.

Katz, B., and Miledi, R. The Statistical Nature of the Acetylcholine Potential and Its Molecular Components. *J. Physiol. (London)* 224:665–669, 1972.

Loewenstein, W. R., ed. *Handbook of Sensory Physiology.* Vol. I. New York: Springer-Verlag, 1971.

Maturana, H. R., Lettvin, J. Y., McCullock, W. S., and Pitts, W. H. Anatomy and Physiology of Vision in the Frog. *J. Gen. Physiol.* 43:129–175, 1960.

Mountcastle, V. B., ed. *Medical Physiology.* Vol II. 13th ed. St. Louis: Mosby, 1974.

Penfield, W. The Role of Temporal Cortex in Recall of Past Experience and Interpretation of the Present. *Neurological Basis of Behavior.* Ciba Foundation Symposium. London: Churchill, 1958.

Ramon y Cajal, S. Histologie du Système Nerveux de l'Homme et des Vertébrés. *Consiejo Superior de Investigaciones Cientificas.* Madrid: Instituto Ramon y Cajal, 1952–1955.

Roeder, K. *Nerve Cells and Insect Behavior.* Cambridge, Mass.: Harvard University Press, 1963.

Ruch, T. C., and Patton, H. D., eds. *The Nervous System.* Vol. I of *Physiology and Biophysics.* Philadelphia: Saunders, 1976.

Schmitt, F. O., ed. *The Neurosciences Study Programs.* Vols. 1 and 2. New York: Rockefeller University Press, 1967, 1970. Vol. 3. Cambridge, Mass.: The M.I.T. Press, 1974.

Shepherd, G. M. *The Synaptic Organization of the Brain.* New York: Oxford University Press, 1974.

(Symposium) Sensory Receptors. *Cold Spring Harbor Symp. Quant. Biol.* Vol. 30. New York: Cold Spring Harbor Laboratory of Quantitative Biology, 1965.

Von Bekesy, G. *Experiments in Hearing.* New York: McGraw-Hill, 1960.

Wells, M. J. *Brain and Behavior in Cephalopods.* London: Heinemann, 1962.

MOTILITY

Robert M. Dowben

All living organisms display movement. In animals, as contrasted with plants, movement is vigorous, specific, and noteworthy for its sheer quantity. Indeed, animals spend a major fraction of the energy made available by the metabolism of food for movement. Unicellular organisms show streaming of their cytoplasm and often possess flagella, which aid in locomotion. Multicellular metazoan organisms have specialized cells derived from the mesenchyme, whose function it is to develop tension and to perform mechanical work in a controlled fashion. Thus muscles are used by animals to move toward food and away from noxious materials, or to move food toward them and aid in the elimination of waste products.

In higher organisms three kinds of muscle are found: *skeletal muscle*, which is primarily used for voluntary motion; *cardiac muscle*, which contracts rhythmically throughout the life of the animal and pumps blood to all parts of the body; and *smooth muscle*, which is found in the walls of hollow organs, blood vessels, and various ducts and is responsible for contraction of the viscera and propulsion of their contents. The energy for the mechanical work produced by muscles comes from ATP, an energy-rich compound produced by the metabolism of food, which is discussed in Chapter 7.

In this chapter we will focus primarily on the structure and operation of muscle and its specialized adaptations, although we will touch on the structure and function of flagella, which are found throughout the animal kingdom, and on cytoplasmic motion, which is universal. Studies done on muscles from widely divergent sources indicate that they are constructed and that they work in basically the same way. Differences in structure and function between various muscles seem to be relatively minor variations of a basic, underlying theme.

Our understanding of muscle physiology has developed to a remarkable degree, both of the muscle as a functioning organ and at a microscopic or even molecular level. In muscular contraction, as in few other physiological systems, it is possible to describe gross physiological events in terms of molecular processes. Muscle is also unique in terms of its position as the effector organ of complex and precise regulatory systems, involving especially the central nervous system. Indeed, muscle function, as a part of bodily feedback systems, offers the opportunity to study such control systems, and abnormal function of the control system may be a first clue to toxic effects of disease or environmental pollution.

1 • STRUCTURE OF VERTEBRATE SKELETAL MUSCLE

Skeletal muscle, which makes up 40 percent or more of the body mass in higher animals, consists of bundles of elongated, cylindrical cells called *muscle fibers,* 50 to 200 μ in diameter, running all or most of the length of the muscle. Like other cells, the muscle fiber is surrounded by a plasma membrane called the *sarcolemma.* The interior of each muscle fiber is occupied by contractile elements called *myofibrils,* numerous parallel threads 1 to 3 μ in diameter that run the entire length of the muscle fiber. Each myofibril is surrounded by an organized, internal membrane system called *sarcoplasmic reticulum.* Squeezed between the myofibrils and the sarcolemma is a bit of cytoplasm, the *sarcoplasm,* containing multiple nuclei, numerous mitochondria (*sarcosomes*), lipid droplets, glycogen granules, and other intracellular organelles.

The bundles of muscle fibers, each called a *fasciculus,* are surrounded by a covering of connective tissue, the *endomysium* (Figure 4.1). A muscle consists of a number of fasciculi encased in a heavy outer sheath of connective tissue called the *perimysium.* At each end there is a connective tissue *tendon,* by means of which the muscle is attached to the bony skeleton. Muscles can only shorten. Movements of the bony skeleton back and forth are caused by sets of opposing muscles. The nerve which stimulates the muscle to contract makes contact at a *motor end plate.*

The *cross striations* in skeletal and cardiac muscle were discovered by light microscopy in the beginning of the nineteenth century. As early as 1840, Sir William Bowman suggested that the cross striations resulted from repeating units of material that differed in the refractive index to light. The band that usually appears darker staining is birefringent or *anisotropic* to light; that is, the refractive index of of this band is slightly higher when viewed parallel to the length of the fiber than if viewed perpendicular to the long axis. The dark appearance is accentuated when the microscope is slightly out of focus. The dark anisotropic band is called the A band, whereas the light band is *isotropic* to light and is called the I band.

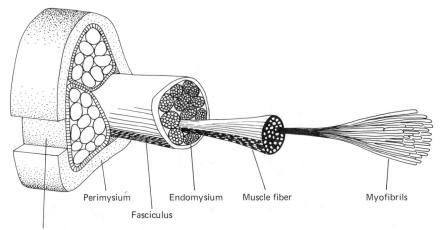

Perimysium   Endomysium   Muscle fiber   Myofibrils

Fasciculus

Cross section of muscle

FIGURE 4.1 A schematic representation of a cross section of skeletal muscle showing its organizational structure.

2 • ULTRASTRUCTURE OF VERTEBRATE SKELETAL MUSCLE

Studies of skeletal muscle ultrastructure by electron microscopy have revealed a rich and complex submicroscopic architecture (Figure 4.2). Each muscle fiber consists of numerous *myofibrils*, 1 to 3 μ in diameter. The A and I bands in adjacent myofibrils are in register. The I band is bisected by a thin, dark line, the Z line. The contractile unit or *sarcomere* is defined as extending from one Z line to the next Z line.

Inside the myofibril are a multitude of longitudinal filaments of two kinds, *thick filaments* and *thin filaments*. The thick filaments are about 120 Å in diameter and about 1.6 μ long; they are located in the center of the sarcomere, arranged in a hexagonal array about 450 Å apart (Figure 4.3). The thick filaments are responsible for the appearance of the A band. The thin filaments, which are about 40 Å in diameter, seem to anchor into the Z line and extend toward the center of the sarcomere through the I band and partly into the A band. The central portion of the A band, called the H zone, is devoid of thin filaments and therefore appears somewhat lighter. The H zone is bisected by a dark M line.

Cross-sectional electron micrographs show various patterns, depending upon the location of the section. Sections through the H zone show only thick filaments arranged in a hexagonal array. Sections through the I band show only thin filaments. Sections through the dense portion of the A band show both thick and thin filaments. In vertebrate muscle there are twice as many thin as thick filaments; the thin filaments are inter-

FIGURE 4.2   A low magnification electron micrograph of rat skeletal muscle showing the cross striations in the myofibrils which are in register. Mitochondria, dark glycogen granules, and bits of sarcoplasmic reticulum can be seen between the myofibrils.

spersed in the array so that each thick filament is surrounded by six thin filaments (Figure 4.4). Insect muscle may have three or four times as many thin filaments, but each thick filament is still surrounded by six thin filaments.

Each thick filament has two projections from opposite sides at about 143 Å intervals along its length. These *cross bridges* (Figure 4.5) project from the thick filament to touch one of the adjacent thin filaments. Projections of successive cross bridges are rotated 120° around the axis of the thick filaments. Each thin filament receives cross bridges from the three adjacent thick filaments. At any given cross-sectional level, the direction of the cross bridges is not in register, but rather, cross bridges from the next nearest thick filaments are rotated by ± 120°. Thus the cross bridges reaching one-half of the thin filaments are coplanar every 430 Å, while on the other half of the thin filaments the cross bridges touch in sequence every 143 Å and make a helical pattern.

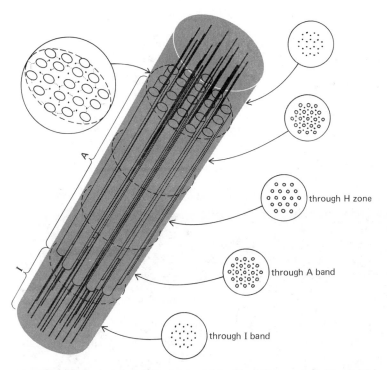

through H zone

through A band

through I band

FIGURE 4.3 A schematic representation of the filament arrangement in skeletal muscle. The thin filaments extend from the Z lines through the I band and into the A band as far as the H zone. The thick filaments extend from one edge of the A band to the other. The arrays in cross section at various levels of the sarcomere are shown in the insets. (From R. M. Dowben, *Cell Biology.* New York: Harper & Row, 1971.)

During shortening, the two sets of interdigitating filaments slide with respect to each other. Thus during shortening the A band remains a constant width, but as the thin filaments move in toward the center of the sarcomere and the distance between Z lines decreases, the I bands and the H zone become narrower. To reiterate, the width of the A band remains constant whether a muscle fiber is relaxed, contracted, or stretched. On the other hand, the I band and H zone become narrower during shortening and wider during stretching. When a muscle is made to shorten very markedly, the thin filaments from opposite Z lines may actually cross over and overlap. With enormous degrees of shortening, the A bands reach the Z lines, causing deformation of the ends of the thick filaments, giving rise to *contraction bands* at the Z lines. This process is frequently referred to as the *sliding filament model of muscle contraction.*

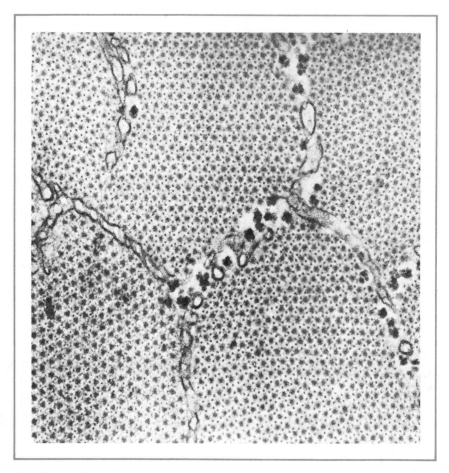

FIGURE 4.4   Electron micrograph of a cross section of rabbit muscle through the dense portion of the A band showing thick filaments in a hexagonal array, each thick filament surrounded by six thin filaments. (Courtesy of H. E. Huxley)

Skeletal muscle has a well-developed endoplasmic reticulum, which in muscle is called the *sarcoplasmic reticulum*. The sarcoplasmic reticulum forms an extensive membranous system within the cytoplasm around each myofibril. The structure of the sarcoplasmic reticulum differs from the endoplasmic reticulum of other cells in that it shows a characteristic pattern that is repeated at the same relative position in every sarcomere. Figure 4.6 shows a schematic representation of components of this internal membranous system. The sarcoplasmic reticulum bulges out at regular intervals to form cisternae on either side of a deep invagination of the sarcolemma (muscle cell membrane). The invagination of the sarcolemma, called a *t-tubule* or transverse tubule, together with the lateral cis-

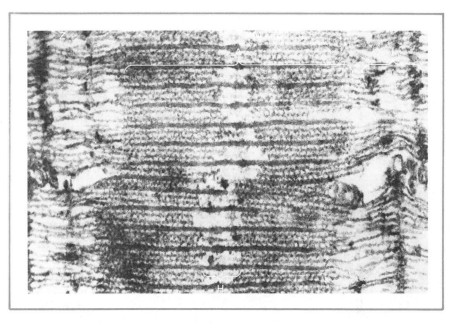

FIGURE 4.5 Electron micrograph of rabbit muscle cut longitudinally showing cross bridges in the dense portion of the A band, and the M line bisecting the H zone. (Courtesy of H. E. Huxley)

ternae of the endoplasmic reticulum on either side, constitutes the *triad* (Figure 4.6). In frog muscle the triads occur at the level of the Z lines, whereas in mammalian muscle they are found at the junction of the A and I bands. The triads, as will be explained later, play an important role in coupling the excitation of muscle with the development of the active, contractile state.

3 • SMOOTH MUSCLE

Unlike skeletal and cardiac muscle, smooth muscle does not show any cross striations. Smooth muscle is found in the walls of blood vessels, in the walls of hollow organs, in the ducts of various glands, and in the uterus. Individual smooth muscle cells tend to be spindle-shaped, with one or a few nuclei located in the center of the cell. The cells are connected to each other by connective tissue. Smooth muscle is innervated by the autonomic nervous system and is not under the same sort of voluntary control as skeletal muscle, so it is often called *involuntary muscle*.

Owing to artifacts that arose during preparation of specimens, it only recently was shown that smooth muscle also contains thick and thin filaments (Figure 4.7). Unlike skeletal muscle, however, the filaments are

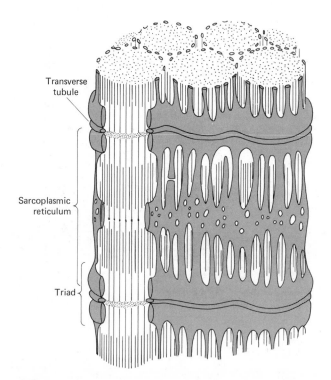

FIGURE 4.6 A diagrammatic reconstruction of the sarcoplasmic reticulum associated with several myofibrils, taken from frog sartorius muscle. Note that the transverse tubules are continuous and at the level of the Z lines. In mammalian muscle the transverse tubules occur at the junction of the A and I bands. (From L. D. Peachey, *J. Cell Biol.* 25:222, 1965.)

not arranged in a highly organized, regular pattern. Although thick filaments are surrounded by thin filaments with which they are associated, the filament structure is more random. The filaments tend to be longer, thick filaments often reaching lengths of 5 μ or more. Smooth muscle, compared to skeletal muscle, contains very little sarcoplasmic reticulum. There seems to be considerable variation in different smooth muscles as to the extent of development of sarcoplasmic reticulum.

4 • EVENTS IN MUSCLE CONTRACTION

Events leading to muscular contraction can be divided into a number of distinct processes:

1. Stimulation of the motor nerve gives rise to an action potential that is conducted

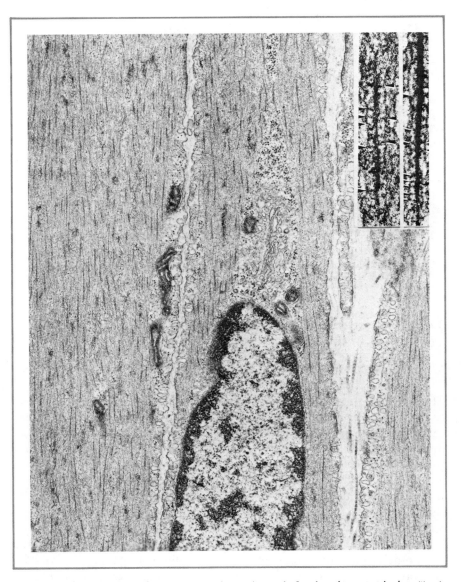

FIGURE 4.7 Longitudinal-oblique sections of smooth muscle. Incubated in a stretched position in Krebs solution and fixed with trialdehyde before sectioning for electron micrographs. The filament structural details are shown at higher magnification. (Courtesy of A. P. Somlyo)

to terminal arborization, where a chemical transmitter substance, usually acetylcholine, is liberated.

2. Acetylcholine produces depolarization of the sarcolemma at the motor end plate (Figure 4.8), which in turn initiates a propagated action potential along the muscle fiber sarcolemma.

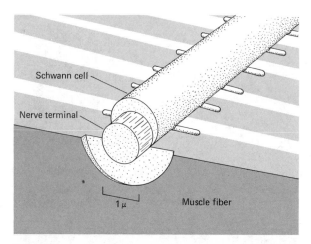

FIGURE 4.8   A diagram of the motor nerve endings and motor end-plate at the myoneural junction. (Modified from R. Birks, H. E. Huxley, and B. Katz, *J. Physiol.* 150:136, 1960.)

3. The action potential invades sarcolemnal invaginations called the t-tubular system and, by an as yet undefined process, initiates the release of calcium from the triadic component of sarcoplasmic reticulum.
4. The released Ca^{2+} reaches the filament system and contraction is initiated.

The temporal relation between action potential and contraction is shown in Figure 4.9. In a typical skeletal muscle fiber the action potential caused by depolarization of the membrane reaches a maximum approximately 1.5 msec after the muscle is stimulated. There is a gradual repolarization of the sarcolemma to the resting potential within about 5 msec of electrical stimulation. With regard to contraction, there is a *latent period* of 3 to 5 msec after stimulation before any tension begins to develop. Tension develops slowly, reaching a maximum between 50 and 120 msec after stimulation. Thus the electrical state of the sarcolemma has returned to the resting state before any appreciable tension is developed and long before the point of maximum tension is reached. From the time of maximum tension, the decay of force is very gradual, taking place over 100 to 500 msec.

The muscle cell membrane can be depolarized a second time after repolarization but before tension development has run its course. Such a second stimulus will cause another twitch, which adds onto the first (Figure 4.9) and increases the tension further. If the muscle is stimulated repeatedly at a frequency of 15 to 120 stimuli per second, individual twitches cannot be detected; instead there is a strong, maintained, and smooth

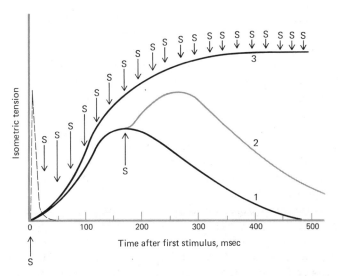

FIGURE 4.9 A diagram of the duration of a muscle action potential (dashed line) and the time course of tension development during (1) a single twitch, (2) two twitches in which tension is summated, and (3) a tetanus. S signifies a stimulus.

contraction called a *tetanus* (Figure 4.9, curve 3). Single twitches, or tetanic tension, can be obtained in single muscle fibers as well as in the whole muscle. The tension generated by the individual muscle fibers is summated and transmitted through the connective tissue and tendons to produce the force developed by the whole muscle. Relaxation of the muscle is thought to be enhanced by the extent of passive tension generated in the elastic components of the muscle (for example, connective tissue and tendons) during the active contraction. Such a mechanism is consistent with the observation that muscles with a greater percentage of connective tissue show a greater tendency to return to rest length after contraction has ceased. However, it must be remembered that true relaxation can occur only if Ca^{2+} activity around the myofilament is reduced. The removal of calcium from the sarcoplasm during the relaxation phase of contraction is mediated by sarcoplasmic reticulum, and it is known to require energy.

5 • MECHANICS OF CONTRACTION

Resting muscle is elastic. Most of the elasticity resides in the connective tissue of whole muscle, although the muscle fibers themselves contribute to the elasticity. Muscle may be regarded as consisting of a contractile element that generates tension when activated and two springs—one in series with the contractile element (*series elastic component*) and one in parallel (*parallel elastic component*). When the contractile element is activated and generates tension, it stretches the series elastic component and compresses the parallel elastic component. The elastic components do not have linear stress-strain properties or linear internal frictional properties; that is, they do not obey either Hooke's law or Newton's law for viscous elements.

The mechanics of muscle contraction are studied experimentally using an apparatus such as that shown in Figure 4.10. The muscle is stimulated by a series of electrodes along its entire length so that all of the contractile elements will be activated completely and simultaneously. The muscle is fixed at one end and connected to the lever system by a wire attached to the tendon at the free end. For one type of experiment the muscle is prevented from shortening by a stop that fixes the lever at various positions. A contraction under these circumstances where muscle length is maintained constant is called an *isometric* contraction; the tension developed in an isometric contraction is measured by a strain gauge connected to the lever arm as a function of muscle length.

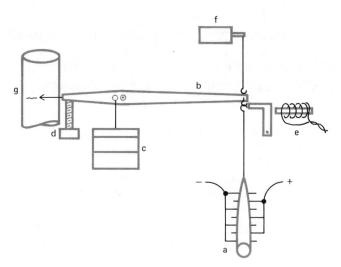

FIGURE 4.10 A diagram of the apparatus used for studying the mechanics of muscle contraction. The muscle is stimulated by a series of electrodes (a). The tendon is attached to a lever system (b) on which may be placed various weights (c). The lever system may be immobilized by an adjustable stop (d) or quickly released by means of a solenoid-activated stop (e). Tensions are measured by means of a strain gauge (f), and movement is recorded by means of a pen attached to the lever (g).

In another type of experiment the muscle is allowed to shorten while lifting a weight attached to the lever arm. This type of contraction is called *isotonic* because the tension developed by the muscle is determined by the weight and remains constant during the shortening; the velocity of shortening is measured during an isotonic contraction as the tension is varied.

5.1 Force-Velocity Relation

Figure 4.11 depicts a series of shortening curves for several loads. Several observations should be made. The apparent latent period before shortening begins is equal to the true latent period plus the time required for the muscle contractile elements to stretch the series elastic component (mainly connective tissue) to develop tension equal to that of the load. Before the tension developed exceeds the load, the muscle cannot lift the weight. As the load increases, the apparent latent period lengthens. As the load increases, the initial velocity of shortening is less. For any given load, the velocity of shortening remains approximately constant or decreases slightly during most of the contraction. During the constant-velocity portion of the contraction, the velocity is less as the load is increased. Last, the maximum amount of shortening diminishes as the load is increased.

If we plot the velocity developed for different loads in a series of individual isotonic contractions, we obtain a hyperbolic curve such as that

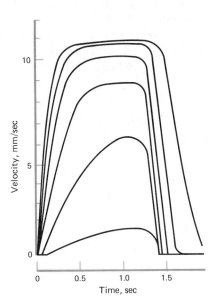

FIGURE 4.11 A family of isotonic contractions for various loads. From top to bottom, loads are 1, 2.5, 5, 10, 20, and 30 g. (From D. R. Wilkie, *Brit. Med. Bull.* 12:179, 1956.)

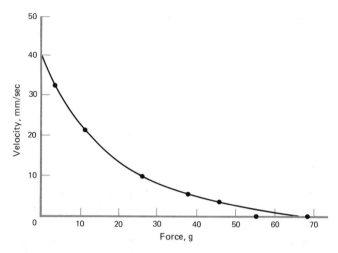

FIGURE 4.12 Force-velocity curve calculated from Hill's equation for a family of isotonic contractions using frog sartorius muscle at 0°C. (From D. R. Wilkie, *Brit. Med. Bull.* 12:179, 1956.)

shown in Figure 4.12. The *force-velocity curve* fits a relation derived by A. V. Hill:

$$(P + a)V = (P_0 - P)b \tag{4-1}$$

where P is the tension developed in a particular shortening, P_0 the maximum tension that the muscle can develop in an isometric contraction, V the velocity of shortening, and a and b constants. The constant a is related to the constant α, the coefficient of shortening heat in the Hill heat equation discussed in Section 7. The constant b depends upon the intrinsic speed of the muscle and has characteristic values for red, mixed, white, or smooth muscle.

5.2 Length-Tension Relation

In an isometric contraction the tension developed during tetanus depends upon the length of the muscle. Figure 4.13 shows a plot of the maximum tension developed by a muscle as a function of length. It should be noted that tension is required to stretch a relaxed muscle beyond rest length. For muscle stretched beyond rest length the tension developed by the contractile elements is equal to the total tension of active muscle minus the tension due to the elastic components, which can be measured when the muscle is not stimulated. Note that the maximum tension is developed by the contractile units when the muscle is approximately at rest length. Different striated muscles will show different length-tension relations de-

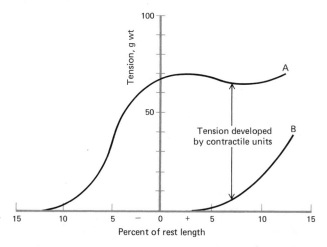

FIGURE 4.13 The length-tension curve for a muscle stimulated tetanically (curve A) and the stress-strain curve (resistance to stretching) for the same muscle in the resting state (curve B). For stretched muscle the tension developed by the contractile units is equal to the difference between curve A and curve B. (Modified from D. R. Wilkie, *Brit. Med. Bull.* 12:178, 1956.)

pending upon the type of muscle, the anatomic arrangement of muscle fibers, and the amount of connective tissue.

5.3 Studies in Single Muscle Fibers

Connective tissue limits the extent to which whole muscle can be stretched to about 10 to 15 percent of rest length. This limitation can be overcome by studying the tension developed by single muscle fibers from which almost all connective tissue has been removed. These experiments are performed using an elegant electromechanical feedback system to maintain constant length during the contraction. The tension developed can be compared to the relative position of the filaments by obtaining electron micrographs of the muscle fibers fixed in situ, great care being taken to avoid artifacts due to shrinkage.

The results of such an experiment carried out by A. F. Huxley and his collaborators on single muscle fibers are shown in Figure 4.14. Little or no tension is developed if the muscle is stretched to so great an extent that there is no overlap of the thin and thick filaments (sarcomere length = 3.6 μ or more). For smaller degrees of stretching, the tension developed increases linearly as the overlap length between the two sets of filaments increases until all of the cross bridges extending from the thick filaments are covered (at a sarcomere length of about 2.25 μ). The central region of the

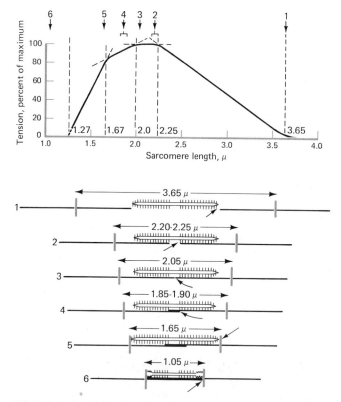

FIGURE 4.14 Tension developed by a single muscle fiber as a function of length compared to the degree of overlap of the filaments in the sarcomere. From A, where thin and thick filaments just overlap to B, where overlap is sufficient to cover all of the cross bridges, tension developed is proportional to the degree of overlap. (From A. M. Gordon, A. F. Huxley, and F. J. Julian, *J. Physiol.* 184:185, 1966.)

thick filaments is devoid of cross bridges; the tension remains constant as greater overlap occurs in this central region (from 2.0 to 2.25 μ). For more extensive overlap of the filaments, the tension developed decreases. At very short lengths the thick filaments are compressed by the Z disks and appear to limit mechanically the development of tension (below 1.67 μ sarcomere length).

6 • EXCITATION-CONTRACTION COUPLING

Excitation-contraction coupling is the process by which depolarization of the muscle cell membrane following stimulation by its motor nerve leads to activation of the contractile units. Immersion of a muscle fiber in KCl solution results in contraction even though the sarcolemma is depolarized everywhere at the same time and longitudinal currents are not generated. Therefore longitudinal currents within the muscle fiber are not responsible for coupling depolarization to contraction. Several experiments indicate that depolarization of the t-tubules, the invaginations of the sarcolemma just described, plays an important role in excitation-contraction coupling.

In a series of elegant experiments using surface microelectrodes on single myofibrils, Huxley and his collaborators showed that localized depolarization of the sarcolemma produced localized contractions when the electrode was placed at the t-tubule sites at the Z lines in frog skeletal muscle. Muscle fibers treated with hypertonic glycerol solution, where the t-tubules are disconnected from the sarcolemma, show normal action potential but no contraction. From a variety of such experiments, it was concluded that depolarization of t-tubular membrane is a crucial step in coupling the electrical activity of muscle to the contractile response. The depolarization of the t-tubular system initiates calcium release from the triad of sarcoplasmic reticulum by a yet undefined process. The specific increase in Ca^{2+} concentration of sarcoplasm following depolarization of the t-tubules was demonstrated by Ashley, using a chemoluminescent protein from jellyfish that requires Ca^{2+} (Figure 4.15). When the Ca^{2+} concentration in the sarcoplasm exceeds about 10^{-6} M, interaction with the thin filament protein troponin initiates the active state. A very small reduction in tension, the *latency relaxation*, and the production of a small amount of heat, the *latency heat*, which precede the development of any

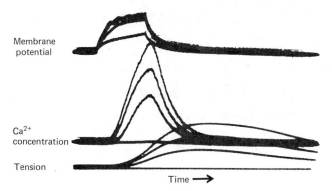

Membrane potential

Ca^{2+} concentration

Tension

Time →

FIGURE 4.15 Aequorin, a luminous protein secreted by certain jellyfish, was injected into giant muscle fibers from barnacles. The oscilloscope traces show (top) the time course of the membrane potential changes, (middle) the time course of intracellular Ca^{2+} concentration as indicated by light production from the injected aequorin, and (bottom) the time course of tension production. (From C. C. Ashley, *Am. Zool.* 7:653, 1967.)

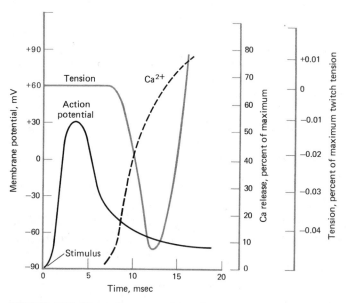

FIGURE 4.16 The early portion of a single twitch showing the simultaneous increase in sarcoplasmic Ca^{2+} concentration and transient decrease in tension (*latency relaxation*) that occurs after the action potential has almost died away. (From J. R. Bendall, *Muscles, Molecules, and Movement.* New York: Elsevier, 1969.)

tension, are coincident with the release of Ca^{2+} ions into the sarcoplasm (Figure 4.16) from the sarcoplasmic reticulum and t-tubule triads.

The Ca^{2+} in the sarcoplasm is quickly removed by an active transport system in the membranes of the sarcoplasmic reticulum. This active transport system, which is coupled to ATP hydrolysis, lowers the Ca^{2+} in the sarcoplasm to less than 10^{-8} M, and relaxation of the contractile units ensues. Ca^{2+} ions accumulate in the sarcoplasmic reticulum, particularly in the lateral cisternae of the triads. From the lateral cisternae, Ca^{2+} ions probably diffuse passively into the cavities of the t-tubules. The active transport system of the sarcoplasmic reticulum is extremely efficient, maintaining the Ca^{2+} of the sarcoplasm well below 10^{-8} M during the resting state and accumulating Ca^{2+} in the lateral cisternae against enormous electrochemical gradients.

7 • ENERGETICS OF CONTRACTION

The energy used in contraction comes from glycolysis under anaerobic conditions (in the absence of oxygen) or by a combination of glycolysis, β-oxidation of fatty acids, and respiration in the presence of oxygen. (These processes are described in detail in Chapter 7.) All of these metabolic processes generate ATP. The hydrolysis of ATP is the immediate source of energy for contraction. The amount of ATP hydrolyzed is related to the amount of mechanical work performed by the muscle.

In an ingenious series of experiments, the concentrations of free fatty acids and glucose in the arterial and venous blood to a limb were measured. By comparing the concentrations it was possible to show that skeletal muscle in the limbs utilized mainly fatty acids rather than glucose as a source of metabolic material. Measurements of O_2 utilized and CO_2 produced gave a respiratory quotient (see Chapter 5, Section 3.2) of about 0.76, which is additional evidence that mainly fats are utilized.

Under optimal conditions the mechanical efficiency (mechanical work performed)/(work + heat) of contracting muscle may be as high as 35 percent, and in specialized muscles like insect flight muscle the mechanical efficiency may exceed 80 percent. The energy liberated by metabolism but not utilized for mechanical work appears as heat. Resting muscle generates a small amount of heat from basal metabolism amounting to about 2×10^{-3} cal/g/min in frog muscle at 20°C. Contracting muscle generates a large amount of additional heat over and above the basal heat. After stimulation, heat is generated even before shortening can be detected. Under aerobic conditions the heat produced by a contraction, whether it is a single twitch or a tetanus, can be divided into two parts: the *initial heat* and the heat of recovery, or *delayed heat*.

In a tetanus, heat production rises rapidly, reaching a level of about 1.2 cal/g/min or more than 500 times the basal heat; this burst of heat is the initial heat. After the tetanus has stopped, heat production continues at about 5 to 10 times the basal level and constitutes the delayed heat. The delayed heat production persists for such a long time that in terms of total calories produced, the initial heat and delayed heat are about equal. In the absence of oxygen (under anaerobic conditions) the initial heat has about the same magnitude as the initial heat under aerobic conditions. However, the delayed heat is largely absent under anaerobic conditions, and lactic acid accumulation is greater. When oxygen is readmitted to the muscle, the delayed heat is then generated.

It has been shown by A. V. Hill, the British Nobel Laureate whose lifework was devoted to studies of the energetics and mechanics of contraction, that energy changes during a single muscle twitch can be described by the following equation:

$$U = A + \alpha \, \Delta x + W + h \qquad (4\text{-}2)$$

where U is the total energy, A the heat of activation of the contractile units, W the mechanical work performed by the contraction, and $\alpha \, \Delta x$ the heat of shortening, which is proportional to Δx, the change in muscle length. The constant α has the dimensions of a force and depends upon the load; α is related to the constant a, which appears in Equation 4-1, the Hill force-velocity equation. The term h expresses the difference in heat output attributable to the decay of the active state at the beginning of relaxation.

The hydrolysis of ATP provides the immediate source of energy for muscular contraction, and the contractile protein myosin, which is an ATPase, is undoubtedly involved in the process. Muscle contains only small concentrations of ATP, about 2 to 4 μmoles/g, which is enough for about eight contractions. Muscle also contains large amounts of phosphocreatine (PC) and abundant quantities of the enzyme creatine phosphokinase (CPK), which catalyzes the interconversion of ATP and PC:

$$ADP + PC \overset{CPK}{\rightleftharpoons} ATP + C \qquad (4\text{-}3)$$

Thus PC acts as a reservoir of energy-rich phosphate bonds that quickly resynthesize ATP. A muscle in which respiration is inhibited by lack of oxygen or by cyanide and glycolysis is poisoned by iodoacetate (thereby preventing the synthesis of ATP) will contract many times. Initially the ATP is resynthesized from PC and ADP. After most of the PC is hydrolyzed, ATP is no longer regenerated, and creatine and inorganic phosphate accumulate; when the ATP is exhausted, the muscle will go into *rigor*, a state of sustained contracture. Rigor does not involve the ordinary process of contraction but rather a different kind of interaction between the contractile proteins, and it is followed by irreversible denaturation of the contractile proteins.

8 • CONTRACTILE PROTEINS

The contractile or myofibrillar proteins, which make up the myofibrillar filaments, are responsible for contraction. The myofibrillar proteins can be solubilized by concentrated salt solutions (0.6 M KCl); their dissolution is accompanied by the disruption and disappearance of the filaments as seen in electron micrographs.

The thick filaments are composed mainly of *myosin*, a threadlike protein molecule about 20 Å in diameter and 1600 Å long, possessing a globular head about 50 Å in diameter and 210 Å long. The myosin molecules have two hinge regions (areas of relatively great flexibility), one about 800 Å from the tail end and one immediately behind the head (Figure 4.17). Myosin has a tendency to aggregate into long threads as the salt concentration or pH of a solution is lowered. At first two molecules come

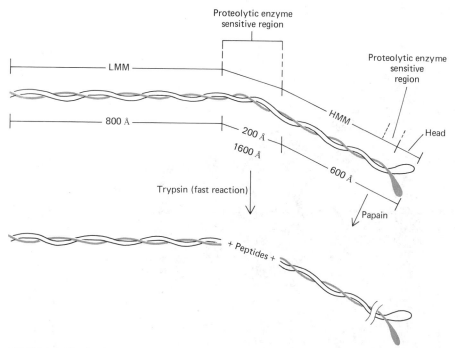

FIGURE 4.17 A schematic representation of the myosin molecule showing the two identical large subunits wound in a superhelix. The small subunits are located in the globular head. Partial proteolysis by trypsin cleaves the molecule at the hinge region into a heavy meromyosin (HMM) portion and a light meromyosin (LMM) portion, each containing parts of the original large subunits.

together tail to tail, with the heads pointing in opposite directions. Gradually more molecules attach, and the thread grows in both directions (Figure 4.18). The body of the thread contains the tail portions of the myosin molecules as far as the first hinge region, whereas the remainder of the tail and the head project outward to form the cross bridges. The thick filaments also contain some minor proteins that have not yet been completely purified and whose functions are not known.

Purified myosin requires strong salt solutions to dissolve; it also has divalent cation-stimulated ATPase enzyme sites, where it is assumed that the ATP that yields the energy utilized for contraction is hydrolyzed. Myosin can be cleaved at the first hinge region by limited proteolysis using the enzymes trypsin and chymotrypsin into a tail region (light meromyosin) and a region containing the head (heavy meromyosin). The heavy meromyosin can be cleaved at the hinge region behind the head by the proteolytic enzyme papain into a head and a cross bridge. The solubility properties of myosin are determined by the tail portion, whereas the ATPase activity is located in the heads (Figure 4.17).

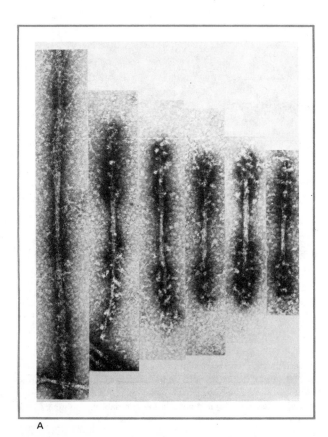

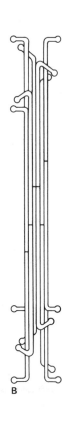

FIGURE 4.18 Arrangement of myosin molecules in the thick filaments. (**A**) An electron micrograph of aggregation of myosin molecules into threads with projections near the ends and a thinner region in the middle. (Negative stain with phosphotungstic acid; electron micrograph courtesy of H. E. Huxley.) (**B**) A diagram showing the arrangement of molecules with the tails oriented toward the center of the thread and the heads projecting toward the ends.

The myosin molecule has a molecular weight of about 500,000 daltons. It is composed of two identical large subunits of about 225,000 daltons that run the entire length of the molecule and are intertwined in a right-handed superhelix in the tail portion and three nonidentical small subunits of 21,000, 19,000, and 17,000 daltons located in the heads only. Different small subunits are found in embryonic, white, red, cardiac, and smooth muscle, and they are believed to determine the different ATPase properties of these various types of muscle.

Actin, tropomyosin, and *troponin* are the major proteins that make up the thin filaments. Actin is a 47,000-dalton globular protein, containing one nucleotide (ATP or ADP) and one divalent cation. Globular actin (G-actin) is soluble only in very weakly buffered distilled water containing a little ATP. Living muscle contains only negligible amounts of G-actin. Globular actin salts out, or aggregates, at very low salt concentrations (1 mM $MgCl_2$). It forms a two-stranded, right-handed superhelix called fibrous actin (F-actin) with a pitch of 700 Å, containing about 13 G-actin units per strand per turn. The double-stranded F-actin is very stable and forms the core of the thin filaments. During the aggregation of G-actin to F-actin, the ATP attached to each actin monomer is hydrolyzed to ADP and inorganic phosphate. Actin has the ability to interact or associate with myosin and markedly stimulates its Mg^{2+}-activated ATPase activity.

$$n \text{ G-actin-ATP} \underset{\text{ATP}}{\overset{\text{salt}}{\rightleftharpoons}} (\text{F-actin-ADP})_n + n\text{P}_i \qquad (4\text{-}4)$$

Tropomysin is a long, slender molecule of 68,000 daltons, composed of two identical subunits, which is soluble only in strong salt solutions. In the thin filaments the tropomyosin molecules lie end to end in the major grooves of the actin superhelix (Figure 4.19). The gross dimensions of the molecules are such that there is one tropomyosin molecule for about every seven actin molecules. Tropomyosin inhibits the actin stimulation of myosin Mg^{2+}-activated ATPase activity. That is, in the presence of both tropomyosin and actin, myosin Mg^{2+}-activated ATPase activity is low.

Troponin is a 78,000-dalton spherical protein, which is sttached to one end of the tropomyosin molecule. Troponin avidly binds calcium. In the presence of Ca^{2+}, troponin lifts the tropomyosin inhibition of the actin stimulation of myosin Mg^{2+}-activated ATPase activity. Thus total thin filaments containing actin + tropomyosin + troponin stimulate myosin Mg^{2+}-activated ATPase activity in the presence of greater than 10^{-6} M Ca^{2+}

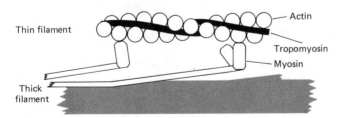

FIGURE 4.19   A diagram showing the organization of a thin filament and its relation to an adjacent thick filament. Actin monomers are arranged to form a two-stranded right-handed superhelix. Tropomyosin molecules lie in the large grooves of the superhelix. There is one tropomyosin molecule for approximately seven actin monomers.

but not in the absence of Ca^{2+}. Troponin is composed of three subunits: TN-C, the calcium-binding subunit (molecular weight 18,000 daltons); TN-T, the subunit that binds the troponin complex to tropomyosin (molecular weight 37,000 daltons); and TN-I, the inhibitory subunit (molecular weight 23,000 daltons) that turns off the actin stimulation of Mg^{2+}-activated myosin directly in the absence of tropomyosin and more strongly in the presence of tropomyosin. Recent experiments indicate that TN-I can be phosphorylated and that it may act as a regulatory subunit shifting the Ca^{2+} curve of interaction with troponin to higher or lower Ca^{2+} concentrations.

In summary Mg^{2+}-activated myosin ATPase activity is required for contraction to make energy available from ATP hydrolysis. Mg^{2+}-activated myosin ATPase activity is markedly stimulated by actin. In the absence of Ca^{2+} the actin of the thin filaments is turned off by the tropomyosin and troponin. In the presence of $> 10^{-6}$ M Ca^{2+} the actin inhibition is lifted and active contraction takes place.

In scallops and other invertebrates pure actin is not sufficient to activate the myosin ATPase activity, but Ca^{2+} ions are directly required. Scallop and other myosins that require Ca^{2+} directly possess a special regulatory subunit to which the Ca^{2+} ions bind.

9 • MECHANISM OF MUSCLE CONTRACTION

The nature of the motor or machine that makes the two sets of interdigitating filaments actually slide to produce tension and shortening is presently unknown; understanding of the mechanism of muscle contraction is one of the most important problems facing cell biologists. Any theory of muscle contraction must satisfy the observations already mentioned:

1. Shortening occurs as a result of sliding of the sets of interdigitating filaments.
2. Muscle is activated by the release of Ca^{2+} into the sarcoplasm and relaxes when the Ca^{2+} is removed.
3. The maximum tension developed (isometric contraction) is proportional to the amount of filament overlap.
4. During the major part of an isotonic contraction, the velocity of shortening is constant or decreases slightly.
5. The energy used increases as the mechanical work performed increases.
6. The heat of shortening is independent of the velocity of shortening.
7. ATP is hydrolyzed.

The theoretical models that have been proposed, none of which may be regarded as proved, may be divided into two groups: those models that require movement of the cross bridges to produce tension, and those models in which cross-bridge movement is passive.

The theoretical models that require cross-bridge movement to produce tension all suppose that, upon activation, the cross bridges undergo repeated cycles of movement that involve attachment of the myosin head to a binding site on an actin molecule, a structural change in the cross bridge and head, causing movement of the thin filament, detachment of the myosin head from the thin filament, and, finally, movement of the myosin head and cross bridge back to a comparable starting position farther down the filament. During this sequence of steps a molecule of ATP is hydrolyzed to provide energy for the structural changes in the myosin head and cross bridge.

Of the theoretical models in which cross-bridge movement is passive, the most attractive proposes an electrostatic mechanism for force production in muscle. ATP hydrolysis produces a negative surface charge on the myosin heads. The thin filaments have an axially oriented polarizability in the relaxed state; in the active state, the polarizable groups reorient perpendicular to the long axis. The electrostatic field caused by the negative surface potential on the myosin head causes movement of the thin filaments toward the center of the sarcomere. This mechanism is the electrostatic equivalent of the solenoid and is sometimes called the solenoid model.

10 • RED AND WHITE MUSCLE

Two types of muscle fibers are found in vertebrate skeletal muscle: red and white muscle fibers. The two types of muscle fibers are histochemically and functionally distinct and derive their names from gross appearance corresponding to dark and light meat, respectively, in chicken and turkey. Many muscles contain both types of fibers, which can be distinguished by various histochemical stains.

Red (slow or Type I) muscle fibers contain many sarcosomes (mitochondria) and large amounts of the oxygen-carrying protein myoglobin. White (fast or Type II) muscle fibers, on the other hand, contain large amounts of phosphorylase and glycolytic enzymes and large deposits of glycogen. Red muscle fibers respond to a stimulus with a slow twitch but are capable of sustained activity for long periods of time during which they derive energy predominantly from respiration. Fast muscle fibers react to a stimulus with a rapid twitch; they are capable of bursts of great tension production, during which glycolysis and lactate production are prominent.

The ATPase activity of myosin isolated from red muscles is less than white muscle myosin ATPase activity. The differences in ATPase activity and other physical properties of myosin from red and white muscle are due to the presence of different small subunits (approximately

20,000 daltons) in the myosin molecule. Mixed reconstituted myosin molecules made up of large subunits from red muscle and small subunits from white muscle have the high ATPase activity characteristic of white muscle. Contrariwise, mixed reconstituted myosin made of large subunits from white muscle and small subunits from red muscle has the lower ATPase activity characteristic of red muscle.

White and red muscle fibers have a different type of innervation. White muscle fibers are innervated by large nerve fibers, which terminate in plaquelike motor end plates. On the other hand, red muscles are innervated by small nerve fibers, which end in grapelike clusters on the muscle fiber. A single muscle fiber never has both types of innervation.

The characteristics of muscle appear to be determined by the motor nerve, but the mechanism is not understood at present. If two motor nerves to a white and red muscle are cut and reversed, so that upon regeneration of the nerves the fast motor nerve innervates the red muscle and the slow motor nerve the white muscle, after a few months the muscles reverse their characteristics. The former red muscle, now innervated by a fast motor nerve, becomes a white muscle, losing myoglobin and sarcosomes, increasing glycogen and glycolytic enzymes, and changing the ATPase activity of the myosin. The opposite changes take place in the other cross-innervated muscle.

Many muscles are mixed muscles, containing a mixture of red and white muscle fibers. Cardiac muscle has intermediate properties but is closer to red skeletal muscle than white muscle.

11 • CATCH MUSCLE

It is generally appreciated that oysters, clams, and other invertebrates can maintain enormous tensions with some of their muscles for very long periods of time. In 1912 the German zoologist von Uexkuell performed a very clever experiment: he inserted a wooden block as a scallop closed its shells. When the contraction closing the shells was fully developed, he removed the block. The shells remained in a slightly open position and resisted efforts to close as well as open them. Thus the powerful adductor muscle that closes the shells of mollusks has a *catch* mechanism that fixes the contractile units for days or even months after a contraction has occurred with only occasional neural stimulation (Figure 4.20). The catch mechanism permits the maintenance of tension by means quite different from a tetanus. Subsequent experiments have shown that the catch mechanism can maintain tension without the large increase of metabolic energy production that occurs during the active contractile state.

Relaxation of the catch mechanism requires stimulation of special nerves to the catch muscles from the visceral ganglia. Experiments have

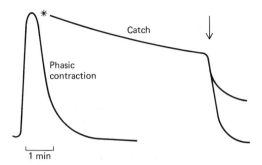

FIGURE 4.20 Time course of tension production during an ordinary twitch (phasic contraction) and its continuation by the catch mechanism. The catch is terminated at the arrow by stimulation of special release nerves from the visceral ganglia. (From B. M. Twarog and Y. Muneoka, *Cold Spring Harbor Symp. Quant. Biol.* 37:490, 1972.)

shown that acetylcholine is the chemical transmitter at the nerve endings that produce a contraction, whereas 5-hydroxytryptamine is the chemical transmitter at the nerve endings that cause relaxation of the catch mechanism. Although the relaxing nerves cause dissipation of the catch mechanism, they do not inhibit the active state of an ordinary contraction.

Molluscan, annelid, and other invertebrate muscles that possess the catch mechanism contain a unique protein, paramyosin, which may amount to as much as 30 percent of the total protein. Paramyosin has a molecular weight of about 220,000 daltons. It is composed of two identical, almost entirely α-helical subunits would in a right-handed superhelix to form a rodlike molecule 20 Å in diameter and 1275 Å long. The primary structure (amino acid composition and sequence) of paramyosin is unrelated to myosin or other contractile proteins. Paramyosin aggregates into thick, long filaments with triangular holes at 725 Å intervals. In the presence of Ca^{2+} ions, paramyosin undergoes a change of state and becomes very rigid.

The ultrastructure of muscles possessing the catch mechanism shows certain distinctive features. The thick and thin filaments are somewhat more randomly oriented, owing to a lack of Z bands and an organized sarcomere architecture. The thin filaments are clustered into *dense bodies* that are analogous to the Z lines of striated muscle. The thin filaments contain actin and tropomyosin, but troponin seems to be absent. The thick filaments are much larger in length and diameter than thick filaments from other muscles, the diameter varying from 300 to 1200 Å in different catch muscles from various species. The thick filaments in catch muscles have a core composed of paramyosin surrounded by an outer layer of myosin. Because of the fact that myosin molecules bind more strongly to paramyosin than myosin molecules bind to each other, the organization of myosin molecules in the thick filaments of catch muscles is determined by the structure of the underlying paramyosin core.

The active state of catch muscles also follows movement of Ca^{2+} ions into the sarcoplasm. However, owing to the absence of troponin in catch muscles, a different activation mechanism operates. The myosin of catch muscles contains unique small regulatory subunits that sensitize the myosin to Ca^{2+} ions. The regulatory light chains do not bind Ca^{2+} but promote the interaction of myosin with thin filaments in the presence of greater than 10^{-6} M Ca^{2+} (molluscan myosin cannot interact with actin in the absence of Ca^{2+} ions). Small amounts of paramyosin inhibit the actin-activated myosin ATPase; this inhibition is overcome by Ca^{2+} ions.

Whether a muscle undergoes just an ordinary twitchlike contraction or enters the catch state depends upon the nature of the nerve stimulation and the presence of drugs. If the muscle goes into the catch state, the intracellular Ca^{2+} level remains high. The paramyosin is thought to undergo a phase transition to a more rigid state; the rigidity is transmitted to the myosin molecules on the outer surface of the thick filaments. The myosin heads are "locked" into position with respect to the actin of the thin filaments. In some ways the catch mechanism resembles rigor in striated muscle.

12 • INSECT FLIGHT MUSCLE

If the beating of the wings of some insects, such as flies, bees, and wasps, produces audible sound, it is evident that these wings must beat several hundred times per second. High-frequency oscillation is a property also found in the timbal muscle of the sound-producing organ in certain insects such as the cicada. Muscles capable of such high-frequency contraction are of a type known as *asynchronous,* or fibrillar, and are restricted to the insect orders Diptera, Coleoptera, Hymenoptera, and to some of the Hemiptera. In such muscles contractions occur more frequently than impulses received from the motor nerve. These impulses exercise only general control over contractile activity, which is maintained by an intrinsic oscillatory mechanism. By this means the duration of a twitch in an asynchronous muscle is very much shorter than that seen in vertebrate striated muscle (about 150 msec). Thus in the blowfly, whose flight muscles contract at about 120 cycles per second, each cycle has a duration of 8.3 msec, while the muscle depolarization frequency is 3/sec. Synchronous flight muscle closely resembles vertebrate striated muscle in that the muscle fibers are stimulated by motor nerves during each wing-beat cycle.

This type of muscle is found in such insects as grasshoppers, dragonflies, moths, and butterflies, which have a relatively low wing-beat frequency.

Synchronous and asynchronous muscles are very similar in structure, but the latter have the important property that when they are extended beyond their resting length, after a short delay, they develop an increase in tension and ATPase activity. This stretch-activated tension develops only when the Ca^{2+} ion concentration is between 10^{-7} and 10^{-6} M which is achieved as a result of low-frequency nervous stimulation. As the muscle contracts, it deforms the elastic exoskeleton. As the exoskeleton returns to its former shape, it stretches the muscle, which responds after a brief delay by increasing the tension produced and contracting once more. Provided that the raised Ca^{2+} ion concentration is maintained in the muscle, this cycle will repeat itself, and the muscle will oscillate at the resonant frequency of the thorax.

Insect flight muscle possesses a number of interesting structural features. First of all, there is remarkably little connective tissue that might damp the oscillatory activity. The ratio of thin filaments to thick filaments in insect flight muscle is 3:1 or more, in contrast to the 2:1 ratio found in most vertebrate striated muscle. The periodicity of cross bridges arising from the thick filaments and the pitch of the thin filament helix are

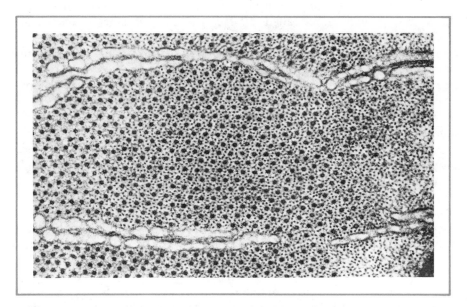

FIGURE 4.21 Electron micrograph of a transverse section of flight muscle from a Belostomatid water bug, *Lethocerus cordofanus,* showing the arrangement of filaments. Note that the ratio of thin to thick filaments is 3:1. (Courtesy of B. M. Luke/Agricultural Research Council Unit)

approximately in register in insect flight muscle, being 385 and 770 Å, respectively. In rabbit psoas muscle the myosin pitch is 430 Å, and the thin filament helix pitch is 770 Å. These features give insect flight muscle a distinctive appearance in electron micrographs (Figure 4.21).

Only a few examples of the great diversity of structural organization and function found in the animal kingdom have been described; an enormous variety of specializations of muscle have evolved to permit animals to deal effectively with various needs. This diversity demonstrates the great adaptive capacity of living systems.

13 • MOTILITY IN SYSTEMS OTHER THAN MUSCLE

Motility is characteristic of all living systems, and particularly of metazoan animals. The most general form of motility is called *protoplasmic streaming;* it is the churning or movement of cytoplasm found to some extent in almost all types of cells. Protoplasmic streaming is particularly prominent in cells that are free to move, such as phagocytes that display *ameboid motion.* Movement of the cytoplasm is appreciated by watching the movement of intracellular particles. Sometimes the intracellular particles can be observed to undergo sudden jumping movements over distances as long as 30 μ. This phenomenon, called *saltatory particle streaming,* is related to protoplasmic streaming.

The processes of cytoplasmic motility require active metabolism and are decreased or absent during oxygen deprivation or in the presence of respiratory poisons such as cyanide. Protoplasmic streaming also stops when glycolysis is inhibited by iodoacetic acid or fluoride. It appears that ATP hydolysis or the hydrolysis of another nucleotide triphosphate is the immediate energy source.

Proteins that resemble muscle actin and myosin have been isolated from many types of cells other than muscle. The nonmuscle actinlike and myosinlike proteins resemble muscle actin and myosin very closely in amino acid sequence and structure. The nonmuscle proteins form actomyosins, which have divalent cation-stimulated ATPase activity, display supercontraction, and resemble contractile proteins from muscle in other ways. It is reasonable to believe, therefore, that the actinlike and myosinlike proteins are involved in cytoplasmic motility.

Microtubules, slender filaments about 200 Å in length and as much as several microns in length, are found in many types of cells and are especially prominent in cells displaying vigorous cytoplasmic streaming. For example, microtubules are very prominent in the axonal processes of nerve cells, where they are thought both to provide some mechanical rigidity and to be involved in the prominent axonal flow by which intracellular matter moves from the body of the nerve cell along the axon to the most peripherally located fiber ends, at rates of approximately 0.5 m/24 hr. Microtubules also are involved in the movement of chromosomes during cell division and in cilia and flagella.

Many microtubules appear to be made up of globular subunits about 45 Å in diameter, and in appearance they have some resemblance to actin monomers in the thin filaments of muscle. Microtubules have been shown to contain two closely related forms of the protein *tubulin*, which has many similarities to muscle actin. It is thought that the force responsible for cytoplasmic motion is generated between the relatively rigid and fixed microtubules and the viscous cytoplasm containing a myosinlike protein.

Cilia and *flagella* are hairlike processes that project outward from the surface of some cells. They are very uniform in diameter, about 0.2 μ, but vary in length from 2 μ to several millimeters. A cross-sectional electron micrograph of cilia is shown in Figure 4.22. Regardless of the cell type or species, cilia and flagella show a similar appearance, namely, nine fibers arranged in a circle around two central fibers. This 9 + 2 structure is called the *axial filament*. Each fiber consists of two or three tubular fibrils,

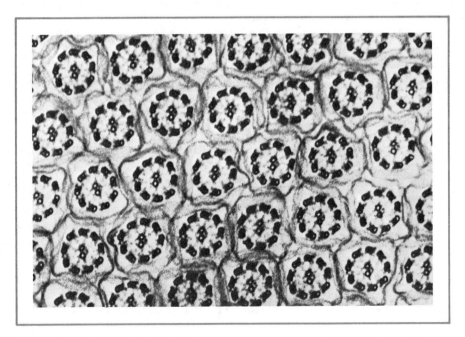

FIGURE 4.22 Lateral cilia from the gill of a mussel, *Anodonta cataracta*, showing the arrangement of fibrils in the *axial filaments*, two central fibrils surrounded by a ring of nine fibrils. (From I. R. Gibbons, *J. Biophys. Biochem. Cytol.* 11:183, 1961.)

approximately 240 Å in diameter, having a dense rim and transparent core, each fibril appearing to resemble other kinds of microtubules. Cilia and flagella also contain actinlike and myosinlike proteins and may use GTP hydrolysis for energy. Thus in nonmuscle systems motility appears to be produced by very similar systems.

REFERENCES

Bendall, J. R. *Muscles, Molecules, and Movement.* New York: Elsevier, 1969.

Gergely, J. *Biochemistry of Muscle Contraction.* Boston: Little, Brown, 1964.

Hill, A. V. *Trails and Trials in Physiology.* Baltimore: Williams & Wilkins, 1966.

Huddart, H. *Comparative Structure and Function of Muscle.* New York: Pergamon Press, 1974.

Huxley, A. F. The Croonian Lecture, 1967. The Activation of Striated Muscle and Its Mechanical Response. *Proc. Roy. Soc. (London)* 178:1–27, 1971.

Huxley, A. F., and Simmons, R. M. Proposed Mechanism of Force Generation in Striated Muscle. *Nature* 233:533, 1971.

Huxley, A. F., and Taylor, R. E. Local Activation of Striated Muscle Fibers. *J. Physiol.* 144:426–441, 1958.

Huxley, H. E. *The Cell.* J. Brachet and A. E. Mirsky, eds. Vol. 4. New York: Academic Press, 1961.

———. The Croonian Lecture, 1970. The Structural Basis of Muscular Contraction. *Proc. Roy. Soc. (London)* 178:131–149, 1971.

Mommaerts, W. F. H. M. Energetics of Muscular Contraction. *Physiol. Rev.* 49:427–508, 1969.

Podolsky, R. J., ed. *Contractility of Muscle Cells and Related Processes.* Englewood Cliffs, N.J.: Prentice-Hall, 1971.

Pringle, J. W. S. The Contractile Mechanism of Insect Fibrillar Muscle. *Progr. Biophys.* 17:1, 1967.

Stracher, A., ed. *The Contractile Process.* Boston: Little, Brown, 1967.

Szent-Györgyi, A. *Chemistry of Muscle Contraction.* New York: Academic Press, 1951.

White, D. C. S., and Thorson, J. The Kinetics of Muscle Contraction. *Progr. Biophys.* 27:175, 1973.

Wilkie, D. R. *Muscle.* New York: St. Martin's Press, 1968.

Yu, L. C., Dowben, R. M., and Kornacker, K. The Molecular Mechanism of Force Generation in Striated Muscle. *Proc. Nat. Acad. Sci. U.S.* 66:1199, 1970.

Zachor, J. *Electrogenesis and Contractility in Skeletal Muscle Cells.* Baltimore: University Park Press, 1971.

CIRCULATION AND RESPIRATION

Stephen C. Wood
Claude J. M. Lenfant

Physiological processes depend on two types of transport. One is information transport (for example, neuronal transmission), which was dealt with in Chapter 3. The second is material transport, which is the subject of this chapter. All living cells must have delivery of substrates and oxygen and removal of metabolic waste products and carbon dioxide. Circulation is the main component of material transport, but this system depends in turn on the *exchange* of materials between the organism and its environment. This exchange of materials requires the coordinated action of the respiratory sytem for gas exchange, the digestive system for food uptake, and the renal system (and other osmoregulatory tissues) for water and electrolyte exchange and waste removal.

Although material transport is involved in the control of body functions, this process is in turn regulated by the systems it is supplying. For example, the circulation regulates the metabolism in local areas of the body by the rate of delivery of oxygen and other substrates to that area. On the other hand, metabolic processes such as CO_2 production have profound effects on local blood flow. Several examples of this type of interaction (that is, feedback control) will be seen in this chapter.

In contrast to the treatment given in most other textbooks, respiration and circulation are discussed together in this chapter. By the presentation of these interrelated topics together, the subject can be discussed from the environmental level all the way down to the cellular and molecular level without dividing the material and presenting it in different parts of the text. Hopefully the approach will aid the reader to understand how the functions of the two systems are integrated.

1 • INTRODUCTION

A prerequisite for life in vertebrates is the acquisition, transportation, and utilization of molecular oxygen. The most successful species (in an evolutionary sense) are those that can deliver the most oxygen to the cells over the widest range of energy requirements and oxygen availability. This achievement requires the activity of two closely integrated systems: (1) the *respiratory system*, which renews the respiratory medium (environmental air or water) and provides a permeable surface for diffusion of gases between it and the blood; and (2) the *cardiovascular system*, which renews the blood on the inner side of the gas exchanger and transports it to the cells, where oxygen is utilized. These same renewal processes, which are recognized as *ventilation* and *circulation*, are needed to eliminate the carbon dioxide produced by metabolizing cells.

The overall result of these processes, measurable externally as oxygen consumption and carbon dioxide production, is achieved through four linked steps:

1. *Ventilation*. Convective transport of respiratory medium to and from the exchange surface.
2. *External gas exchange*. Diffusion of O_2 and CO_2 between environment and blood.
3. *Circulation*. Convective mass transport of blood.
4. *Tissue gas exchange*. Diffusion of O_2 and CO_2 between tissue capillary and cells.

Figure 5.1 illustrates the basic relationship between the active processes (ventilation and circulation) and passive diffusion (external and tissue gas exchange). The same basic pattern of oxygen tensions applies to all vertebrates; that is, whether it is air or water, lungs or gills, there is a progressive decrease in oxygen pressure from environment to cell.

In this chapter we will examine each of the four steps of gas transport, how they vary among classes of vertebrates and among species, and how they vary in relation to each other and to the availability and demand for oxygen. Special emphasis will be placed on the ventilatory and circulatory processes.

When modern vertebrates are compared, there is a seemingly chaotic variety in the morphology and physiology of the gas transport process. This variety becomes indeed remarkable but much less chaotic when each of the four steps in oxygen transport is viewed with respect to the other steps and to the total environmental physiology of each species. In fact, by viewing these four steps with the premise that each one is capable of adaptations that have survival value, a great deal of "evolutionary design" becomes apparent. The interactions and relative importance of

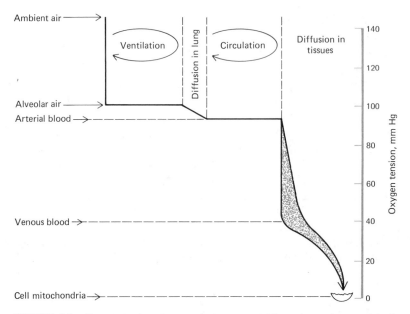

FIGURE 5.1 Oxygen tensions in man during transport from the environment to the cell. (From C. Lenfant, *Preservation of Red Cells,* Publication ISBN 0-309-02135-0, H. Chaplin, Jr., ed. Washington, D.C.: Division of Medical Sciences, National Academy of Sciences–National Research Council, 1973.)

each step in the oxygen transport in extant aquatic, amphibious, and terrestrial vertebrates tell us a great deal about the evolution and adaptability of the respiratory and cardiovascular functions.

2 • VENTILATORY MASS TRANSPORT

Ventilation refers to the convective movement of the respiratory medium over the external side of the gas exchange surface. If the respiratory medium is water, the following types of ventilation are possible: the gills are actively moved past the water (as in amphibian larvae); water is pumped over the gills (as in most fishes); or water is rammed through the gills during fast swimming, a process known as *ram-jet ventilation* (as in some oceanic fishes). If the respiratory medium is air, ventilation actively transports the gas molecules between the ambient air and the lungs. Auxiliary gas exchange surfaces, such as skin in amphibians, are ventilated passively when the respiratory medium or the animal is moving. Because the respiratory medium significantly influences ventilatory mass transport, it is appropriate, before comparing the mechanisms of ventilation, to compare the relevant physical properties of air and water.

TABLE 5.1   Solubility Coefficients for Respiratory Gases as a Function of Temperature and Salinity

| | Gas Solubility, ml/liter/mm Hg | | | |
| | *Fresh Water* | | *Seawater** | |
Temperature	O_2	CO_2	O_2	CO_2
0°C	0.0647	2.254	0.0497	1.90
5°C	0.0567	1.874	0.0436	1.57
10°C	0.0505	1.571	0.0390	1.34
15°C	0.0455	1.341	0.0353	1.16
20°C	0.0414	1.155	0.0324	1.00
25°C	0.0381	0.999	0.0300	0.88
30°C	0.0351	0.875	0.0281	0.77
37°C	0.0322	0.683	0.0256	—

* Salinity at 34°C to 35°C. CO_2 solubility will vary with nature of seawater.
Source: D. J. Randall, *Fish Physiology*, vol. 4, W. S. Hoar and D. J. Randall, eds. New York: Academic Press, 1970.

In terms of their effect on the mechanics of ventilation, the most relevant properties of the respiratory medium are density and viscosity. A liter of water weighs about 1000 times as much as a liter of air and is 100 times as viscous. In terms of its influence on the rate and volume of ventilation, the relevant property of the respiratory medium is oxygen content. A liter of aerated water at 0°C contains 10 ml of O_2, whereas air contains 21 times more oxygen (209.5 ml O_2, STPD). As Table 5.1 shows, the relative paucity of oxygen in water becomes more pronounced as temperature and salinity increase above zero. The water breather, in addition to working much harder to ventilate a given volume, must also ventilate a much greater volume than an air breather in order to obtain the same volume of oxygen per minute. The effect of water breathing on temperature regulation is discussed in Chapter 10, Section 5.1.

2.1   Mechanics of Ventilation

Most gills and all lungs are located internally, an obvious advantage considering their fragility and vascularity, and communicate with the environment through channels or tubes through which the respiratory medium is circulated, usually by rhythmic pumping. The volume and the rate of pumping combine to produce the ventilation volume per unit time. Its magnitude is a function of the respiratory medium and, as may be predicted, is about 28 times higher in fishes than in air breathers.

Gill Ventilation.   The mechanics of pumping water past gills is similar for *elasmobranch* and *teleost* fishes. Water flows through the mouth (buccal cavity) over the gill filaments and out through the opercular cavity (Figure 5.2). Several studies using sensitive electronic flow and pressure transducers have established that the water movement depends on the interaction of two pumps: a positive pressure pump in the buccal cavity and a suction pump in the opercular cavity (parabranchial cavity in elasmobranchs). The general arrangement of this double-pumping mechanism is shown in Figure 5.3. During most of the respiratory cycle the action of the two pumps results in a continuous pressure gradient, thus a continuous flow (with cyclic velocity) of water past the gills. In bottom-dwelling (benthic) species, this continuous flow depends heavily on the opercular suction pump whereas many free-swimming (pelagic) species rely on ram-jet ventilation. As described later in the chapter, the direction of blood flow through the secondary lamellae of the gills is counter to the water current. The continuous flow of water is important to the efficacy of the "countercurrent" exchange of gases at the gills.

In *cyclostome* fishes the mechanics of ventilation differ from those in elasmobranchs and teleosts. In lampreys water enters through the mouth and is propelled by muscular action of the gill pouches past the gills and to the exterior through multiple gill openings. When they are

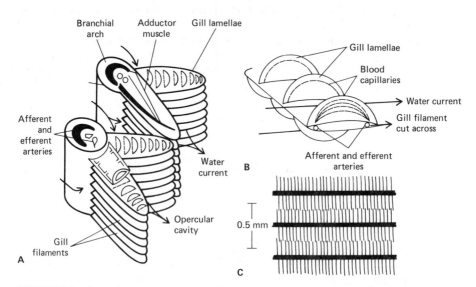

FIGURE 5.2   General structure and flow relations in teleost gills. (**A**) Anatomy of gill arches. (**B**) Magnified structure of a single gill filament. (**C**) Sievelike arrangement of filaments and lamellae. (From G. M. Hughes, *New Scientist* 11:347, 1961. This article first appeared in *New Scientist*, London, the weekly review of science and technology.)

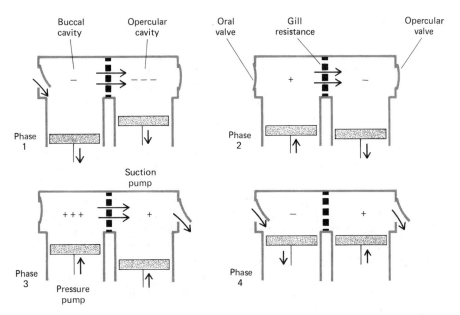

FIGURE 5.3 Mechanical representation of the double-pumping mechanism providing gill water flow. Water flow is almost continuous because phase 1, in which the opercular suction pumps are active, and phase 3, when the buccal pressure pump is active, occupy about 90 percent of the entire cycle. The transition phases, 2 and 4, are so brief that only an insignificant interruption of flow occurs. (From G. M. Hughes, *New Scientist* 11:347, 1961.)

feeding, pharyngeal inflow is not possible, and a unique tidal ventilation is necessary whereby water is inhaled and exhaled through the same external openings. In hagfishes there is a functional dorsal nostril containing a specialized pumping apparatus for water entry and a single branchial opening for water exit; further, the gill ducts are equipped with sphincters to facilitate unidirectional flow.

DEAD SPACE. In water-breathing species (as in air-breathing species) some of the "inhaled" volume of water (or air) does not participate in gas exchange. This volume is called the *dead space*. The magnitude of the dead space is determined by two factors: anatomic and physiologic. The anatomic factor is represented by the volume of air or water that does not come in contact with the gas exchange surface (lamellae in gills and alveoli in lungs). The physiologic factor is determined by lack of sufficient time to reach equilibrium or by a distance that is too great for diffusion or both. This results in a failure to establish gas tension equilibrium between the blood and the volume of water (or air) that is in contact with the gas exchange surface. In fishes the total dead space varies greatly, and it may be as high as 60 percent of the ventilation volume.

The cost of ventilation may be estimated by measuring the oxygen consumption of the respiratory system in relation to the volume of air or water that is pumped. Schumann and Piiper (1966) found that ventilation in the teleost (*Tinca tinca*) required 0.7 to 1.8 ml O_2 per liter of water pumped. In the range of ventilation volumes studied (1 to 3 times resting ventilation) the cost of breathing in fishes is 3.5 to 9 times higher than in man. Considering the highly unfavorable physical properties of water as a respiratory medium, this difference is not large. This is explained by the inherently greater efficiency of gas exchange in gills compared with lungs due to the countercurrent flow of blood and water. In terms of overall metabolic rate, however, the cost of breathing in fishes represents a high percentage of the total oxygen consumption. For *Tinca*, ventilation uses 32 percent of the total oxygen uptake at rest and 49 to 69 percent of total O_2 uptake at 3 times resting. ventilation.[1] Other fishes (for example, trout) are not taxed quite so heavily, but even here about 18 percent of total O_2 uptake is used for breathing.

Lung Ventilation.   Lung ventilation occurs in all terrestrial and in some aquatic vertebrate classes, but there is great diversity in the comparative morphology of vertebrate lungs and in basic ventilatory mechanics. There are two basic patterns of lung ventilation. The first involves positive pressure, which inflates the lungs like a balloon. This type is found in air-breathing fishes and in amphibians. The other basic pattern is negative pressure, which enlarges the cavity containing the lungs and results in aspiration or "sucking" of air into the lungs. This difference is due to the absence or presence of a rib cage and related muscular apparatus. This section briefly describes some aspects of the mechanics of ventilation in these two types of pumps.

In air-breathing fishes and amphibians ventilation is mechanically similar to that of gill ventilation, but the degree of structural modifications for air breathing is proportional to the dependence on air for overall gas exchange. In the three genera of lungfishes the buccal cavity provides the positive pressure pump for filling the lungs. Buccal pressure is also used by the garfish (*Lepisosteus*) to inflate the swimbladder, which is used as a functional lung during facultative air breathing.

Ventilation of amphibian lungs is also affected by buccal action (buccopharyngeal pump). Figure 5.4 describes the mechanics in anuran amphibians. First, the entrance to the lungs (glottis) is kept closed, the nostrils are open, and the lowering of the floor of the mouth permits the filling of the buccopharyngeal cavity with fresh air. Then the nostrils are

---

[1] This is in contrast to only about 5 percent of the total oxygen consumption in man at 5 times the resting level of ventilation.

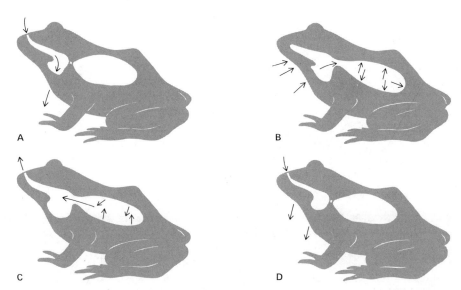

FIGURE 5.4 Mechanical movements and patterns of air flow in the bullfrog, *Rana catesbeiana*. (**A**) Lowering of buccal cavity with glottis closed and nostrils open. (**B**) Nostrils close and contraction of buccal cavity forces stored "fresh" air into the lungs. (**C**) Glottis opens, elastic recoil of lungs forces "used" air out through nostrils. (**D**) Glottis closes, nostrils open, and elastic forces cause lowering of buccal cavity. (From C. Gans et al., *Science* 163:1223, 1969. Copyright © 1969 by the American Association for the Advancement of Science.)

closed, the glottis is open, the floor of the buccal cavity is elevated, and the resulting pressure inflates the lungs. Expiration is largely passive by elastic recoil of the inflated lungs. The relative role of ventilation in total gas exchange varies tremendously among amphibians, ranging from nil in lungless salamanders to very high in terrestrial toads.

Mammals, reptiles, and birds depend on negative instead of positive pumping for ventilation. In *mammals* the negative pressure pump used to inflate the lungs consists of the diaphragm and external intercostal muscles, which, upon contracting, enlarge the thorax, reducing the intrathoracic, intrapleural, and intraalveolar pressure to subatmospheric.

It is reasonable to ask how the contraction of muscles connecting one rib to another can result in an upward and outward movement of the chest during inspiration. As Figure 5.5 illustrates, the geometry of the intercostals determines which of the two ribs (upper or lower) will be moved most by an equal but opposite force acting on both ends of the muscle. Because the external intercostals slope obliquely downward and forward from the upper to lower rib, the lower insertion is more distant from the point of rotation (vertebral articulations) than the upper insertion. Consequently more torque is applied to the lower rib than to the upper rib, so

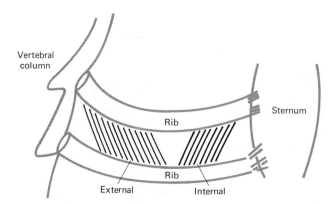

FIGURE 5.5 Geometric relationship of external and internal intercostal muscles. Contraction of external muscles elevates lower rib (inspiration); contraction of internal muscle depresses upper rib (expiration).

the net movement is elevation of the ribs ("bucket handle" movement). This increases the transverse diameter of the thorax. At the same time, the sternum is moved forward by this rib movement, thereby increasing the anteroposterior thorax diameter. The internal intercostals (internal interosseus) slope obliquely downward and backward, so the upper insertion is farther from the point of rotation and contraction pulls the upper ribs downward.

The pressure gradient for air flow, and the volume of air taken into the lungs, depends on the strength of contraction of these muscles and accessory muscles (scalene and sternomastoids), which are used for maximal breathing effort.

Alveolar pressure equals atmospheric pressure at the end of the inspiration and exceeds atmospheric pressure during expiration. At rest (eupnea), expiration is the passive result of relaxation of the inspiratory muscles. Expiratory muscles (internal intercostals and abdominals) are used during maximal breathing effort, that is, increased ventilation volumes or in cases of breathing with mechanical obstruction.

In marine mammals the pattern of breathing is greatly different. Because the breathing frequency may be less than one time a minute, whales and other diving mammals utilize as much as 90 percent of their total lung volume for tidal volume (see Figure 5.6 for definition of lung volumes and capacities). Each expiration ("blow") results in almost total collapse of the lungs. Olsen and colleagues (1969) found that a captive pilot whale could expire almost 40 liters of air in half a second (88 percent of total lung volume). Although this suggests active expiration pressure, measurements show that this is not the case, since there is no increase in esophageal

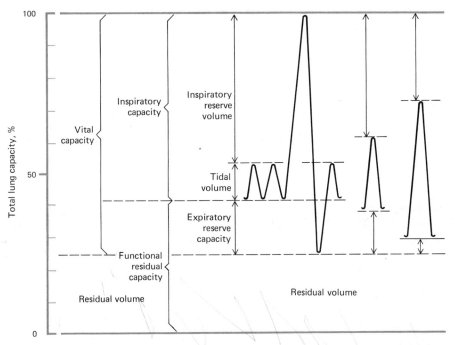

FIGURE 5.6 Volumes and capacities of normal human lungs. Tidal volume is the volume of gas that enters and leaves with inspiration and expiration. Inspiratory reserve volume is the maximum volume of gas that can be inspired beyond the resting tidal inspiration. Residual volume is the volume of gas remaining in lungs after maximum expiration. Expiratory reserve capacity is the maximum volume of gas that can be expired beyond resting tidal expiration. Functional residual capacity is the resting volume of lungs, that is, the volume of gas in lungs when respiratory muscles are relaxed. Vital capacity is the maximum volume of gas that can be expired after maximum inspiration.

pressure during the expiration. Hence it appears that the driving force consists solely of the lung elastic recoil. The whale, in contrast to man, can increase ventilation up to 30 times the resting volume by increasing frequency alone and without use of expiratory muscles. Man makes use of his expiratory reserve when ventilatory volume exceeds 6 to 8 times the resting value; larger increases of the ventilatory effort can increase the tidal volume up to the vital capacity (Figure 5.6). In the whale the normal tidal volume is so large that there is almost no reserve volume. Thus the only way for a whale to increase minute volume is to increase breathing rate (See Table 5.2).

In man the anatomical dead space has a volume of about 150 ml, which is about one-fortieth of the total lung capacity. In many diving mammals the anatomical dead space is a much greater fraction of the total lung volume. This is an important adaptation to prolonged diving because

TABLE 5.2 Breathing Rate, Tidal Volume, and Minute Volume in Man and
Other Vertebrates

Species	Condition	Respiratory Frequency, breaths/min	Tidal Volume, ml	Minute Volume, liters/min
Man	Resting	11.70	750	7.43
	Heavy work	21.20	2030	42.90
	Maximum exercise	40.00	3050	111.00
Marmot				
(*Marmota marmota*)	Awake	8.00	22	0.1740
	Hibernating	0.68	13	0.0089
Killer whale				
(*Orcinus orca*)	Beached	1.11	46200	51.30
Mouse				
(*Mus muscelus*)	Resting	1.63	0.15	0.024
Diamondback terrapin				
(*Malaclemys centrata*)		3.70	14	0.051
Australian lungfish				
(*Neoceratodus fosteri*)	In water	28.00	15	0.420

hydrostatic pressure compresses the lungs and forces air into the large airways where almost no gas diffuses into the blood. This prevents excessive nitrogen from dissolving in plasma and eliminates the risk of "bends," which are caused by nitrogen coming out of solution as bubbles if ascent to the surface is too rapid. In the fin whale the anatomical dead space (150 to 200 liters) is about one-tenth of the total lung volume.

The energy expended by the respiratory muscles can be determined by measuring total oxygen consumption at different ventilation volumes. The increase in total oxygen consumption caused by voluntary hyperventilation in a resting individual is due to the oxygen consumed by the respiratory muscles. Data for man are shown in Figure 5.7. At rest the oxygen needed to ventilate 1 liter of air is about 0.5 ml. At 50 liters/min ventilation this increases to about 1 ml O_2/liter, and at 100 liters/min it is about 3.5 ml/liter. During active exercise requiring 100 liters/min ventilation (which corresponds to a total oxygen consumption of about 3.5 liters/min) the work of the respiratory muscles accounts for about 10 percent of the total oxygen consumed. The critical level of ventilation occurs at about 150 liters/min; beyond this, all of the additional oxygen would be required for the respiratory muscles. This critical value, rarely obtained under normal conditions, is much less when ventilation is mechanically obstructed or lung elasticity is reduced. The curve for humans with obstructive lung disease is much the same as the pattern for fishes.

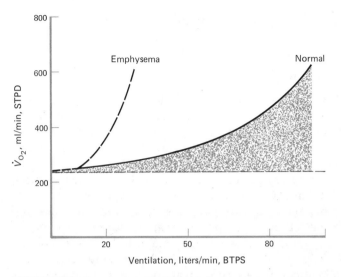

FIGURE 5.7 Oxygen consumption of the ventilatory muscles in normal and emphysematous humans. Hatched area represents extra energy expended per liter of air ventilated in the normal subjects. (From P. Dejours, *Physiologie*, C. Kayser, ed. Paris: Flammarion, 1963.)

Reptiles utilize buccopharyngeal ventilation, but mainly for olfaction ("sniffing") and not for lung inflation. Some lizards use this pumping for behavioral displays or to inflate their bodies as a defense against being dislodged from a crevice. The lungs of snakes are usually elongated and asymmetric, the right lung being reduced or vestigial. The mechanics of reptilian ventilation are not completely understood, but apparently exhalation is active, brought about by rib movement or body compression by transverse musculature. In crocodilians the volume of the pleural cavity is affected mainly by movements of the liver. Contraction of abdominal muscles pushes the liver cephalad and decreases pleural volume. Contraction of the diaphragmaticus muscle pulls the liver toward the pelvic girdle and increases pleural volume.

Birds possess an unusual system comprised of lungs and air sacs, representing a pattern of gas exchange intermediate between the countercurrent system of fish gills and the tidal system of other lungs. (The lungs constitute only 9 to 17 percent of the total respiratory system volume in birds.) The basic ventilatory mechanism is similar to mammals for inspiration (active inspiration by enlargement of the thoracoabdominal cavity); however, because of the low compliance of the avian lungs, expiration requires participation of expiratory muscles. These antagonistic sets of muscles operate what amounts to a bellows for ventilating the parabranchial gas exchange capillaries. The inflation and deflation of air sacs thoroughly flush the entire system with fresh air during each respiratory cycle. The movements of the sternum by flight muscles are not necessarily correlated with respiratory movements.

2.2   Regulation of Ventilation

Rate and depth of breathing are determined both by the environment and the metabolic rate. The range of values of these parameters and the degree of species diversity are considerable as illustrated by the data shown in Table 5.2. In any one species the minute volume can vary in response to various stresses (exercise, temperature, change in environment, and the like). If the normal respiratory pattern is regular or rhythmic (as in man), both frequency and depth of breathing can vary to change ventilatory volume. If the respiratory pattern is irregular (as in marine mammals and most poikilothermic vertebrates), the period of apnea-interrupting ventilatory periods can also vary. In either case the magnitude of tidal volume and the respiratory frequency combine to provide what is important to oxygen transport, that is, the minute volume of air or water most appropriate for maintaining relative homeostasis.

Three levels of control operate to adjust minute volume to appropriate values. These levels may be classified as central, neurogenic, and humoral.

Central Control.   The basic respiratory cycle of all vertebrates requires an intact brain stem and medulla. The localization of the central control areas is poorly defined in many species, but their function is clear.

In mammals, changes in respiration follow stimulation of several regions of the cortex and of the pons and the medulla. Electrical stimulation of the cortex has been used to identify areas or respiratory inhibition and respiratory facilitation. The respiratory response is often accompanied by changes in other visceral functions and in somatomotor activity. Although it is clear that the cortex is the center of voluntary and emotion-associated changes in respiratory activity, there is no evidence that it is involved in the basic coordination and integration of normal rhythmic breathing. The respiratory responses to stimulation of the pons and of the medulla are greater and more discrete than those following cortical stimulation. The medullary center—often called the respiratory center—is the site of rhythmic respiratory activity. This activity is, however, modulated by centers located in the pontine area, and it becomes erratic when the medullary center is isolated from the pons by transsection. The pons contains two centers: the *apneustic* center, located in the lower pons, and the *pneumotaxic* center, situated high in the pons. The former causes cessation of ventilation after inspiration when it is separated from the other centers, but it is normally inhibited by the pneumotaxic center, which also has a facilitory function, as shown by a slowing of respiration after its ablation. In lower vertebrates the central control of respiration is not documented as well as in mammals except for the medullary centers, which appear to be universal.

In mammals the coordinated activity of these centers produces basic rhythmic respiration as follows (Figure 5.8): Activity of the inspiratory center in the medulla is transmitted to the pneumotaxic center and by

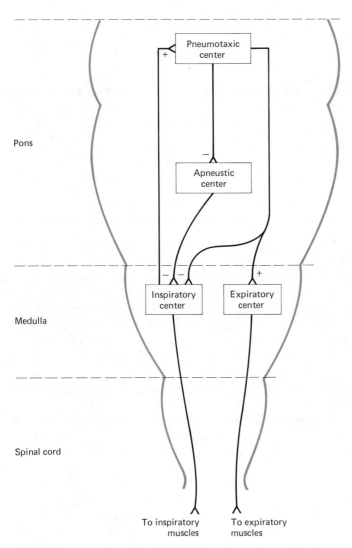

FIGURE 5.8 Diagrammatic representation of the central control of respiration. Plus sign indicates stimulatory influence; minus sign indicates inhibitory influence.

phrenic nerves to the inspiratory motor neurons. In turn sufficient excitation causes the pneumotaxic center to excite the medullary expiratory center, which inhibits the inspiratory center. Release from this inhibition starts the next cycle. Overall respiratory activity results from integration by these centers of the ventilatory stimuli of neural and humoral origin.

Neurogenic Stimuli. Neurogenic stimuli may be central or peripheral in origin. In higher vertebrates the central stimuli include temperature receptors in the hypothalamus, which may induce panting; emotion "receptors" in the cerebrum, which cause hyperpnea; chemoreceptors in the medullary surface, which, in response to increased H^+ ion activity, induce hyperpnea. Peripheral neurogenic stimuli, especially mechanoreceptors (see Chapter 3, Section 11.3), play a primary role in ventilatory control of mammals and probably also in lower vertebrates, including fishes. A familiar example is the ventilatory response to exercise in mammals, in which "ergoreceptors" of skeletal muscle stimulate breathing before there is a significant change in blood gases or pH (humoral stimuli). Other peripheral input to the respiratory center comes from pain receptors, carotid sinus baroreceptors, nasal and laryngeal irritant receptors, and lung inflation receptors. The latter are the basis of the Hering-Breuer reflex, in which lung inflation causes reflex inhibition of the inspiratory center.

Humoral Stimuli. Humoral stimuli—P_{CO_2} (partial pressure of carbon dioxide), P_{O_2}, pH—act on both central and peripheral chemoreceptors (see Chapter 3, Section 11.4, for a general discussion of chemoreceptors). The peripheral chemoreceptors are located primarily in the carotid and aortic bodies. They directly monitor arterial blood gases and pH, and they transmit tonic impulses to the respiratory center via the glossopharyngeal and vagus nerves. The central chemoreceptors are located on the surface of, or in, the medulla; they monitor pH in the cerebrospinal fluid, which surrounds the central nervous system.

In the evolutionary transition from aquatic to aerial respiration the central chemoreceptors have assumed a dominant role in regulating respiration. The reasons for this are that (1) these receptors are sensitive primarily to P_{CO_2}-induced changes in H^+, whereas the peripheral receptors are more sensitive to P_{O_2}, and (2) in the evolution of air breathing, P_{CO_2} replaces P_{O_2} as the primary respiratory stimulant. In aquatic species the dominant respiratory stimulus is P_{O_2} (water or blood), which is monitored by O_2-sensitive tissue in branchial arch structures. As discussed in Section 3.2 of this chapter, the greater solubility and diffusivity of CO_2 in water minimizes CO_2 retention; that is, CO_2 is voided so easily and rapidly that it could not serve as a useful control of respiration. As air breathing becomes obligatory, CO_2 retention occurs (because of relative hypoventilation), and the respiratory response to water P_{CO_2} increases, although it remains variable in intensity. This has been well demonstrated by numerous studies in facultative and obligatory air-breathing fishes.

In higher vertebrates the effectiveness of CO_2 as a respiratory stimulant is well established. For instance, man's ventilation may increase by 10 liters/min in response to as little as 1 mm Hg increase of P_{CO_2} in his arterial blood. There are, however, other species, such as some diving mammals and some birds, that exhibit a very low sensitivity to CO_2, as shown by little or no respiratory response to large changes in arterial P_{CO_2}. In these species arterial P_{O_2} becomes the major respiratory stimulant. For example, in birds flying at altitudes of nearly 8000 m (reported by the Mt. Everest expedition in 1953) the P_{O_2} in the "alveolar" regions of the lungs ranges from 0 to 35 mm Hg, depending inversely on "alveolar" P_{CO_2}, or P_{ACO_2}. If, by hyperventilating, the P_{CO_2} is decreased to 10 mm Hg, the arterial P_{O_2}, or P_{aCO_2}, increases, and the arterial blood becomes more than 50 percent saturated with oxygen. Decreased P_{O_2} may evoke a ventilatory response in other species which are also very responsive to P_{CO_2} changes, as shown in Figure 5.9. However, these data also show that the magnitude of the ventilatory response to the same stimulus (P_{O_2} in alveolar air

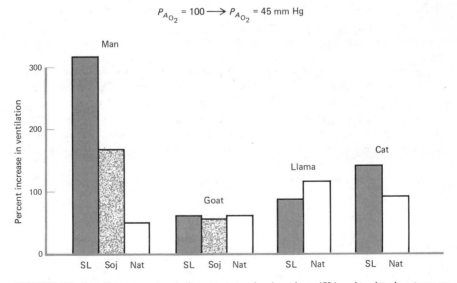

FIGURE 5.9 Ventilatory response to hypoxia in sea-level residents (SL) and in altitude sojourners (Soj) and natives (Nat) when alveolar P_{AO_2} is lowered from 100 to 45 mm Hg. Note the difference between species in regard to the magnitude of the response and the effect of adaptation. (From C. Lenfant, *Am. Zool.* 13:449, 1973.)

lowered from 100 to 45 mm Hg) is extremely variable among these species, and that in some species, such as man, it can become attenuated after prolonged exposure to decreased P_{O_2}.

3 • EXTERNAL GAS EXCHANGE

The old hypothesis that oxygen was actively secreted by lung tissue into the blood has long been rejected.[2] It is now well established that gas molecules move passively—by diffusion—across the respiratory epithelium.

The factors limiting external gas exchange are therefore those affecting the rate of diffusion, which in turn is greatly determined by the difference in gas partial pressure across the gas exchange surface and by the type of gas exchange organ as illustrated in Figure 5.10. Hence, factors such as ventilation, metabolic rate, circulation, and blood composition are of extreme importance. In air-breathing animals, gas exchange is also a function of barometric pressure, inspired gas composition, and the gas exchange ratio. As will be discussed in this section, these factors, especially the latter three, impose completely different limitations in water breathers. The general principles governing diffusion are described in Chapter 2, but it is useful in the present context to compare some special features of diffusion of gases in air and in water.

3.1 Mechanism of Gas Exchange (Diffusion)

The rate of diffusion of gas (dq/dt) from an area of higher to lower partial pressure is directly related to the pressure gradient (ΔP), the cross-sectional area in which diffusion occurs (A), and the diffusion coefficient (K), which depends on the properties of the gas and also reflects the permeability of the membrane. The rate of diffusion is inversely related to the thickness of the membrane (L). Hence the rate of diffusion is given by

$$\frac{dq}{dt} = \frac{KA\,\Delta P}{L} \tag{5-1}$$

---

[2] J. B. S. Haldane and Christian Bohr were among the physiologists in the early 1900s who favored the hypothesis that O_2 diffusion would be inadequate during exercise and O_2 secretion by alveolar epithelium must occur. One of Bohr's students, August Krogh, disagreed with this concept, and with his wife, Marie, provided unequivocal data showing that "the absorption of oxygen and the elimination of carbon dioxide in the lungs takes place by diffusion and by diffusion alone." (Krogh and Krogh, 1910)

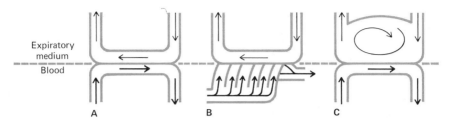

FIGURE 5.10 Models illustrating types of gas exchange in vertebrates. (**A**) Countercurrent model for fish gills. Water (in the interlamellar space) and blood (in the secondary lamellae) flow in opposite directions. Expired water equilibrates with incoming venous blood, and effluent arterial blood equilibrates with inspired water. (**B**) Cross-current model for avian lungs. In avian lungs the parabronchi are surrounded by a sheath of air capillary/blood capillary meshwork in which gas exchange takes place. This is represented in the model by blood vessels having gas exchange contact with parabronchial gas inside a single cross-sectional element of the parabronchus only. Thus blood flows effectively at a right angle to the air tube axis. The arterial blood originates from a mixture of effluent blood whose partial pressure values vary along the length of the air tube. (**C**) Uniform pool model for mammalian lungs. Capillary blood equilibrates with an external medium pool (alveolar space) in which partial pressures are uniform (functionally achieved by stirring or by absence of significant diffusion resistance). Pulmonary gas exchange in reptiles and amphibians is another example of this type of model, but the pool is not as uniform; that is, more heterogeneity exists in terms of nonventilated and/or nonperfused lung areas. (From J. Piiper and P. Scheid, *Resp. Physiol.* 14:86, 1972.)

Although the surface area (A) and thickness (L) of a respiratory surface are anatomically determined, the effective diffusion area and effective thickness may be less than this because of the physiological dead space (see Section 2.1). The latter includes alveoli that are unperfused and, in water breathers, unstirred fluid layers that increase the diffusion path by an amount that varies with the velocity of fluid flow in contact with the respiratory membrane. This has been well demonstrated in fishes, wherein the calculated rate of diffusion for oxygen (calculated from morphological measurements) is always more than that rate of diffusion measured because of the resistance of the water, mucus, and tissue of secondary lamellae, plasma, and red cells to oxygen diffusion.

The partial pressure difference for oxygen (ΔP_{O_2}) is usually difficult to measure directly. It may be estimated from samples of expired gas or water and efferent blood, but these values are often falsely elevated because of the admixture between exchange ventilation volume and dead-space volume (producing high estimates of external P_{O_2}) or admixture between afferent and efferent blood flow (producing low estimates of internal P_{O_2}).

The diffusion coefficient (also called Krogh's permeability coefficient) is a multifarious property of the respiratory membrane. Consequently it is difficult to measure directly and must be calculated by measuring the other variables of the diffusion equation. Table 5.3 shows the value of Krogh's permeability coefficient in air, water, gelatine, and non-

TABLE 5.3 Permeability Coefficients of O_2 at 20°C

Material	K (ml O_2/cm²/min/atm)
Air	11
Water	34×10^{-6}
Gelatin	28×10^{-6}
Muscle	14×10^{-6}
Connective tissue	11.5×10^{-6}
Chitin	1.3×10^{-6}

Source: from Krogh, 1919.

living tissue. The value of K for oxygen in human lungs can be estimated from measurements of diffusion capacity for oxygen, lung area, and diffusion distance. Diffusion capacity is the ratio of the rate of oxygen uptake[3] to ΔP_{O_2} (that is, $D_{O_2} = \dot{V}_{O_2}/\Delta P_{O_2}$), and in resting human lungs $D_{O_2} \simeq 15$ ml/min/mm Hg. Since $\dot{V}_{O_2} = KA\,\Delta P_{O_2}/L$, the above equation can be rewritten as $D_{O_2} = KA/L$, and, using approximate values $A = 70$ m² and $L = 1\,\mu$, we can calculate the approximate value of K as

$$K = \frac{15 \text{ ml/min/mm Hg} \times 1\,\mu}{70 \text{ m}^2}$$

or

$$K = 1.63 \times 10^{-6} \text{ ml } O_2/\text{cm}^2/\text{min/atm}$$

These calculations show that for human lungs, the effective permeability coefficient in vivo is lower than most of the in vitro values found by Krogh for various tissue preparations.

For gases other than oxygen, the relative rate of diffusion depends on their solubility coefficient and molecular weight. Hence since CO_2 is approximately 28 times more soluble than O_2, at 20°C its diffusion rate is about 26 times faster than O_2.

3.2 Gas Exchange Principles: Water Breathing versus Air Breathing

The molar ratio of carbon dioxide production to oxygen consumption ($\dot{V}_{CO_2}/\dot{V}_{O_2}$) is called the gas exchange ratio (R), which in the steady state is equal to the metabolic respiratory quotient (RQ). For carbohydrates $R = 1$; that is, there is an equal volume exchange of O_2 and CO_2. Gas ex-

---

[3] O_2 uptake $\dot{V}_{O_2}$ is equal to rate of diffusion dq/dt or volume of oxygen diffusing across the respiratory surface in 1 min.

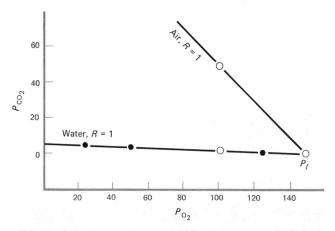

FIGURE 5.11 O_2-CO_2 diagram for air breathers and water breathers. P_I is the inspired partial pressure of O_2 and CO_2. See text for explanations. (From H. Rahn, *Resp. Physiol.* 1:5, 1966.)

change in air- and water-breathing animals can be graphically described by the "O_2-CO_2 diagram," which shows the relationship between P_{O_2} and P_{CO_2} in the respiratory fluid (air or water) and in the blood.

The O_2-CO_2 Diagram.   The example in Figure 5.11 (Rahn and Fenn, 1955; Rahn, 1966) shows two lines corresponding to a gas exchange ratio of 1 for an air breather and a water breather, respectively. Since equal volume of CO_2 and O_2 have the same partial pressure in air,[4] the slope of the line $R = 1$ in air is 1.0. Since the coefficient of solubility (α) of O_2 and CO_2 is not the same in water, equal volumes of O_2 and CO_2 do not have the same partial pressure; in this case the slope of the line $R = 1$ is determined by the relative solubility of the two gases.

The usefulness of this diagram is that, whatever the volume of ventilation and the amount of gas exchanged, the gas tensions in the expired air or water and at the respiratory surface (in air or water and in blood) will lie somewhere on the R line, provided that $R = 1$. It is, of course, possible to draw different lines for any value of R. The equation for gas exchange is given by

$$\frac{P_{E(CO_2)}}{P_{I(O_2)} - P_{E(O_2)}} = \frac{\alpha_{O_2}}{\alpha_{CO_2}} R \tag{5-2}$$

[4] In a volume of air with total pressure $= 760$ mm Hg, each 1/760 volume exerts a partial pressure of 1 mm Hg; thus for any gas 1.32 ml/liter exerts a partial pressure of 1 mm Hg at 0°C.

where I is inspired and E expired gas tensions in water or air. If inspired P_{O_2} = 150 mm Hg and expired P_{O_2} = 100 mm Hg, then the expired P_{CO_2} for a fish will be

$$P_{CO_2} = \frac{0.04}{1.15} (1)(150 - 100) = 1.74 \text{ mm Hg}$$

where 0.04 and 1.15 are values of the coefficient of solubility at 20° C (see Table 5.1). For an air breather with the same inspired and expired P_{O_2}, the lung P_{CO_2} will be

$$P_{CO_2} = (1)(150 - 100) = 50 \text{ mm Hg}$$

These equations provide the physical bases for CO_2 retention in air breathers and the fact, alluded to earlier, that P_{CO_2} becomes the main humoral stimulus of respiration. The ventilation ($\dot{V}$) required in ml/min for every milliliter of O_2 uptake or CO_2 output can be calculated from the equation

$$\dot{V} = \frac{1}{(P_I - P_E)\alpha} \tag{5-3}$$

Hence in the example shown in Figure 5.11 the minute ventilation for each milliliter of oxygen uptake from the water is

$$\dot{V}_{water} = \frac{1 \text{ ml } O_2/\text{min}}{(150 - 100)(0.041)} = 480 \text{ ml/min}$$

To remove 1 ml O_2 from air at the same expired P_{O_2}, the ventilation is

$$\dot{V}_{air} = \frac{1 \text{ ml } O_2/\text{min}}{(150 - 100)(1.18)} = 17 \text{ ml/min}$$

where 1.18 is the solubility coefficient for O_2 in air; that is, 1.18 ml O_2/liter air/mm Hg at 20°C.

These calculations exemplify the great difference in ventilatory requirements between air- and water-breathing species. They further show that as the efficiency of gas exchange rises, there is a decrease in the ventilatory requirement per unit O_2 consumed.

3.3   Transitional Species: Bimodal Gas Exchange

Much of the theory in the preceding section assumes unimodal (that is, gills or lungs) gas exchange and does not apply to transitional species (air-breathing fishes and amphibians) that have bimodal gas exchange. An evolutionary analysis of gas exchange clearly shows gradual transition between strictly aquatic fishes using gills to the purely terrestrial mammal using lungs. Indeed, there are many species that can be both water and air breathers; in these species the relative importance of aerial to aquatic exchange is influenced mainly by the ratio of available to required oxygen.

Ambient temperature is a major determinant of this ratio in ectothermic aquatic vertebrates. Increasing temperature decreases this ratio for two reasons: it increases metabolic rate (demand) and decreases O_2 solubility in water (availability). This is exemplified by the respiratory response of the bowfin *Amia calva* to temperature and oxygen tension of the inspired water (Figure 5.12). At 10°C gill breathing is the primary mode of gas exchange, even in hypoxic water. However, with increasing temperature, increased air breathing and use of the air bladder for exchange become apparent, and at 27°C the gills are relatively unimportant, even in well-oxygenated water. In other species with bimodal gas exchange a similar effect of temperature is evident. In the salamander *Taricha granulosa*, the skin contributes more to gas exchange than the lungs at low temperature, but the increase in total $\dot{V}_{O_2}$ as temperature increases is all due to an increase in O_2 uptake by the lung, with skin $\dot{V}_{O_2}$ remaining constant.

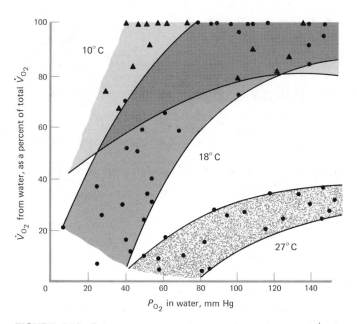

FIGURE 5.12 Relative importance of aquatic O_2 uptake to total $\dot{V}_{O_2}$ in *Amia calva* as a function of water P_{O_2} and temperature. (From K. Johansen, D. Hanson, and C. Lenfant, *Resp. Physiol.* 9:167, 1970.)

Modern dipnoi show a clear transition in the relative contributions of aerial and aquatic gas exchange. The Australian lungfish, *Neoceratodus*, has efficient gills, which, at rest, account for 100 percent of the O_2 and CO_2 exchange. In contrast the African lungfish, *Protopterus*, is more dependent on aerial gas exchange, and the lung accounts for 88 percent of O_2 uptake and about 55 percent of CO_2 elimination. Even more dependent is the South American lungfish, *Lepidosiren*, where the lung accounts for 96 percent of O_2 uptake and about 55 percent of CO_2 elimination. If these lungfishes are taken out of water and forced to utilize only aerial exchange, *Lepidosiren* and *Protopterus* show only a slight reduction in total oxygen consumption and no decrease in arterial blood oxygenation. *Neoceratodus*, however, in spite of vigorous attempts to breathe air, suffers a rapid depletion of arterial P_{O_2} and a 60 percent reduction in O_2 uptake. An order of amphibians, the salamanders, provide another example of gradual transi-

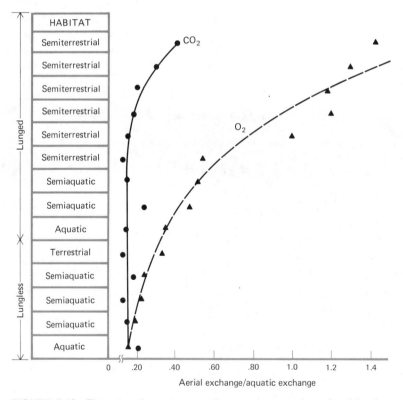

FIGURE 5.13 The ratios of aquatic to aerial gas exchange in lunged and lungless (Plethodontid) salamanders at 15°C. Lunged salamanders select more terrestrial habitats and show a greater dependence on aerial O_2 uptake. The high solubility and diffusibility of CO_2 allow the skin to be the major site for CO_2 elimination for most species. (Reprinted from C. Lenfant, K. Johansen, and D. Hanson, *Fed. Proc.* 29:1127, 1970.)

tion between aquatic and aerial breathing. In these animals the mode of gas exchange is closely related to morphological development and to the habitat, as seen in Figure 5.13. The compilation presented in this figure shows that the exchange of CO_2 remains preferentially aquatic (through gills and skin), although the lung—as attested to by increased aerial O_2 exchange—becomes more functional. This is again to be credited to the fact that carbon dioxide has greater solubility than oxygen in water.

At the phylum level, representative members of present-day vertebrates also show a progressive increase in the contribution of aerial gas exchange (Figure 5.14). Again, the salient feature is that oxygen uptake becomes independent of water much earlier than carbon dioxide elimination, whose extrapulmonary component is detectable even at the mammalian (man) stage.

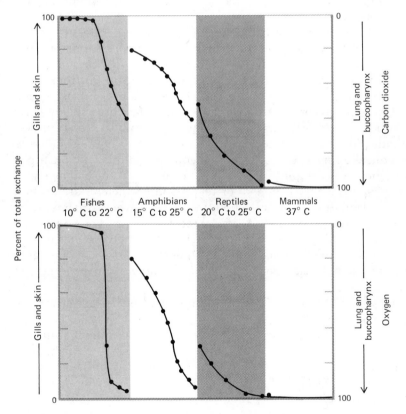

FIGURE 5.14 A summary of the relative importance of aquatic gas exchange (gills and skin) and aerial gas exchange (lungs and buccopharynx) in four classes of vertebrates. The area above each curve indicates the aerial contribution to total gas exchange. (Reprinted from C. Lenfant, K. Johansen, and D. Hanson, *Fed. Proc.* 29:1127, 1970.)

3.4 Functional Morphology of Exchange Organs

A comparison of the morphological features of vertebrate gas exchange organs reveals distinct and often unusual adaptations favoring adequate diffusion. The principal limits to diffusion are discussed in Section 3.1 of this chapter and represented in Equation 5-1. Surface area is the component of this equation showing the greatest correlation with oxygen requirement. In this section the diverse gas exchangers are described with respect to adaptations that match diffusion rate and metabolic rate.

Gills. If an animal depends mainly on gills for external gas exchange, the degree of gill development and vascularity correlates with the availability of and demand for oxygen. When the gills are external, as in salamander larvae, the morphological changes are quite striking; for instance, as shown by Bond (1960), hypoxic water induces a rapid growth of gill size and vascularity.

When the gills are internal (fishes), the correlation of development with environmental oxygen and the demand for it is not so apparent, but quantitative measurements leave no doubt that the correlation exists. Individual and species differences in oxygen demand result from many factors, including the effect of body size. Hence it is not surprising that both the total oxygen consumption and the gill surface area are similarly related to body weight. Although there is much variability among species, a logarithmic plot of gill surface area as a function of body weight (Figure 5.15A) produces a linear relationship. The equation of this relationship is $A = aW^b$, where A is the gill surface area, a a constant, W the body weight, and b the proportionality exponent or regression coefficient (see Section 5.1). A similar equation relates metabolic rate to body weight; that is, $\dot{V}_{O_2} = aW^b$, where $\dot{V}_{O_2}$ represents the O_2 consumed in ml/min. For both equations the average value of b is approximately 0.80, implying that indeed gill surface area is directly proportional to O_2 demand. As shown in Figure 5.15, the very active pelagic species such as *Thunnus* have the largest gill areas.

Hughes (1970) has also shown that the diffusion distance (water to blood) appears to decrease with increasing gill area with tuna, having a diffusion distance of less than 1 μ. On the other hand, a thick diffusion barrier is characteristic not only of some sluggish species (flatfish, bullhead, and so on), but also of species with atrophied gills, as in air-breathing fishes. In *Anabas*, which obtains most of its oxygen by air breathing, the water-blood pathway is about 15 μ thick (Figure 5.15B).

Lungs. In air-breathing species the lung volume is directly related to body weight; in mammals and reptiles the proportionality exponent is

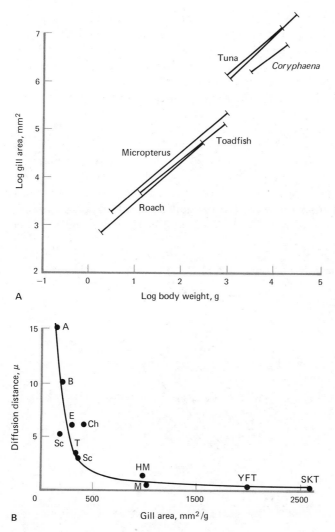

FIGURE 5.15 Relationship between body weight and gill area in teleost fish (upper graph) and the relationship between gill area and diffusion distance in a number of elasmobranch and teleost fishes (lower graph). Code: A—*Anabas;* B—bullhead; Ch—icefish; E—eel; HM—horse mackerel; M—mackerel; Sc—*Scyliorhinus;* T—trout; SKT—skipjack tuna; YFT—yellow fin tuna. (From G. M. Hughes, *Separatum Experientia* 26:117, 1970.)

1.00, whereas in amphibians it is about 0.75. The respiratory surface area in amphibians (lungs plus skin) and in reptiles and mammals is also linearly related, on a log scale, to body weight, with an exponent approximately equal to 0.80. Since oxygen uptake is also proportional to body weight for many lunged vertebrates, there is in air-breathing animals, as in fishes, a direct relationship between lung surface area and metabolic rate. However, at least in mammals (in contrast to fishes) there is little change in the thickness of the diffusion barrier with changing body weight (Figure 5.16).

In amphibians and reptiles the direct relationship between the metabolic requirements and lung surface area is achieved primarily by enlargement of the volume of saclike lungs (which in many salamanders and in all snakes occupy almost the entire length of the body). In homeotherms the basal oxygen demand is 10 to 100 times greater for any given body weight than the standard oxygen demand of poikilotherms. This demand for larger lung surface area is met in birds by the extensive

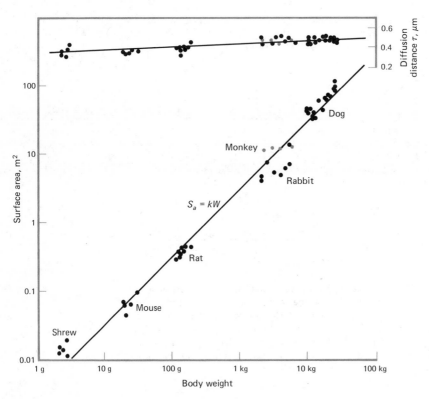

FIGURE 5.16 Alveolar surface area (S_a) and diffusion distance (τ) as a function of body weight in mammals. (From E. R. Weibel, *Resp. Physiol.* 14:35, 1972.)

air capillary system and in mammals by subdivision of the lung into alveoli. It is interesting to compare these quite distinct evolutionary solutions to meeting the pulmonary surface area requirement.

Bird lungs are unique by being divided into a gas-exchanging portion, the parabronchi of the lungs, and a ventilating portion, the air sacs, which serve as bellows and circulate inhaled air to the air capillaries of the parabronchi. Figure 5.17 shows the morphological relationships between these two portions of the avian respiratory system. The pathway of air through this system has long been a controversial subject. Recent experiments by Bretz and Schmidt-Nielson (1970) and Scheid and Piiper (1970) were the first attempts at *direct* measurements of airflow through bird lungs. In both studies it was observed that airflow through the bronchial system is unidirectional; that is, during both inspiration and expiration air flows from the mesobronchus to the parabronchi. The absence of anatomical valves presented the problem of accounting for this unidirectional flow. The answer seems to lie in what have been called "aerodynamic valves" in which the flow is directed by resistance features of the architecture of this system. As Schmidt-Nielsen (1972) points out, an increase in the diameter of the parabronchi by a factor of 2.3 (from epinephrine) will result in a 29-fold decrease in resistance to flow (see Section 4.2).

The air capillaries provide the respiratory surface area and are closely intertwined with blood capillaries. The diameter of the air capillaries is 3 to 10 μ, and thus a much higher ratio of surface area to volume is provided than is possible in mammals, where the smallest alveoli are 35 μ in diameter. The air capillaries, in contrast to alveoli, are rigid. This is essential in preventing collapse, which, in spite of the presence of a surface ac-

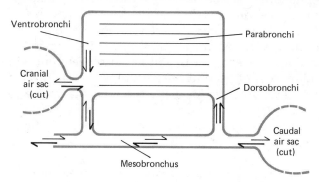

FIGURE 5.17 Mechanisms of unidirectional flow in parabronchi of avian lungs; measurements in duck lung preparations. Heavy arrows—inspiration; light arrows—expiration. (From P. Scheid, H. Slama, and J. Piiper, *Resp. Physiol.* 14:83–95, 1972.)

tive substance (surfactant), would result from the extremely high surface tension in such a small cylinder. Duncker (1972) provides some data on specific surface area of the air capillaries that ranges from 17.9 cm²/g body weight for the domestic goose (which has 10-μ air capillaries) to 40.3 cm²/g body weight for the pigeon (which has 3- to 5-μ air capillaries).

A similar relationship of body weight and relative metabolic requirement to specific surface area occurs in alveolar lungs. In humans, for example, the relationship between body weight and specific surface area is about 10 cm²/g body weight, whereas in the shrew it is about 40 cm²/g body weight. The diameter of the alveoli also decreases with increasing specific area from 197 μ in man to 32 μ in the shrew. This indicates that animals with a high relative metabolic rate (such as the shrew) have the smaller alveolar diameter but the larger relative surface area.

ROLE OF SURFACTANT: STABILITY OF ALVEOLI.   As the alveoli are gas pockets lined with a liquid-and-air interface, they are subjected to a surface tension that increases the pressure inside them. Because this internal pressure is inversely related to the radius (Laplace's law, $P = 2T/r$), the smaller alveoli would empty into the larger ones; that is, they would collapse if it were not for the presence of a surface active material—the *surfactant*—lowering the surface tension. The absence of surfactant leads to the collapse of the lung at the end of each expiration. Collapsed alveoli, like collapsed balloons, require much more pressure to inflate than partially inflated alveoli.

The role of surfactant has been mostly studied in man, whereas studies in other species are still very sparse. A direct relationship has been reported between the amount of surface active material present in different species and the pulmonary surface area. Also, some recent data indicate that species with smaller alveoli (shrews) synthesize more surfactant than those with larger alveoli (sloths). These findings are consistent with the role of surfactant in humans and are supportive of the notion of unity in biological function. However, surfactant has also been found in air-breathing fishes and in frog, turtle, and chicken lungs; its role in such nonalveolar lungs is not clear. In a phylogenetic sense it is tempting to speculate that biochemical preparedness for lung evolution began quite early, but a more likely explanation is that we do not yet know the function of surfactant in the lower vertebrates.

Skin.   Some gas exchange through the body surface occurs in all animals, but its relative contribution to total gas exchange ranges from almost zero in birds, mammals, and terrestrial reptiles to 100 percent in lungless salamanders. Cutaneous gas exchange (skin and buccopharyngeal) is not very important in fishes, except for eels during land excursions when the skin is a major respiratory surface. Measurements of gas exchange across the skin of salamanders, frogs, and toads have established that the skin is most important for CO_2 elimination in all amphibians. In terrestrial frogs the lungs account for most O_2 uptake, whereas in aquatic anurans the skin is equally important. In salamanders the skin is dominant for both O_2 uptake and CO_2 elimination.

Anatomical studies on the vascularity of the skin have complemented the physiological studies. The capillary surface area is about equal in the skin and lungs of most anurans. In an aquatic salamander *Triton* the skin contains about 75 percent of the respiratory capillaries, whereas in the tree frog *Hyla* the lungs and skin have the same fraction. In lungless salamanders the buccopharyngeal cavity has 5 to 10 percent of the respiratory capillaries; the balance is in the skin. The buccal cavity is richly vascularized and is an important gas exchange surface in the electric eel *Electrophorus*, which has atrophied gills and is an obligate air breather. The role of the skin compared to the gills or the lung often changes between the developmental stages of an animal. A good example are the salamanders, which display a remarkable diversity of respiratory behavior: their gas exchange is via gills in the neotenic forms, but it becomes cutaneous at the adult stage.

4 • CIRCULATORY MASS TRANSPORT

Gas exchange through the body surface imposes limitations on body size and activity. Furthermore, to be effective as a gas exchanger the body surface must sacrifice mechanical protection and osmotic control. These limits favored the evolution of specialized respiratory organs that allow the general body surface (except in amphibians) to become impermeable to water, protective, and insulative. Successful evolution of terrestrial vertebrates depended on "lungs and legs." In fact skin breathing was probably not important, and according to Romer (1972) reptiles and higher vertebrates evolved directly from fishes, with amphibians being a side branch.

Development of specialized respiratory organs, however, cannot improve upon oxygen uptake by general surface diffusion unless there is concomitant development of the cardiovascular system to provide internal convection between the exchange organ and the cells. As described by Johansen (1972), the progressive development of cardiovascular function in vertebrates is centered around four primary requirements. (1) The cardiovascular system must provide adequately for an immense range of cellular demand for oxygen and metabolic substrates. (2) It must perfuse the gas exchange organ at a flow rate optimum for maximum diffusion equilibrium—that is, matched with ventilation—and with a perfusion pressure low enough to minimize nongaseous exchange. (3) It must perfuse systemic tissues at higher pressures to provide filtration transport and with a flow rate proportional to the O_2 uptake of the tissue but not wastefully fast. (4) It must maximize separation of oxy- and deoxygenated blood.

The structural and functional properties of vertebrate cardiovascular systems are adapted to mode of gas exchange, behavior (activity level), metabolic rate, and body temperature. But before these anatomical and physiological adaptations of the system itself are considered, it is useful to examine properties related to gas transport and adaptability of the convective medium, that is, the blood.

4.1   Respiratory Functions of the Blood

The respiratory properties of blood are defined by the chemical interactions of gas and blood. In vertebrates these properties derive from the presence in red cells of hemoglobin (HHb),[5] which combines reversibly with O_2 and CO_2. Without such a respiratory pigment, circulation mass transport would be limited by the amounts of gas physically dissolved in plasma. This gas volume depends on the partial pressure and solubility coefficient of each gas (Henry's law). The solubility coefficients are almost the same for gases in plasma and fresh water, so the values in Table 5.1 are useful for estimating blood gases in physical solution. At partial pressures equivalent to those in alveoli, 100 ml of human blood (37°C) will contain 0.30 ml of O_2 and 2.7 ml of CO_2 in physical solution. In 100 ml of blood at the same partial pressures the actual content of these gases is about 20 ml of O_2 and 45 ml of CO_2. These differences represent the amount of O_2 in chemical combination with HHb and the amount of CO_2 in the form of bicarbonate. Only in rare cases where dissolved oxygen is high and its uptake by the tissues is low (as in the ice fish, which has no respiratory pigment) can vertebrates get along on dissolved O_2. If human blood contained no HHb and O_2 transport depended on physical solution alone, the perfusion requirement (cardiac output) at rest would need to be about 20 times higher than the actual value.

Oxygen-carrying Capacity.   In most vertebrates the hemoglobin molecule is tetrameric and its reversible combination with oxygen is described by the equation $HHb_4 + 4O_2 \leftrightarrow HHb_4(O_2)_4$. Since each monomer of HHb has a molecular weight of approximately 16,000 g, and 1 mole of HHb combines with 1 mole of O_2 (22.4 liters STPD), each gram of HHb can combine with about 1.38 ml O_2/g HHb. Hence the concentration of functional hemoglobin determines the oxygen capacity of the blood. However, in

---

[5] HHb indicates nonionized hemoglobin, which has acid properties.

many instances the in vivo oxygen capacity is less than the value predicted from hemoglobin concentration because of the presence of nonfunctional hemoglobin derivatives such as carboxyhemoglobin or methemoglobin, which do not combine with oxygen. The discrepancy between theoretical and in vivo O_2 capacity is often pronounced in lower vertebrates, where very high levels of methemoglobin are common. In human red cells methemoglobin—the oxidized (Fe^{3+}) form of hemoglobin—is rapidly reduced to hemoglobin (Fe^{2+}), so the normal level of methemoglobin is less than 1 percent of the total hemoglobin. This cellular reductase system is deficient in poikilothermic vertebrates; for example, levels of methemoglobin of 10 percent or more are common in fishes, and otherwise healthy turtles have been found with more than 90 percent methemoglobin. The tolerance to such high levels, especially in turtles, may be related to the high capacity of these animals for anaerobic metabolism.

Oxygen Transport and Delivery.   Theoretically the amount of oxygen available to the tissues is a function of the arterial oxygen content (Ca_{O_2}) and the rate of blood flow ($\dot{Q}$)through the tissue; that is, O_2 transport $= \dot{Q}$ liters/min $\times$ Ca_{O_2} ml/liter. But the oxygen actually delivered (equal to oxygen uptake, $\dot{V}_{O_2}$) is less than the amount transported. Not all of the O_2 transported is utilized by tissues, so the actual O_2 delivery is given by the equation (Fick equation) $\dot{V}_{O_2} = \dot{Q} \times (Ca_{O_2} - C\bar{v}_{O_2})$, where $C\bar{v}_{O_2}$ is the oxygen content of mixed venous blood. The range of obtainable $\dot{V}_{O_2}$ depends, therefore, on both the range of cardiac output and the range of arteriovenous oxygen difference ($Ca_{O_2} - C\bar{v}_{O_2}$). An increase in arteriovenous O_2 difference is the cheapest way (in terms of energy) to meet increased $\dot{V}_{O_2}$. The only limitation to increasing O_2 delivery in this way is the fact that tissue P_{O_2} must be maintained at a sufficient level in order for oxygen diffusion into the cells to occur. It is the oxygen dissociation curve of hemoglobin that enables ($Ca_{O_2} - C\bar{v}_{O_2}$) to make a large contribution to the $\dot{V}_{O_2}$ equation.

Oxyhemoglobin Dissociation Curve.   The relationship between oxygen tension in the blood and the hemoglobin oxygen saturation (S) is called the oxyhemoglobin dissociation curve. If the hemoglobin is monomeric (for example, lamprey hemoglobin), the slope of this curve is hyperbolic. When more than one monomer is involved, however, there is interaction between the oxygen-binding sites (hemes), resulting in a sigmoid dissociation curve (S-shaped). The degree of heme-heme interaction shows great variability among vertebrates. It is quantified most often by plotting the oxygen dissociation curve as $\log [S/(100 - S)]$ versus $\log P_{O_2}$, which, within limits of 10 percent and 90 percent saturation, results in a straight line. The slope of this line (called n) is 1 if there is no heme-heme interaction (hyperbolic curves) and becomes increasingly greater as the curve becomes more sigmoidal. For human hemoglobin $n \simeq 2.7$; that is, there are at least 2.7 interacting oxygen-binding sites. The position and shape of the oxyhemoglobin dissociation curve have great physiological significance in regard to the delivery of oxygen into the tissues. The ideal dissociation curve is positioned so that arterial blood P_{O_2} results in full saturation and is shaped so that a large amount of oxygen can be given up to the tissues with a limited fall in P_{O_2}. Figure 5.18 illustrates these features and their relationship to $\dot{V}_{O_2}$ (Fick equation). Delivery of oxygen to the tissues can be increased (in the steep portion of the curve) with a minimal decrease in mixed venous P_{O_2} ($P\bar{v}_{O_2}$). This is essential since final oxygen uptake by the mitochondria and other subcellular fractions depends on passive O_2 diffusion. The graphic solution to the Fick equation emphasizes a central point; that is, increased $\dot{V}_{O_2}$ can be provided by increased flow and by increased arteriovenous O_2 difference, and there is a large reserve in both factors. Factors that control the position of the curve in relation to the P_{O_2} axis can be highly advantageous to meeting increased tissue oxygen requirements. Consideration of the curve shows why: as long as the upper portion of the curve is unaffected, a right displacement of the central steep portion of the curve will allow increased arteriovenous oxygen content difference without lowering venous P_{O_2}. Thus an increased $\dot{V}_{O_2}$ can be effected

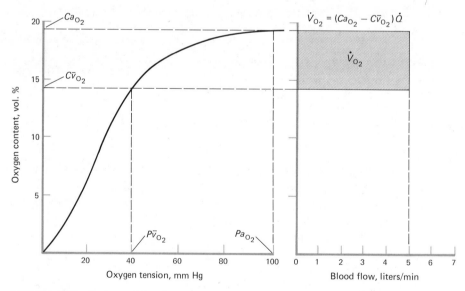

FIGURE 5.18 Oxyhemoglobin dissociation curve for human blood (pH 7.4, 37°C) and graphic representation of the Fick equation. (From C. Lenfant, *Preservation of Red Cells*, Publication ISBN 0-309-02135-0, H. Chaplin, Jr., ed. Washington, D.C.: Division of Medical Sciences, National Academy of Sciences–National Research Council, 1973.)

without increased blood flow and without sacrificing the final P_{O_2} diffusion gradient. As we will describe here, such factors are numerous and operate on an immense variety of vertebrate hemoglobins.

FACTORS INFLUENCING HEMOGLOBIN AFFINITY FOR OXYGEN. The position of the oxyhemoglobin dissociation curve in relation to the P_{O_2} axis is an expression of the affinity of the hemoglobin for oxygen. This is usually quantified by the P_{50}, the partial pressure of oxygen required for 50 percent of the hemoglobin to be saturated with oxygen. A displacement of the curve to the right (right shift) corresponds to an increase of P_{50} and to a decrease of the hemoglobin affinity for oxygen, since a higher P_{O_2} is required to maintain 50 percent hemoglobin saturation.

A survey of the available data for vertebrates reveals a seemingly infinite variety of oxyhemoglobin dissociation curves. The basic O_2 equilibrium properties of HHb are genetically determined, but many cellular factors determine the phenotypic expression. Paramount among these factors are temperature and ionic strength of the erythrocyte cytoplasm and the activity of ligands such as H^+, CO_2, and certain organic molecules that can combine with HHb, usually in competition with O_2. The effect of an increase in any of these factors is a decrease in the hemoglobin affinity for oxygen—that is, a shift of the curve to the right.

Temperature. Because of the general effect of temperature on the rate of a chemical reaction (Arrhenius theory), an increase in blood temperature decreases the hemoglobin affinity for oxygen, whereas a decrease in the temperature has an opposite effect. The magnitude of the temperature effect on the affinity for oxygen is related to the heat of combination of 1 mole of oxygen with hemoglobin (van't Hoff equation). In mammals this results in a temperature coefficient ($\Delta \log P_{50}/°C$) of about 0.024; that is, P_{50} changes by about 1 mm Hg/°C. It is noteworthy that during exercise, which causes the temperature to increase, the temperature-dependent right shift of the dissociation curve favors an increase of the arteriovenous oxygen content and is therefore adaptive to the increased $\dot{V}_{O_2}$.

In vertebrates whose body temperature varies with ambient temperature the temperature coefficient of HbO_2 binding is obviously very important. Comparative data are limited, but the available evidence suggests that numerous adaptations allow these ectothermic vertebrates to escape from, or alter, the temperature-dependence of their oxyhemoglobin dissociation curve. Indeed, species that experience very diverse ambient temperatures may be adversely affected by marked or rapid changes in hemoglobin affinity for oxygen. Two solutions to this problem have evolved. Fishes and frogs exposed to prolonged temperature change have an alteration in the basic position of the oxyhemoglobin dissociation curve. Thus at any given temperature the curve for cold-adapted individuals lies to the right of those adapted to a warmer temperature. An alternate solution is provided by hemoglobins with reduced or no temperature sensitivities. This may be common in diving lizards, which invariably have a rapid decrease in body temperature upon submergence. Iguanas, for example, behaviorally regulate body temperature to about 35°C by basking in the sun. They often dive into 25°C rivers and, because of the high specific heat of water, suffer a rapid 10°C drop in body temperature. A decrease in P_{50} of 10 mm Hg would occur if the temperature sensitivity of iguana hemoglobin were equal to that in mammals. This would seriously handicap oxygen delivery and active metabolism.

The fact that the position of the dissociation curve for iguanas and other diving reptiles is less temperature-sensitive than that of mammals is a definite adaptation to their life style. Tuna hemoglobin is perhaps the least sensitive to temperature, which may also be a molecular adaptation to pronounced temperature changes attending the rapid seasonal migration from tropical to northern Atlantic water. The Australian lungfish, *Neoceratodus*, which likewise experiences large temperature fluctuations, also has HHb with a low temperature effect. Stenothermal species—for example, the ice fish, *Trematomus*—characteristically show normal temperature sensitivity of HbO_2 equilibrium. The bowfin, *Amia*, on the other hand, is a eurythermal species but has HHb with a high temperature sensitivity. For the bowfin, however, the pronounced shift of its oxyhemoglobin dissociation curve could be advantageous because high temperature induces air breathing, which is best served by a hemoglobin with a relatively low O_2 affinity.

pH and CO₂. The pronounced effect of plasma pH on the affinity of hemoglobin for oxygen was first demonstrated by Bohr, Hasselbalch, and Krogh (1904). Figure 5.19 shows a reproduction of their original graph. They actually studied CO_2 effects on hemoglobin affinity for oxygen and assumed, as have others until recently, that the CO_2 effect was solely due to the effect of CO_2 on plasma pH. The chain of events resulting in the release of O_2 from oxyhemoglobin when CO_2 is added to plasma may be summarized as follows:

$$CO_2 + H_2O \longleftrightarrow H_2CO_3^- + H^+$$
$$H^+ + HbO_2 \longleftrightarrow HHb + O_2$$

An increase in H^+ (decrease in pH) from CO_2 or a metabolic acid produces an immediate rightward shift of the dissociation curve, and an increase in pH has the opposite effect. The magnitude of this effect (Bohr effect) is given by measuring the change in log P_{50} for a unit change in pH (Δlog P_{50}/ΔpH). For human hemoglobin this equals -0.48 if titrated with CO_2

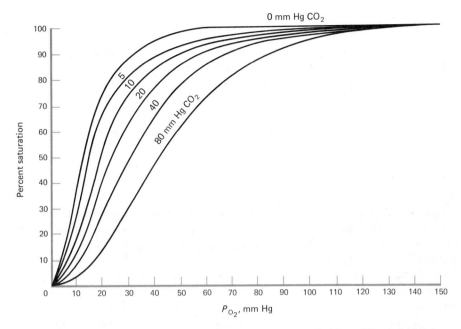

FIGURE 5.19 Reproduction of the original graph by Bohr, Hasselbalch, and Krogh (1904) demonstrating the effect of P_{CO_2} upon the hemoglobin affinity for oxygen of human blood (Bohr effect).

and -0.40 if titrated with fixed acid or base. The difference between these values (-0.08) is due to a direct decrease in affinity brought about by CO_2 binding to N-terminal amino acids of hemoglobin. In vivo pH and P_{CO_2} vary simultaneously, so to estimate the P_{50} change with changing pH of human hemoglobin the value of -0.48 should be used.

Although almost all vertebrate hemoglobins are pH-sensitive, the magnitude of the Bohr effect is quite variable. In hemoblogin solutions from different mammals $\Delta\log P_{50}/\Delta pH$ increases with decreasing body weight, but there is no clear trend in whole blood for the Bohr effect to vary with body weight. However, when the "effective" Bohr effect is considered, the relationshp to body weight disappears completely. The "effective" Bohr effect is defined as the extra volume of oxygen released when the blood is acidified by 0.1 pH unit. It is a function of the O_2 capacity of the blood and of the shape of the dissociation curve; in terms of oxygen delivery it is most important to the animal.

In order for the Bohr effect to occur in vivo, the CO_2 produced by metabolizing cells must be rapidly hydrated. In the red cells, which are in capillaries for very brief periods, the enzyme carbonic anhydrase performs this essential job and may be the limiting factor in the magnitude of the in vivo Bohr effect. Larimer and Schmidt-Nielsen (1960) showed that the magnitude of the Bohr effect in mammals is directly proportional to the concentration of carbonic anhydrase in the red cells.

In some fishes P_{CO_2} and pH have an effect on O_2 capacity as well as affinity (Figure 5.20). This is called the Root effect. At first glance this effect seems to have no survival value since, in effect, it produces anemia. However, in vivo the Root effect is the key to the function of the swim bladder, an organ used by fishes to control their buoyancy.

In some deep-water species the swim bladder may contain oxygen at a partial pressure of 76,000 mm Hg (100 atm). This concentration of oxygen is possible because acidification of arterial blood entering the "gas gland" of the bladder results in the release of oxygen into physical solution. Gases in physical solution obey Henry's law; that is, the amount of gas physically dissolved equals the solubility coefficient times the partial pressure. For oxygen, if the arterial blood upon acidification releases 3 ml O_2/100 ml blood, dissolved O_2 would increase from 0.3 to 3 ml/100 ml, and, since the solubility coefficient is constant, the arterial P_{O_2} would increase by a factor of 10. Countercurrent flow of blood entering and leaving the gas glands allows this P_{O_2} increase to be continuously multiplied until the bladder is gassed to the appropriate pressure for any given depth.

Organic Phosphates. The position of the oxygen dissociation curve of intact red cells is known to be determined by the concentration of intracellular organic phosphates independent of temperature, pH, and CO_2.

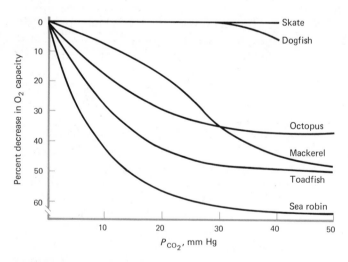

FIGURE 5.20 The effect of P_{CO_2} on the oxygen capacity of blood (Root effect) in fishes and the octopus. (From C. Lenfant and K. Johansen, *Resp. Physiol.* 1:21, 1966.)

In mammalian red cells the principal organic phosphate is 2,3-diphosphoglycerate (DPG); in the red cells of birds and turtles inositolhexosephosphate (IHP) or inositolpentosephosphate (IPP) is the major organic phosphate; in amphibians and reptiles, ATP predominates. Fish red cells have both ATP and guanosine triphosphate (GTP) as hemoglobin modulators. Human hemoglobin with all organic phosphates removed (by column chromatography) has about the same affinity for oxygen as myoglobin. Similar increases in affinity for oxygen occur with all vertebrate hemoglobins when they are "stripped" of organic phosphate. The presence of high concentrations of DPG in mammalian red cells had been known for decades, but no one knew what possible role DPG could play until the effect on the HbO_2 affinity was demonstrated.

In man and many other mammals hypoxia of varied physiological and pathological origins evokes an increase in the level of DPG and consequently a rightward shift of the dissociation curve. The feedback control operates primarily through changes in red cell pH and works in the following manner. Hypoxia produces an increase of deoxyhemoglobin, which is less acid than oxyhemoglobin, so red cell pH is increased. Red cell pH also increases in acute hypoxia because hyperventilation raises the plasma pH. The increased red cell pH has two effects, both leading to an increase in DPG concentration: (1) activation of DPG mutase (the enzyme that converts 1,3-DPG into 2,3-DPG) and an increase in the rate of glycolysis, and (2) inhibition of DPG phosphatase, the enzyme that converts 2,3-DPG into 3-PG. It must be emphasized that the direct effect of increased intracellular pH (Bohr effect) is to shift the curve to the left, whereas the indirect influence of increased DPG concentration is to shift the curve to the right. Thus the in vivo dissociation curve may not change at all when these two influences are antagonistic.

The regulation of the intracellular organic phosphates has not been studied in other species of vertebrates, but their effects on HHb function are similar to those of DPG.

PHYSIOLOGICAL SIGNIFICANCE OF THE OXYHEMOGLOBIN DISSOCIATION CURVE. The regulation of hemoglobin affinity for oxygen in mammals is extremely intricate because of the complexity of factors controlling organic phosphate metabolism and the interaction of organic phosphates with H^+ ions. In lower vertebrates the situation is even more confusing because of lack of homeostatic control of temperature and pH. Also, in contrast to mammals, the nucleated red cells of fishes, amphibians, reptiles, and birds have their own active aerobic metabolism. This becomes a very important factor during hypoxia if the major organic phosphate is ATP, as in the case of fishes, amphibians and some reptiles. Since oxidative phosphorylation is the primary source of ATP, hypoxia should cause a decrease in red cell ATP and, if the hemoglobin affinity for oxygen is ATP-sensitive, a shift to the left in the dissociation curve.

It is significant that hypoxia, at the red cell level, evokes an opposite response in lower vertebrates and mammals, and it seems unlikely that these opposite responses could both be adaptive in the sense of increasing the tolerance to hypoxia. This sort of paradoxical situation explains why the question of whether the position of the oxyhemoglobin dissociation curve influences the delivery of oxygen to the tissues remains unanswered. Nonetheless, there are numerous examples, some of which are discussed below, that substantiate the hypothesis that it is a parameter of great physiological importance.

Some of the most striking evidence comes from studies of ontogenetic changes in the position of the dissociation curve. In all mammals (except the cat) fetal blood has higher affinity for oxygen than maternal blood. The physiological significance of this is quite clear considering the fact that P_{O_2} of arterialized fetal blood is usually less than 30 mm Hg, equivalent to the P_{O_2} of air at an altitude of 39,000 feet. The increased affinity of fetal hemoglobin means that O_2 released from the maternal hemoglobin will be taken up by the fetal hemoglobin. The Bohr effect assists the O_2 transfer since the maternal curve shifts to the right and the fetal curve shifts to the left during CO_2 transfer.

Similar ontogenetic changes in hemoglobin affinity for oxygen occur in all vertebrates. Often these changes are correlated with a transition from aquatic to aerial gas exchange and an increase in metabolic rate. A marked decrease in affinity for oxygen follows birth, or hatching, in most fishes, amphibians, reptiles, birds, and mammals. In the chicken P_{50} increases by 13 mm Hg within 24 hours after hatching. Slower but equally dramatic increases in P_{50} occur during the next month of postnatal development, and the increase in P_{50} is closely correlated with an increase in red cell organic phosphate (IHP). In amphibians the increase in P_{50} following frog metamorphosis is well known. At 20°C and pH 7.6 the P_{50} increases from about 5 mm Hg in tadpoles to 40 mm Hg in adult bullfrogs (Johansen and Lenfant, 1972). In anuran metamorphosis the functional changes seem to result from changes in primary structure of the hemoglobin molecule itself. However, recent studies indicate that in other cases the differences between fetal and maternal red cells in hemoglobin affinity for oxygen disappears when the respective hemoglobins are purified. In mammals there is a tendency for P_{50} to be inversely related to the body weight, which in turn is inversely related to the relative oxygen requirement (Figure 5.21). This indicates that P_{50} and relative oxygen requirement may be dependent variables, a fact of some importance since it seems to maintain a large pressure gradient between blood and tissues in those species with the highest metabolic rate.

Other evidence for the physiological significance of the oxyhemo-

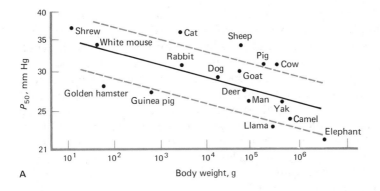

A

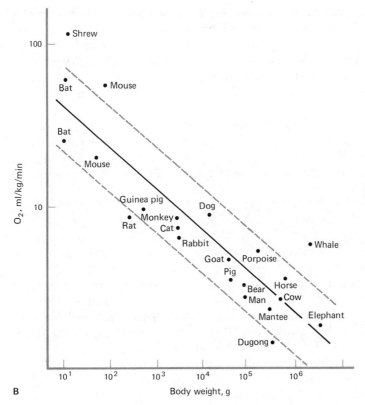

B

FIGURE 5.21 (**A**) P_{50} as a function of body weight. (**B**) The inverse relationship between the lag of metabolic rate and body weight. Dotted lines indicate plus or minus one standard deviation. (From H. Bartels, *Lancet* 9:600, 1964.)

globin dissociation curve is found in the extensive data correlating the curve with environmental gas levels. The adaptability of hemoglobin affinity for oxygen is especially apparent in the aquatic environment, where respiratory conditions range from extreme hypoxia in stagnant and polluted water whose P_{O_2} is close to zero to the saturated tide pools where algal photosynthesis produces oxygen tensions of 300 mm Hg or more. The respiratory properties of fish blood show the general pattern of high affinity in species which are sluggish and inhabit oxygen-deficient water and low affinity in active pelagic species. The trend for increased affinity in oxygen-deficient water is surprisingly clear, considering the multitude of factors affecting such interspecies comparisons. Furthermore, this same trend appears in a single species (eels) adapting to hypoxic water, where the left shift of the dissociation curve is mediated by a decrease in red cell ATP (Wood and Johansen, 1972).

As we already stated, the proof is not established that an animal is better off if its oxyhemoglobin dissociation curve is in one position rather than another; it is, however, particularly notable that the adaptations discussed in this section all provide a maximum arteriovenous oxygen content difference with a minimum sacrifice of the capillary oxygen tension. These are the important criteria of a useful dissociation curve.

Role of Blood in CO_2 Transport and Acid-Base Balance.   Mammals and birds have a relatively constant blood pH maintained through renal and respiratory control systems. Blood pH is quantitatively related to P_{CO_2} and to bicarbonate concentration by the Henderson-Hasselbalch equation:

$$pH = pK + \log \frac{[HCO_3^-]}{\alpha[P_{CO_2}]} \tag{5-4}$$

The respiratory center of the brain monitors and regulates arterial P_{CO_2} by controlling alveolar P_{CO_2} through ventilation. At a constant P_{CO_2} a decrease in bicarbonate results in a decreased pH, as shown in Figure 5.22. Physiologically, this type of pH change usually results from the addition to blood of a metabolic acid, such as lactic acid, which lowers pH and removes HCO_3^- by shifting the equilibrium $H^+ + HCO_3^- \longleftrightarrow H_2CO_3$ to the right. This pH change can be compensated by ventilatory changes in P_{CO_2}. When blood is titrated with CO_2, the H^+ (produced from $CO_2 + H_2O \longleftrightarrow H_2CO_3 \longleftrightarrow H^+ + HCO_3^-$) is buffered by intracellular (mainly hemoglobin) and plasma proteins. The change in pH for a given HCO_3^- increase is less as HHb concentration increases (Figure 5.22). Stated another way, for each H^+ buffered, a bicarbonate is formed. The slope of this titration curve $\Delta HCO_3^-/\Delta pH$ is called the buffer capacity. The specific buffer capacity is this value per gram HHb. A high buffer capacity minimizes pH changes when arterial CO_2 fluctuates.

As discussed earlier, CO_2 retention is associated phylogenetically with occurrence of air breathing, concomitant with keratinization of the

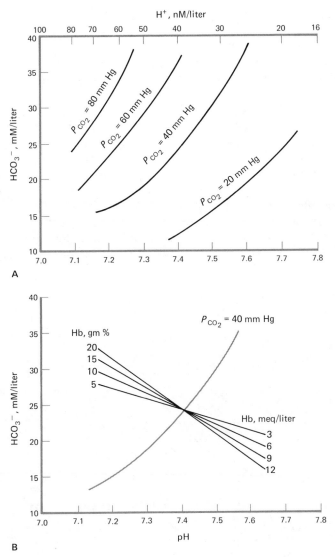

FIGURE 5.22 (**A**) The pH-bicarbonate diagram. For any P_{CO_2} isobar the bicarbonate concentration for any pH is given by the Henderson-Hasselbalch equation. (**B**) Values shown are for human plasma as a function of hemoglobin concentration. (From H. W. Davenport, *The ABC's of Acid-Base Chemistry*. Chicago: University of Chicago Press. © 1969 by The University of Chicago.)

body surface and invagination of the gas exchange organ. The pH reduction from this increased P_{CO_2} is minimized by an appropriate increase in bicarbonate. Data from modern transitional species clearly depict the evolutionary changes in acid-base regulation that occurred during the transition from water breathing to air breathing. Figure 5.23 shows the buffer curves of fishes ranging from obligate air breathers, which, for a given blood pH, have higher P_{CO_2} and bicarbonate. The role of the kidney in acid-base balance is discussed in Chapter 6. Reptiles were the first vertebrates to evolve a "modern" lung to handle both O_2 and CO_2 exchange. This allowed keratinization of the skin and true terrestriality and, by permitting the shunting of blood away from the skin, set the stage for the evolution of temperature regulation (homeothermy). In many aquatic reptiles the trend is toward a high bicarbonate concentration. This could be a great advantage during prolonged diving since more lactic acid could accumulate for a given pH decrease.

In contrast to mammals and birds, blood pH of ectothermic vertebrates is not constant but varies inversely with body temperature (Figure 5.24). As body temperature falls, pH increases due to an increase in the

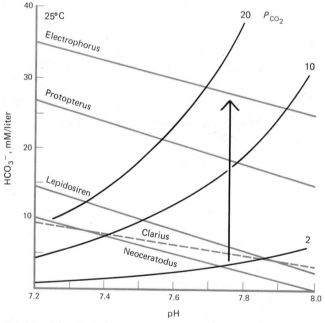

FIGURE 5.23 Buffer curves for blood from water-breathing, transitional, and air-breathing fishes. Arrow represents probable pH at 25°C. Increasing dependence on air breathing from *Neoceratodus* to *Electrophorus* is accompanied by CO_2 retention and increased bicarbonate formation. (Reprinted from B. J. Howell, *Fed. Proc.* 29:1131, 1970.)

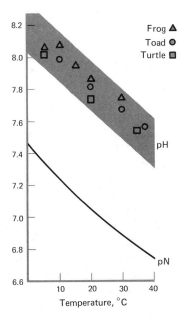

FIGURE 5.24 pH as a function of body temperature in poikilotherms. The solid line pN represents the neutrality of water. (From H. Rahn and F. W. Baumgardner, *Resp. Physiol.* 14:173, 1972.)

HCO_3^-/P_{CO_2} ratio, brought about mainly by a P_{CO_2} decrease. In turn P_{CO_2} decreases because of a reduction in metabolic rate, whereas ventilation remains essentially constant. In other words, at low temperatures there is a relative hyperventilation, producing decreased P_{CO_2} and increased pH. The fact that the blood pH parallels the neutral pH (pN) of water (Figure 5.24) indicates that a constant OH^-/H^+ ratio is maintained at all body temperatures.

4.2 Heart and Circulation

In Section 4.1 it was shown how O_2 delivery to tissues can be adjusted over a wide range of O_2 tensions by modifications in the position of the oxyhemoglobin dissociation curve. In spite of this plasticity in the curve, the maximum arteriovenous oxygen content difference in an individual is only 3 or 4 times the resting value. Oxygen consumption, on the other hand, may increase to 10 to 20 times the resting value during activity. The difference must be made up by an increase in blood flow ($\dot{Q}$).

In this section the role of circulation in meeting tissue O_2 demands and the evolutionary development of the circulatory system will be considered. The selective pressure for more efficient blood flow can be appreciated by comparing the absolute values of O_2 consumption in different vertebrates. In small mammals and birds the resting oxygen consumption can be 100 times as high as that of fishes, amphibians, and reptiles (Brett, 1972). As described by Johansen (1972) in his review of cardiovascular function in vertebrates, these metabolic factors "afforded the most important selection pressure for the progressive development of the cardiovascular system and its functional properties among the air-breathing tetrapods."

Patterns of Vertebrate Circulation. In general those species with higher metabolic rates have more efficient cardiovascular systems, and in a phylogenetic survey there is an obvious trend in several key factors affecting efficiency: (1) separation of systemic and pulmonary circulation and (2) increased arterial blood pressure. The first maximizes arteriovenous oxygen content difference, whereas the second permits increased flow rates and, in terrestrial vertebrates, is necessary to overcome gravitational forces. These two features attain maximum development in mammals and birds, and are in fact essential for homeothermy and high metabolic rates. It is not necessarily true, however, that incomplete separation of systemic and pulmonary circulation is a primitive feature. In some cases this is adaptive to routinely encountered physiological stress by higher vertebrates (for example, diving). Figure 5.25 shows the general architecture of vertebrate cardiovascular systems. In teleost fishes the blood flow to the gills is in series with the systemic flow. In lungfishes, amphibians, and reptiles, where anatomical separation is not complete, the circulation to the lungs is in parallel with systemic blood flow. This permits shunting of blood flow to occur; that is, if the interatrial septum is incomplete, pulmonary and systemic blood returning to the heart can mix. Likewise, an incomplete ventricular septum permits right-to-left or left-to-right shunting, depending on pressure gradients and resistance to flow. In crocodilian reptiles, where anatomical separation of the heart is complete, the right and left aortic arches communicate through an opening (*foramen Panizzae*).

In spite of this anatomical opportunity for shunting in all poikilothermic vertebrates, many species have an amazingly effective double circulation. To appreciate this and other physiological adaptions, it is helpful first to review some general features of the heart and circulation.

Vertebrate Heart. The heart is a muscular pump equipped with unidirectional valves. The mass of heart muscle is proportional to body weight (and therefore to metabolic rate) and, like skeletal muscle, is increased (hypertrophied) by prolonged maximum activity. The muscle cells and areas of specialized muscle tissue (nodes) autonomously depolarize, giving the heart an inherent rhythmic contraction; the area of the heart with the highest rate of depolarization becomes the "pacemaker," setting the inherent heart rate (beats or contractions per minute). In most vertebrates this inherent heart activity is influenced by neural and humoral factors, and both the rate (chronotropic factors) and strength (inotropic factors) of contraction are variable over a wide range.

Neural Factors and Innervation of Vertebrate Heart. All embryonic vertebrate hearts begin beating before receiving a nerve supply, and in the most primitive vertebrates, the hagfishes, this absence of innervation persists through adult life. In all other vertebrates the adult heart receives autonomic nerves.

The autonomic control of the heart and peripheral circulation in mammals is illustrated in Figure 5.26. The sinoatrial (SA) node is normally under tonic control by both sympathetic and parasympathetic (vagal) nerves, so changes in heart rate result from the reciprocal changes in the firing rate of these nerves. Autonomic blockade produces what is called the intrinsic heart rate. In man the intrinsic heart rate (about 105 beats/min in young adults) is higher than the normal heart rate, indicating a dominance of parasympathetic control. Blocking or cutting the sympathetic nerves produces a slight reduction in rate (bradycardia), whereas cutting the vagus (or blocking with atropine) produces pronounced rate increases (tachycardia). The right branch of the vagus mainly affects the SA node, producing sinus bradycardia, whereas the left vagus influences mainly the atrioventricular (AV) node. Sympathetic innervations via the superior, middle, and inferior cardiac nerves are distributed throughout the heart muscle.

Autonomic control of peripheral circulation is mediated mainly by the vasomotor center of the medulla, which supplies tonically active sympathetic innervation to peripheral vessels. Parasympathetic nerves supply a small fraction of the total number of vessels (head, viscera, genitalia, and so on) and consequently do not have a pronounced effect on total vascular resistance. Sympathetic discharge releases norepinephrine, which elicits vasoconstriction except for the sympathetic nerves going to skeletal muscle and cutaneous blood vessels. In the latter two cases the neurons release acetylcholine (cholinergic sympathetic nerves), which elicits vasodilation.

Activity of the vasomotor center is controlled by input of neural impulses from the carotid and aortic baro- and chemoreceptors, the hypothalamus, cerebrum (for example, blushing), skin, and viscera. There is also a direct effect of P_{CO_2} and pH on the center. Increased P_{CO_2} and decreased pH are both stimulants, resulting in increased peripheral resistance. Decreased P_{O_2} also affects the vasomotor center, but in this case the stimulation is mediated through the peripheral chemoreceptors.

Until recently it was thought that fish hearts received only parasympathetic innervation from the vagus nerve and that acceleration of

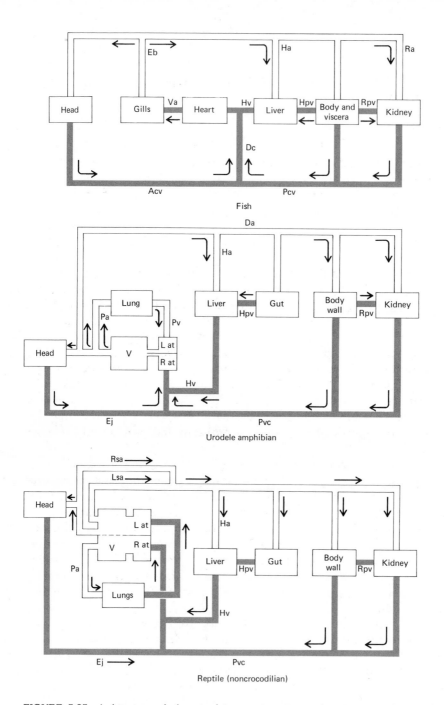

FIGURE 5.25 Architecture of the circulatory system in representative vertebrates. Code: Acv—anterior cardinal vein; Da—dorsal aorta; Dc—ductus Cuvier; Eb—efferent branchial arteries; Ej—external jugular vein; Fp—*foramen Panizzae;* Ha—hepatic artery; Hpv—hepatic portal vein; Hv—hepatic vein; L at—left atrium; Lsa—left systemic arch; Lv—left ventricle; Pa—pulmonary artery; Pcv—posterior cardinal vein; Pv—pulmonary vein; Pvc—posterior vena cava; Ra—renal

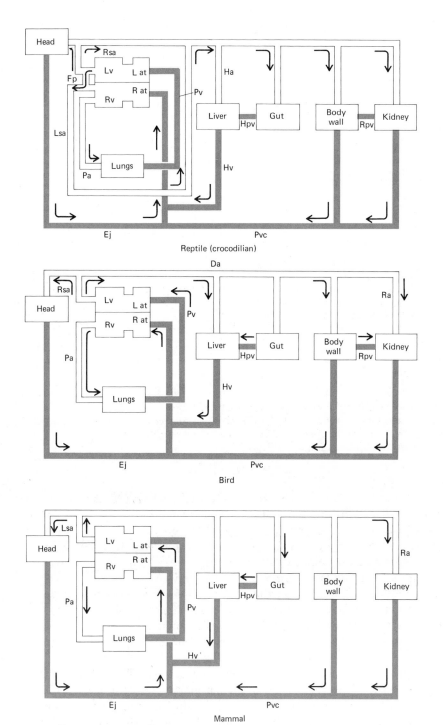

Reptile (crocodilian)

Bird

Mammal

artery; R at—right atrium; Rsa—right systemic arch; Rpv—renal portal vein; Rv—right ventricle; V—ventricle; Va—ventral aorta. (From F. N. White, *Animal Physiology: Principles and Adaptations,* M. S. Gordon, ed. New York: Macmillan, 1972.)

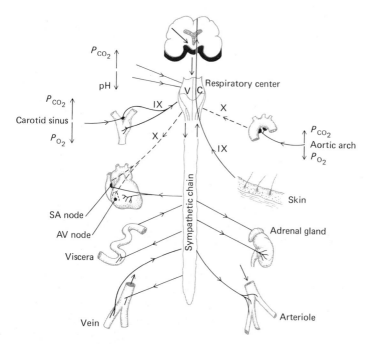

FIGURE 5.26 Schematic diagram illustrating the neural input and output of the vasomotor center (VC). IX—Glossopharyngeal nerve; X—vagus nerve. (From R. M. Berne and M. N. Levy, *Cardiovascular Physiology*, 2d ed. St. Louis: Mosby, 1972.)

heart rate could result only from decreased vagal activity (that is, reduced inhibition). The inhibitory fibers release acetylcholine, which has an inhibitory effect on cardiac muscle (except in lampreys). The mechanism of action of acetylcholine on cardiac muscle is to decrease the permeability of the pacemaker cell membranes to sodium ions, so the rate of sodium "leak" and resulting autonomous depolarization is reduced. Only in lampreys is the response of cardiac muscle the same as skeletal muscle; that is, stimulation of the vagus (or application of acetylcholine) causes an increase in heart rate. Although elasmobranch and teleost hearts appear to receive only vagal nerves, physiological and pharmacological studies indicate the presence of excitatory (adrenergic) fibers. These sympathetic fibers are thought to be interspersed among cholinergic vagal fibers and, for those going to the ventricles, to run along the coronary arteries. Recent reviews by Johansen (1970) and Satchell (1971) point out the considerable controversy that exists regarding both the innervation and reflex control of fish hearts. How, for example, are rate changes mediated in the aneural hagfish heart? Two possibilities have been proposed. First, the hagfish heart has been reported to have granules containing epinephrine or other catecholamines and to possess an intrinsic cardioactive chemical which has properties similar to norepinephrine and epinephrine. A second, alternative explanation is a stretch-induced increase in heart rate. This phenomenon occurs in all vertebrate hearts and appears to be due to mechanical sensitivity of the pacemaker tissue. The stretch response also increases cardiac output by increasing stroke volume (Frank-Starling mechanism), which, in denervated mammalian hearts, accounts for most of the increase in cardiac output during exercise. For instance, greyhound dogs with denervated hearts show a three- to fourfold increase in cardiac output and only a 5 percent reduction in average running speed.

In anuran amphibians sympathetic nerves are in the vagus, but in urodeles there is anatomically defined sympathetic innervation of the heart. The hearts of all other vertebrates, be they from reptiles, mammals, or birds, are abundantly supplied by anatomically distinct vagal and sympathetic nerves. Adrenergic fibers innervate the pacemaker node, atria, and ventricle(s), where they release norepinephrine. The epinephrine-containing cells in the myocardium, as circulating epinephrine, have a positive inotropic effect; that is, they increase the strength of the muscular contraction. This results from activation of the enzyme adenyl cyclase, which causes an increase in the level of cyclic AMP (see Chapter 9, Section 2.1), which in turn increases the myocardial metabolism and mobilizes calcium needed for muscle contraction.

HUMORAL FACTORS AND REFLEX REGULATION OF HEART ACTIVITY. Stimulation of the vagus nerve causes bradycardia in all vertebrates except in the lamprey, where tachycardia occurs. The tachycardia in lampreys can be blocked by curare, indicating the presence of acetylcholine receptors similar to those in skeletal muscle; the bradycardia of other vertebrate hearts can be blocked by the autonomic blocking agent, atropine.

The level of cholinergic stimulation (vagal tone) is reflexively controlled by numerous factors. The most important stimuli eliciting reflex bradycardia are anoxia and increased arterial pressure. In fishes these reflexes are mediated through gill receptors, which are sensitive to water oxygen content, water flow, and arterial pressure. In higher vertebrates these chemoreceptors and pressure receptors are located in the carotid and aortic bodies.

It is informative to compare the reflex regulation of single and double circulatory systems. In fishes it is the requirement for gill perfusion that dominates the regulation of heart rate. Many investigators have noted a direct relationship between heart rate and respiratory rate, which is presumed to provide efficient matching at the gills of blood and water flow. Satchell (1960) found that the heart of the elasmobranchs *Squalus acanthius* and *Mustelus antarcticus* would beat with every second, third, or fourth respiration and that the P wave of the EKG occurred as the mouth opened. In other fishes the heart rate exceeds respiratory rate and the degree of synchrony often depends on water P_{O_2}; that is, as inspired P_{O_2} declines, the heart rate becomes more closely coupled with respiration.

In air-breathing vertebrates with a parallel pulmonary circuit, the systemic circulation requirements dominate heart rate regulation via carotid and other receptors, but the influence of respiration on heart rate persists; this is often called the respiratory–heart-rate response (RHRR), an example of which is shown in Figure 5.27.

ELECTRICAL ACTIVITY: THE ELECTROCARDIOGRAM (EKG).   The electrical activity of cardiac muscle that initiates contraction can be recorded by body surface electrodes. In man, when placement of the surface electrodes is standardized and the EKG is subjected to vector analysis, a wealth of detailed functional information has been accumulated. In all vertebrates, regardless of surface electrode placement, the basic EKG pattern is similar to that recorded in man (Figure 5.28). The P wave represents atrial depolarization; it originates in the sinus venosus in fishes and amphibians and in the sinoatrial node in other vertebrates. It is followed by an interval in which the electrical signal is delayed in nodal tissue (atrioventricular

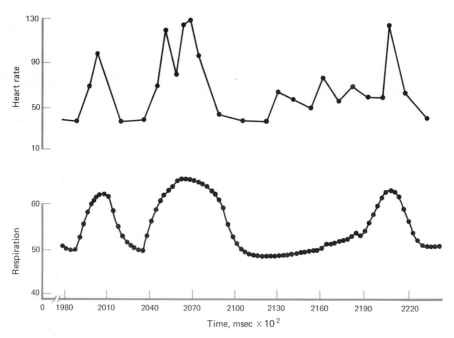

FIGURE 5.27 Relationship between heart rate and respiration during regular breathing (first peak in respiration) and a spontaneous "sighing" breath (second peak) followed by a period of apnea. (From J. F. Amend and H. E. Hoff, *Arch. Int. Pharmacodyn.* 189:34, 1970.)

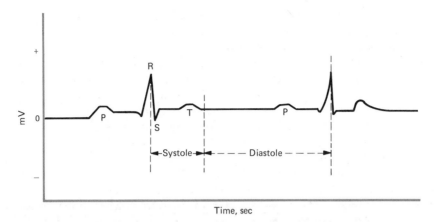

Time, sec

FIGURE 5.28 Main events of the cardiac cycle in humans. The basic pattern applies to most vertebrate hearts.

node) during ventricular diastole. The QRS complex signals ventricular depolarization, whereas the T wave corresponds to repolarization of ventricular muscle. A special conduction pathway—the Purkinje fibers—transmits the action potential to the apex of the ventricle in fishes, birds, and mammals. This specialized conduction system has not been established for amphibians and reptiles. In the hagfish the electrical activity of the auxiliary portal heart is aphasically superimposed on the EKG of the branchial heart. In the lungfishes a B wave is produced when the conus arteriosus depolarizes; it usually occurs between the QRS complex and T wave.

In many species the delay of the action potential at the atrioventricular node (P-R interval) is proportional to the total EKG interval regardless of heart rate; for example, in the horse (40 beats/min) and shrew (600 beats/min) the P-R interval is 30 to 50 percent of the total EKG interval. This permits optimal ventricular filling. The total period for ventricular filling (diastole), however, is the interval between the end of the T wave and the R wave, and in an individual vertebrate increased heart rate causes this interval to decrease much more than the interval of systole. Thus in man, for instance, as heart rate increases from 65 to 200 beats/min, the duration of systole decreases from 0.3 to 0.16 sec, whereas diastole goes from 0.62 to 0.14 sec. This disproportionate decrease in diastole limits the effective maximum heart rate since stroke volume will eventually decrease as filling time shortens. During bradycardia converse changes occur. In reflex bradycardia of diving animals it is the prolongation of diastole that accounts for most of the decreased heart rate. In alligators forced under water, the intervals increased from 0.34 to 4.98 sec for diastole (T-Q) and 0.98 to 1.50 sec for systole (Q-T).

The differential effect of rate on diastole and systole is due to the fact that pacemaker depolarization occurs during the T-P interval, and it is this interval which is influenced by autonomic innervation. As mentioned earlier, the inherent depolarization rate of the pacemaker tissue stems from the permeability of the cell membrane to sodium ions. (To review the basis for the membrane potential, refer to Chapter 2.) During diastole there is an inward "leak" of sodium exceeding an outward "leak" of potassium, so the membrane potential becomes progressively less negative until the threshold voltage (about -55 mV in man) is reached and an action potential initiates the all-or-none wave of depolarization. The rate of leak (and therefore the heart rate) varies directly with temperature, a significant factor for ectothermic vertebrates. Some evidence suggests that stretching pacemaker tissue as during filling may increase the leak rate, providing a mechanism for a stretch-induced tachycardia. In addition to these factors are the dominant influences of autonomic neurotransmitters acetylcholine and norepinephrine and circulating catecholamines.

Acetylcholine released during vagal stimulation decreases the leak rate by selectively increasing potassium conductance (outflow) and thereby slows heart rate (except in lampreys). Norepinephrine and epinephrine have the opposite effect; that is, by decreasing outward conductance of potassium, the inward sodium conductance depolarizes the pacemaker tissue at a faster rate.

Circulation

HEMODYNAMIC PRINCIPLES. The flow of blood through the circulatory system is related to the driving pressure and the resistance to flow. The relationship is analogous to Ohm's law (relating current, voltage, and resistance) and forms the basis of hemodynamics. It is helpful to derive each component in terms of the others; that is,

$$\dot{Q} = \frac{\Delta P}{R} \qquad R = \frac{\Delta P}{\dot{Q}} \qquad \Delta P = \dot{Q}R \qquad (5\text{-}5)$$

where ΔP is the difference in pressure (mm Hg) between the beginning and end of the flow path and R is the resistance. The units of resistance are mm Hg/ml/sec, which is called a peripheral resistance unit (PRU).

The vascular resistance is primarily a function of vessel diameter as shown by the equation

$$R = \frac{8l\eta}{\pi r^4} \qquad (5\text{-}6)$$

where 8 is an integration constant, l vessel length, η viscosity, and r the radius. If a constant flow is maintained, then ΔP must increase as R increases. This occurs mainly in the arterioles. The pressure drop across capillary beds is normally less. Although the resistance of individual capillaries is greater than that of individual arterioles, the total resistance is less because of the much larger number of capillaries and greater total cross-sectional area.

Cardiac output is determined ultimately by factors that affect blood flow to the heart, namely, the venous return. Venous return is augmented by skeletal muscle contraction in all vertebrates, where the presence of venous valves (reported even in fishes) provides one-way flow. In fishes venous return is also augmented by cardiac suction, since the pericardium is somewhat rigid and the intrapericardial pressure is negative. This augments the pressure head, resulting in increased flow from veins to atria. When the heart contracts, blood flows to the gas exchange organ through the central circulation and to the systemic tissues through the peripheral circulation.

CAPILLARY FLUID EXCHANGE. The absorption of interstitial fluid into capillaries tends to occur because the colloid osmotic pressure of plasma (π_{pl}) (see Chapter 2, Section 5) is higher than the colloid osmotic pressure of interstitial fluid (π_{if}). Filtration of plasma out of capillaries into interstitial fluid is favored because capillary hydrostatic pressure (P_c) is greater than interstitial fluid hydrostatic pressure (P_{if}). When these opposing forces are equal—that is, $\pi_{pl} = \pi_{if} = P_c - P_{if}$—in any tissue, an isovolumetric condition, or "Starling equilibrium," exists. As discussed in the subsequent subsection on blood volume, this balance between filtration and absorption of fluid is extremely important for maintenance of a constant blood volume. Figure 5.29 illustrates the fluid exchange parameters in a "typical" capillary. At the arterial end of a capillary in a human finger held at heart level, the P_c is about 30 mm Hg. At the venous end of the capillary P_c is 10 to 15 mm Hg. The plasma colloid osmotic pressure is relatively constant from the arterial to the venous end of the capillary ($\pi_{pl} \simeq 25$ mm Hg). Interstitial colloid osmotic pressure (π_{if}) is variable (2.5 mm Hg in skeletal muscle to 20 mm Hg in liver), but for our "typical" capillary we can assign a value of 5 mm Hg. The interstitial fluid pressure (P_{if}) is zero or, according to some studies, less than zero. As the P_{if} becomes greater than zero, edema (tissue swelling) results. If, for our capillary in Figure 5.28, we consider P_{if} to be zero, we can see that the filtration force at the arteriolar end is $(P_c - P_{if}) - (\pi_{pl} - \pi_{if})$, or $(30) - (25 - 4) = +9$ mm Hg. At the venous end ($P_c = 15$ mm Hg), the net filtration force is $(15) - (25 - 4) = -6$ mm Hg, so absorption occurs. In man, where total systemic blood flow is 8000 to 9000 liters/day, approximately 20 liters of blood are filtered from capillaries, and 16 to 18 liters are reabsorbed. The balance of 2 to 4 liters provides the daily lymph flow.

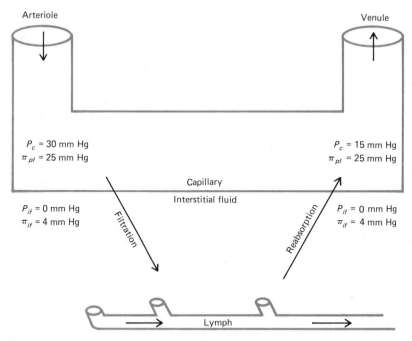

FIGURE 5.29 Fluid exchange in a typical capillary. See text for explanation of symbols and events.

Large deviations from these average pressures occur in many tissues. Pulmonary capillaries, for example, have a P_c of less than 10 mm Hg, an obvious advantage in keeping the lungs dry (glomerular capillaries have a P_c of about 70 mm Hg, allowing filtration of almost 180 liters/day). P_c is also not constant in a given tissue but varies with the ratio of postcapillary to precapillary resistance. Neurogenic arteriolar constriction, for example, causes P_c to fall (see subsection on blood volume control). Conversely, arteriolar dilation such as that occurring in skeletal muscles during exercises causes an increase in P_c and resulting increase in filtration.

In vertebrates other than mammals capillary fluid exchange obeys the same principles (Starling equilibrium), but the absolute pressure values are often far removed from those found in mammals. For example, in freshwater turtles, such as *Pseudemys scripta*, the plasma protein is less than one-half that found in mammals. Consequently the colloid osmotic pressure is only 3 to 8 mm Hg. This results in a functional ascites, that is, accumulation of serous fluid in the coelomic cavity. This fluid is fairly high in bicarbonate concentration and can therefore contribute to lactic acid buffering during submergence. In other reptiles Scholander and his colleagues (1968) found π_{pl} to be in the mammalian range and P_{if} to be negative.

CENTRAL CIRCULATION.   In most fishes the central and peripheral circulations are completely in series; that is, all cardiac output is directed first to the gills, then through the systemic circuits. In birds and mammals these circuits are also arranged in series, but the completely divided heart provides a separate pump for each circuit.[6] In air-breathing fishes, amphibians, and reptiles the central and peripheral circulations are anatomically in parallel. Incomplete separation of the heart or anatomical connection of the aortic arches may allow shunting of blood flow to occur. This may be from the central to the systemic circuit (right to left) or from the systemic to the central circuit (left to right).

At first glance shunting would seem to be in all cases a definite liability since, in terms of transporting O_2 to the tissues, it could be detrimental because of the mixing of oxygenated with venous blood. It is, however, remarkable that functional separation of the central and systemic circuits may exist despite incomplete anatomical separation. Radiological studies and physiological studies, in which oxygen content and pressure of blood entering and leaving the various heart vessels were measured, have combined to show that a complete double circulation is possible in amphibians and reptiles. In the amphibian *Amphiuma tridactylum* separation of pulmonary and systemic venous return is affected by the partial septation (trabecula) of the ventricle. The selective distribution of cardiac output (right atrial venous return to pulmonary arteries, left atrial venous return to systemic arteries) is maintained partially by anatomical direction, owing to the presence of a spiral valve in the bulbis cordes, but depends mainly on vasomotor activity in the systemic and pulmonary circuits; that is, flow distribution from the truncus arteriosus depends on pressures and vascular resistance in the pulmonary artery, carotid arch, and aortic arch. Instead of using a spiral valve to distribute flow to proper aortic arches from the ventral aorta, the reptiles have two aortas and a pulmonary trunk. In crocodilians the ventricles are completely separated, but the right and left aortas are connected by the *foramen Panizzae*. In snakes, lizards, and turtles the ventricle is divided into three distinct chambers (cava) separated by a muscular ridge; this muscular ridge provides for selective distribution of left atrial blood to the right and left systemic aortas and right atrial blood to the pulmonary artery.

Reptiles often use shunting to great advantage. For instance, in croc-

---

[6] As derived for mammalian hearts, cardiac output is the output of only one side of the heart. In species with incomplete separation the gas exchange organ and systemic outputs may vary tremendously, so the only meaningful expression of cardiac output is the sum of these two flows.

odilians the functional separation of pulmonary and systemic venous return is due to the fact that left aortic pressure normally exceeds right ventricular pressure, and therefore venous blood flows only to the pulmonary artery. However, during diving a right-to-left shunt occurs because the pulmonary outflow resistance causes right ventricular pressure to exceed left aortic pressure, so venous blood enters the left aorta; this is associated with bradycardia.

In mammals the pulmonary circulation is in series with the systemic circulation, and functionally it contrasts strikingly with the systemic circulation. Consider the following:

1. It is a low-pressure circulation. The pulmonary arterial pressure oscillates from about 10 mm Hg in diastole to about 25 mm Hg during systole. The mean pressure is approximately 18 mm Hg or one-sixth of that of the systemic circulation.
2. It is a high-blood-flow circulation. Because of its position in series with the systemic circulation, the entire cardiac output perfuses the pulmonary circulation. No other single organ receives as high a blood flow. Although the total flow in the systemic circulation is just as great, it is distributed to many different organs in accordance with their needs.
3. It is a low-resistance circulation. The pulmonary resistance is about seven to eight times less than the systemic resistance. This difference in resistance is due in part to the structural properties of the pulmonary vessels, which allow for their remarkable extensibility.
4. It is greatly affected by the hydrostatic pressure. Because the pulmonary capillary bed is surrounded by air spaces (alveoli) where the pressure is near atmospheric, hydrostatic pressure differences between the upper and the lower parts of the lung influence the distribution of the circulation of blood. At the lower part of the lung the hydrostatic pressure is maximum; thus the vessels are distended and therefore offer the least resistance to the blood flow.
5. It is a blood reservoir. The pulmonary vascular bed is a distensible blood reservoir interposed between the left and right heart. In man it normally contains 10 to 12 percent of the total blood volume, but this amount can change greatly under various conditions. For instance, an increase is seen with increasing blood flow as in exercise, whereas a decrease in pulmonary blood volume is noted on standing up.

These features have progressively evolved from fishes to mammals. Figure 5.30 shows the difference in systemic and gas exchange organ blood pressure (Figure 5.30A) and vascular resistance (Figure 5.30B) in representative vertebrates. In homeotherms, man and duck in Figure 5.30B, the resistance of the pulmonary circuit is about one-tenth that of the systemic circuit; this is necessary to avoid water loss (pulmonary edema) and filtration of substrates other than gases. In obligate gill-breathing fishes it is mandatory for incoming (afferent) gill circulation to have higher pressure than that in systemic circuits, resulting in substantial ion and water exchange. This is countered in fishes by active transport of ions in gill epithelia and by the relatively small surface area and relatively thick diffusion distance. In effect the gill permeability must be a compromise between osmoregulatory and respiratory demands. In amphibians and reptiles the low pulmonary resistance is, as discussed earlier, necessary to minimize right-to-left shunting during air breathing.

Control of Central Circulation.   Circulation to the gas exchange organ is reflexly controlled and responds to local control factors, the participation of which is a universal feature of vertebrate circulatory control. Hypoxia is known to cause branchial vasoconstriction in elasmobranchs and teleosts and pulmonary vasoconstriction in mammals. In gill breathers the vasoconstriction with hypoxia may prevent O_2 loss from blood to gills. In air-breathing fishes such as the bowfin, electric eel, and lungfishes and in amphibians and reptiles, right-to-left shunting results from pulmonary vasoconstriction. In lower vertebrates it provides an adjunct to the inflation-dependent flow changes. As shown in Figure 5.31, changes in pulmonary arterial blood flow are directly related to pulmonary P_{O_2} in the green turtle. There is also direct evidence of increased blood flow during air breaths in air-breathing fishes.

In mammals the survival advantage of pulmonary vasoconstriction has been somewhat controversial. Pulmonary hypertension has been considered maladaptive at high altitude, contributing to "mountain sickness." However, the increase in pulmonary perfusion pressure at high altitude also results in a more uniform distribution of flow through the lungs. At sea level the vasoconstrictor response to hypoxia also minimizes the unevenness of blood flow distribution and to a point enhances the efficiency of gas exchange.

PERIPHERAL CIRCULATION.   The distribution of blood flow between the pulmonary and systemic circuits is, as described earlier, quite variable when these circuits are in parallel. This is not so in fishes, where there is a single source of cardiac output, or in mammals and birds, where both sides of the heart pump equal volumes. However, the regional distribution of systemic blood flow exhibits wide variations between organs and between physiological status. For instance, comparison of the resting and maximum regional flows reveals the tremendous potential for redistribution of blood flow in response to physiological requirements and changes in vascular resistance in the systemic circuits. In man during rest only 18 to 26 percent of the total cardiac output (5.5 liters/min) goes to the skeletal muscles and the skin, but during strenuous exercise these two circuits receive 80 to 85 percent of the total output (25 liters/min). This obviously adaptive redistribution of flow according to tissue oxygen demand and

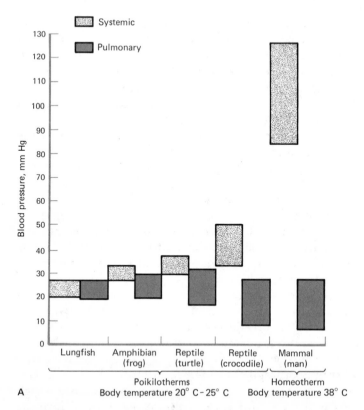

FIGURE 5.30 (**A**) A comparison of systolic and diastolic blood pressures in the pulmonary and systemic circuits of unanesthetized resting vertebrates. (Reprinted from K. Johansen, C. Lenfant, and D. Hanson, *Fed. Proc.* 29:1136, 1970.)

thermoregulatory requirements in man probably has evolved from basic mechanisms that seem to be evident today in many lower vertebrates.

Recent studies have emphasized the fact that cardiac output is partially regulated by local control in peripheral circuits; that is, each organ regulates its own blood flow to maintain homeostasis of O_2 delivery. The mechanism of this autoregulation involves both control of flow and oxygen extraction. Local control is achieved by tissue metabolites such as CO_2 and H^+ ion concentration. The metabolites that are vasoactive provide a complementary system for matching local flow with local tissue activity.

The following describe some examples of systemic circulatory adaptations.

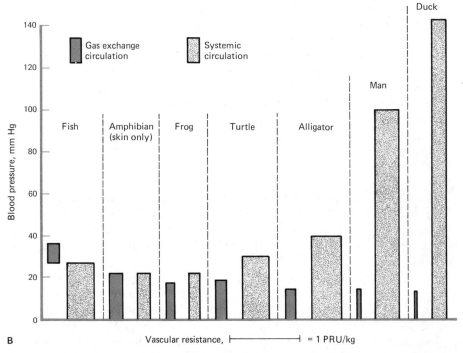

B

Vascular resistance, ⊢———————⊣ = 1 PRU/kg

FIGURE 5.30 (**B**) The perfusion pressure and vascular resistance of the gas exchange and systemic circulation in representation vertebrates. Vascular resistance is expressed by the width of the bars. (From K. Johansen, *Resp. Physiol.* 14:197, 1972.)

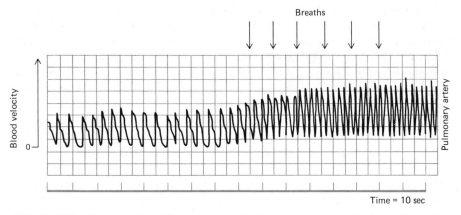

FIGURE 5.31 Pulmonary blood flow in the green turtle during apnea and eupnea. (Reprinted from K. Johansen, C. Lenfant, and D. Hanson, *Fed. Proc.* 29:1140, 1970.)

Cutaneous Blood Flow. Control of cutaneous circulation is directed toward two broad goals: thermoregulation and gas exchange, especially in amphibians. The thermoregulatory function (discussed also in Chapter 10) is present in homeotherms and also in some species of lizards. In the latter the mechanism appears to be a change in total cardiac output and not a selective redistribution of skin blood flow. In man and other scantily insulated homeotherms the adjustments of skin circulation are crucial in preventing overheating, which would occur in warm climates. On the other hand, naked man has relatively little physiological protection against cold stress. For example, total blood flow to the skin during thermal balance (27°C ambient) is about 500 ml/min. Although this can decrease with sympathetic vasoconstriction to 20 ml/min in cold stress, such attempt at insulation is not sufficient, and man must generate heat metabolically (shivering and nonshivering thermogenesis) when environmental temperature drops below about 26°C. However, as the extent of furry or fatty insulation increases, mammals can maintain their basal metabolism at lower environmental temperatures.

During heat stress there is an increase in cutaneous blood flow (hyperemia), which in man may exceed 3000 ml/min. This results not only from a decrease in sympathetic tone but, at least in man, may involve active dilation either by specific nerve fibers or polypeptide metabolites (bradykinin or kallidin) secondary to sweat gland activation. Specific arterial-venous shunts bypass skin capillaries in man and allow more rapid heat transfer. Active neurogenic vasodilatation of skin vessels has been difficult to confirm in spite of good circumstantial evidence of its existence (for example, emotional blushing of man).

Circulation during Diving. Diving elicits cardiovascular reflexes in all vertebrates (including fishes "diving" into air). In vertebrates that do not habitually dive these reflexes are elicited by increased P_{CO_2} and decreased P_{O_2} in arterial blood (asphyxia) and mediated by the peripheral and central chemoreceptors. In habitual divers, such as marine mammals, ducks, and aquatic reptiles, the reflexes are much more pronounced and often elicited more rapidly than in terrestrial vertebrates. The reflexes are often mediated by sensory receptors responding to head immersion and are elicited in the absence of breath holding; for example, in seals, ducks, and man the responses to head immersion are more pronounced than those to breath holding.

The two fundamental reflexes involved are bradycardia and redistribution of blood flow to tissues of low anaerobic tolerance. Although diving bradycardia is quite pronounced in ducks (100 to 14 beats/min), seals (100 to 10 beats/min), and especially in some reptiles (1 beat/5 min in iguanas), the greatest contribution to sustained apnea is provided by the selective distribution of oxygenated blood to the heart and central nervous system. The decrease in cardiac output accompanying bradycardia reduces the oxygen consumption of the heart. Vasoconstriction in skeletal muscles and other tissues results in the maintenance of arterial pressure, and thus even if total systemic flow and pulmonary flow are reduced, the brain flow and heart flow remain at predive levels.

P. F. Scholander and his colleagues made tremendous contributions to our knowledge of diving physiology. In a series of papers beginning in the early 1940s, they demonstrated the intense vasoconstriction of skeletal muscle blood vessels in diving seals. The obvious evidence for this phenomenon was the tremendous increase of blood lactic acid immediately following a dive.

This same pattern of "lactate flush" following diving is observed in birds and in some (but not all) diving reptiles. In addition to the redistribution of systemic flow in most divers, there exists in reptiles a redistribution of cardiac output. In crocodilians the increase in pulmonary resistance causes right ventricular pressure (refer to Figure 5.25) to increase above left aortic pressure resulting in a right-to-left shunt. In noncrocodilian reptiles the increased pulmonary resistance produces an intracardiac right-to-left shunt.

BLOOD VOLUME AND ITS CONTROL. The total volume of vertebrate cardiovascular systems is equal to the volume of cells and plasma contained in the system. In man, blood volume is about 75 to 80 ml/kg body weight, with the red cells making up about 45 percent of this volume. Three main factors determine blood volume: (1) the control of red cell volume, (2) the control of plasma-interstitial fluid exchange, and (3) the control of extracellular fluid volume loss to the environment (kidneys and sweat glands). Chapter 6 presents a detailed discussion of the volume and composition of body fluids.

In mammals individual variations and species differences in red cell and total blood volume are correlated with activity, behavior, and environment; for example, in marine mammals blood volume ranges from 91 to 132 ml/kg body weight; high-altitude residents also develop increased blood volumes as compared to sea-level residents. In lower vertebrates there is no consistent correlation of blood volume with any known factors. For fishes blood volume values range from 18 ml/kg in the bullhead to 140 ml/kg body weight in eels. The same chaotic situation exists for amphibians and reptiles, reflecting perhaps inadequate measurement techniques but more likely the presence of poor regulation of body water and greater tolerance to daily variations between hemodilution and hemoconcentration. In man, for example, a major system for controlling plasma volume is the renin-angiotensin system, which is discussed in Chapter 9, Section 10. Although most other species have a similar system, its function appears to differ and not to have the same fine regulatory capability. Plasma volume (and therefore pressure) is also controlled by fluid exchange between plasma and interstitial fluid. This system, which is closely related to the regulation of systemic blood pressure, has been well studied in mammals, but little comparative data are available. A fall in blood pressure reduces the output of the aortic and carotid stretch receptors. This increases sympathetic vasoconstriction, primarily in the skeletal muscles where the precapillary resistance is increased more than the postcapillary. This causes an imbalance in the hydrostatic and colloid osmotic pressure, so interstitial fluid enters the plasma. A decrease in vascular tone has the opposite effect, namely, filtration of fluid out of the plasma. In exercise, hyperemia, the increase in capillary hydrostatic pressure, results in a loss of up to 15 percent of the plasma volume into the interstitial space of skeletal muscle.

Blood volume is also dependent on red cell volume. Changes in red cell volume can be rapid, by contraction of the spleen, but regulation of red cell volume is primarily geared to long-term adaptations. Hypoxia, an obvious stimulus for feedback control of red cell production, elicits in mammals secretion of the renal hormone, erythropoietin, which increases the rate of red cell production in bone marrow. It is apparently not a uni-

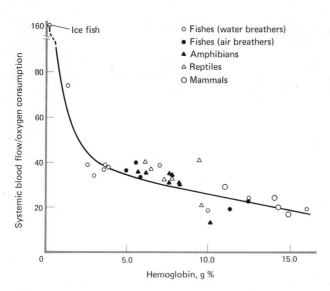

FIGURE 5.32 Relationship between blood flow/metabolic rate and hemoglobin concentration. Note that the blood flow increases markedly only when hemoglobin concentration is lower than 3 to 4 g percent. (Reprinted from C. Lenfant, K. Johansen, and D. Hanson, *Fed. Proc.* 29:1128, 1970.)

versal feature of vertebrates. Comparative data again are meager and conflicting. The erythropoietic response to hypoxia is absent in some reptiles and birds, as is a correlation between O_2 capacity and habitat altitude (Dawson and Poulson, 1962). On the other hand, lizards do appear to have increased erythropoiesis at high altitude (Weathers and White, 1972). In fishes there is a fairly close correlation between the O_2 availability of the habitat and the hemoglobin concentration, and erythropoietic responses to hypoxia do occur. The extreme variability in red cell volume in poikilotherms is a function of the relatively low metabolic rate (compared with homeotherms) and reflects the absence of strong selective pressures for precise red cell volume control. Some antarctic fishes have no red cells and get along nicely in their supercooled habitat merely by having a bigger heart (and greater cardiac output) than most other fishes (Figure 5.32). Other normally red-blooded fishes as well as amphibians and mammals can be made anemic with no severe effects at the resting metabolic rate (Holeton, 1972).

5 • TISSUE GAS EXCHANGE: MECHANISM AND BIOCHEMICAL PROCESSES

The final step in gas transport is the exchange of O_2 and CO_2 between capillaries and cells. The rate of exchange is determined not only by passive diffusion but also by cellular biochemical processes and cellular adaptations.

5.1 Influence of Body Size on Metabolic Rate

When the metabolic rate of mammals is plotted as a function of body weight on log-log graph paper, the data reveal a straight line with a slope of 0.75. Figure 5.33A shows this well-known "mouse-to-elephant" curve, which has the equation $M = aW^b$, where a is the intercept on the metabolism axis and b is the slope of the line (0.75). The weight-specific metabolism (metabolic rate per unit body weight) is derived as follows: $M/W = aW^b/W = aW^bW^{-1} = aW^{b-1}$, so when $b = 0.75$, the slope of the line relating weight-specific metabolism is -0.25, as shown in the figure. The very large range of body weights in mammals (from a 4-g shrew to a 100-ton blue whale) is useful to impress one with the significance of the value of b; that is, if b was equal to 1.0, a blue whale would have to eat half its body weight in food each day, a formidable task at 50 tons of food. Twenty-five million shrews would be equally difficult to feed, as each one

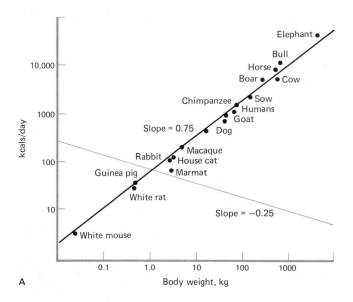

A

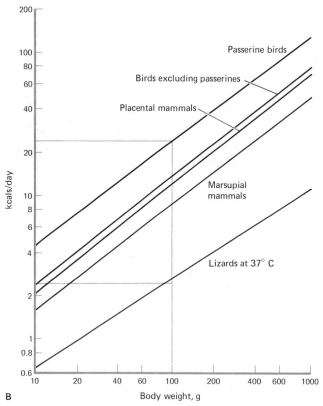

B

FIGURE 5.33 Metabolism as a function of body weight in (**A**) mammals and in (**B**) four groups of vertebrates. (From G. A. Bartholomew, *Animal Physiology: Principles and Adaptations*, M. S. Gordon, ed. New York: Macmillan, 1972.)

consumes about 2 g of food per day. Another popular example to illustrate the importance of b being less than 1.0 involves heat dissipation; that is, if b equaled 1.0, a cow would have to dissipate heat at the same rate as a shrew, but without the advantage of the favorable surface area/volume ratio of shrews. To accomplish the heat dissipation, the cow's body surface would have to have a temperature close to 200°C.

The significance of the constant a is indicated in Figure 5.33B. The value of a for different vertebrate taxa indicates the relative "cost of living." A 100-g lizard, for example, needs a food intake of about 10 kcal per day to sustain resting metabolism, but a 100-g songbird must have 10 times this caloric intake.

What is the implication of the relationship between metabolism and body weight for oxygen transport? The greater O_2 uptake per unit weight of tissue must be provided for by one or both factors of the Fick equation—that is, cardiac output and arteriovenous O_2 content difference—or by adaptations to improve the rate of diffusion.

5.2 Diffusion Process

Diffusion between blood and tissues, and within the tissues, is determined by the same parameters governing external gas exchange,

$$\frac{dq}{dt} = \frac{KA \, \Delta P}{L} \tag{5-7}$$

As in external exchange, the high value of K for CO_2 takes the burden off the other components of this equation. It is O_2 transport that is the problem, and evolutionary solutions to maximize the rate of diffusion involve ΔP, A, and L.

The partial pressure difference (ΔP_{O_2}), which is a function of the oxyhemoglobin dissociation curve, was shown by Parer and Metcalfe (1970) to be maintained at about the same level regardless of body size. This appears to be accomplished by an increase in P_{50} with decreasing size, which in turn results in an increased arteriovenous oxygen content difference but not in a higher capillary P_{O_2}. P_{50} is correlated not only with metabolic rate (Figure 5.21) but also with the critical P_{O_2} (Figure 5.34), that is, the P_{O_2} below which a normal rate of metabolism cannot be sustained. It is important to note that the increase in P_{50} that partially provides for the elevated oxygen requirement in smaller mammals also makes them more susceptible to hypoxia, since any lowering in arterial oxygen tension will result in a much lower oxygen saturation than if the P_{50} were lower. Indeed, because blood O_2 capacity is not related to body weight, the increased arteriovenous O_2 content difference in small mammals depends on a relatively high arterial P_{O_2}.

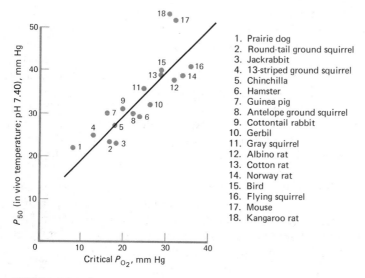

FIGURE 5.34 Oxygen affinity (P_{50}) of mammals as a function of the critical P_{O_2}. (From F. G. Hall, *J. Appl. Physiol.* 21:375–378, 1966.)

The other factors determining the rate of diffusion of O_2 from capillary to cell are the diffusion distance (L) and area for diffusion (A). These factors are a function of capillary density.

Krogh (1919) suggested that small animals, because of their higher metabolic rate, would require a higher capillary density, and his measurements in muscles from horse, dog, and guinea pig showed this to be the case. Others, however, in extending Krogh's study by examining different muscles in a larger number of mammals, have shown that the relationship between capillary density and body size is clear only for mammals smaller than the cat or the rabbit (Figure 5.35A). It is noteworthy that the hyperbolic shape of this relationship is similar to that between metabolic rate and body weight (Figure 5.35B). Indeed, it can be seen in comparing Figure 5.35A and B that between the cow and the rabbit there is only perhaps a twofold increase in the capillary density and in the required rate of O_2 supply to the cells, but in mammals smaller than 1 kg there is up to a hundredfold increase in metabolic rate and in capillary density. Other evidence supporting the role of capillary density is provided by studies during exposure to chronic hypoxia, which show that acclimatization to high altitude is accompanied by increased tissue capillaries in smaller mammals (rabbits, guinea pigs, and rats), but studies on larger mammals are lacking.

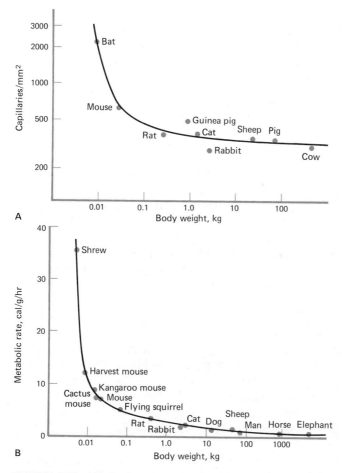

FIGURE 5.35 (**A**) Capillary density of the gastrocnemius muscle and (**B**) metabolic rate as a function of body weight in mammals [compare with (**B**) in Figure 5.21]. When metabolic rate per gram is plotted on an arithmetic scale, as in this figure, instead of a log scale, the tremendous increase of metabolism in the smallest mammals becomes very clear. (From K. Schmidt-Nielson, *How Animals Work*. New York: Cambridge, 1972.)

5.3 Biochemical Processes and Cellular Adaptations

The role of biochemical processes and cellular adaptations can be established by answering two basic questions: (1) What is the ultimate fate of the oxygen molecules that have been so ingeniously ventilated, diffused, circulated, and diffused into each metabolizing cell? (2) What adaptations are possible if the rate of oxygen supply cannot keep up with the demand for energy? (Can the cells themselves adapt to hypoxia? If so, how?

Figure 5.36 shows the main features of aerobic metabolism and partially answers the first question. Oxygen is used in mitochondria as the final electron acceptor in a series of reductions of NAD, flavoprotein, and cytochromes. A mole of glucose provides 686 kcal of free energy and 36 moles of ATP if it is completely oxidized to CO_2 and H_2O. If oxygen is not available, a mole of glucose provides only 48 kcal and 2 moles of ATP plus a toxic metabolic acid (lactate). The oxygen affinity of the cytochrome system is extremely high, and O_2 utilization is independent of P_{O_2} down to levels much less than 1 mm Hg. This explains why active skeletal muscle can extract almost all of the oxygen available in blood with a maximum oxygen consumption that is maintained at very low oxygen pressure. In organs with more complex metabolic and respiratory patterns, the oxygen extraction is not as complete, and much of the total oxygen consumption is dependent on P_{O_2}. The reason for this difference between skeletal and parenchymal tissue is the presence in the latter of subcellular respiratory systems other than mitochondria that utilize molecular oxygen but have much lower oxygen affinities. Two of the systems are the microsomal oxidase system of the endoplasmic reticulum, which has an oxygen affinity of 1 to 5 mm Hg, and the oxidase system of peroxisomes, which has even lower O_2 affinity. These systems are found in all vertebrates and are especially prevalent in the liver, which in mammals requires about 30 percent of total oxygen consumption at rest, more than any other organ.

It is these systems that are the most susceptible to hypoxia and could benefit from high oxygen pressure in the capillary and blood, that is, a higher P_{50}. Also, it is these systems that apparently account for the presence of a critical P_{O_2}.

This subcellular difference in tolerance to hypoxia is manifested at the tissue, organ, individual, and species level. When the O_2 supply is less than the oxygen demand, cells must depend on anaerobic metabolism for energy production. A comparison of the ATP yields in aerobic versus anaerobic glucose metabolism emphasized the potential handicap this imposes. For example, if activity required an energy expenditure of 10 kcal/hr, to supply this by anaerobic glycolysis would use glucose at the rate of about 200 mM/hr, whereas aerobic production of 10 kcal/hr would require only about 14 mM glucose/hr.

The second question posed—that is, whether cells can adapt to hypoxia—has been approached from many experimental bases, and it is now clear that cells can indeed adapt to operate efficiently at lowered levels of O_2. This is especially evident in vertebrate muscle (which is the least favored tissue during redistribution of blood flow during apnea).

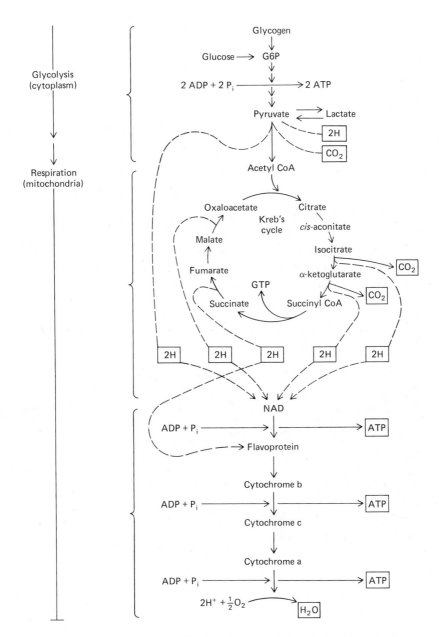

FIGURE 5.36 Major pathways of carbohydrate metabolism in vertebrates. In the absence of oxygen, lactate is the end product; in the presence of oxygen, glucose is oxidized to form CO_2 and H_2O. See text for discussion. (From P. Hochachka and G. Somero, *Strategies of Biochemical Adaptation*. Philadelphia: Saunders, 1973.)

Among vertebrates reptiles are the champions in terms of anaerobic energy production and tolerance to anoxia. The major cellular mechanisms involved in anoxia tolerance in vertebrate muscles are (1) a high capacity for glycolysis (resulting from a high concentration of glycolytic enzymes), (2) the presence of isozymes that favor glycolysis in skeletal muscle and are rapidly activated during anoxia, and (3) tolerance of high lactate concentrations (and concomitant acidosis).

The ability of skeletal muscle to generate anaerobic ATP and tolerate lactic acidemia is an essential part of the diving response, which to be most effective converts the cardiovascular system into a "heart-brain" circuit. The intense vasoconstriction needed to stop oxygen consumption in other circuits is especially essential in exercising skeletal muscle, which would rapidly deplete oxygen down to P_{O_2} levels below critical P_{O_2} of brain and heart.

The extreme development of these enzymatic adaptations and lactate tolerance has been well demonstrated in the diving turtle (Jackson, 1968). Figure 5.37 compares the calorimetric measurement of metabolism with the arterial oxygen available during a 3-hr dive. During the first 30 min availability of oxygen falls precipitously, but there is no reduction in metabolic rate. This is apparently due to a threshold of tolerance (critical

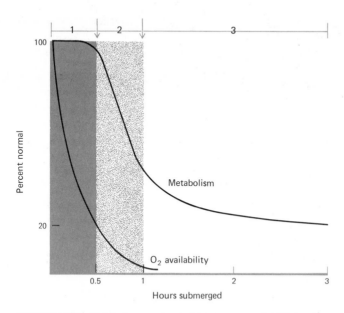

FIGURE 5.37 Metabolism and arterial oxygen availabilities of an aquatic turtle during a 3-hr dive at 24°C. (From P. Hochachka and G. Somero, *Strategies of Biochemical Adaptation*. Philadelphia: Saunders, 1973.)

P_{O_2}) since further reduction in oxygen is accompanied by a sharp decrease in metabolic rate. At 2 hr total anoxia ensues, and metabolic heat production for the remainder of the dive (which at low temperatures may last days in some species) is entirely from anaerobic glycolysis.

Even the hardiest diving turtles must eventually surface and repay their "oxygen debt" by oxidizing lactate back to pyruvate, but some true facultative anaerobes (able to survive indefinitely by anaerobic metabolism) do exist among the teleost fishes. The carp, for instance, survives for months trapped under ice with almost no oxygen.

On the basis of these few examples, it is clear that the evidence for a major role of cell acclimation is convincing; that is, many cases of successful hypoxia adaptation, especially long-term cases, cannot be accounted for by respiratory, circulatory, or erythropoietic changes.

6 • CONCLUSION

In this chapter the respiratory and circulatory functions in vertebrates have been presented and discussed in the context of gas transport from the environment to the cell, and vice versa. Mechanisms and adaptations have been described in a variety of species, aquatic and terrestrial, poikilotherm and homeotherm, to illustrate the complexity and the diversity of these functions. It is clearly evident that both the respiratory and the circulatory functions are extremely versatile and probably well adapted to environment, to ecology and to behavior. Gas transport is achieved through several steps, each with a different relative importance among the different species, so that the most effective system of overall gas transport can be attained.

REFERENCES

Astrand, P. O. Pulmonary Function and Heart Rate during Exercise: Man. *Respiration and Circulation*. P. Altman and D. Dittmer, eds. Bethesda Md.: Federation of American Societies for Experimental Biology, 1970.

Bohr, C., Hasselbalch, K., and Krogh, A. Ueber einen boliogischer beziehung wichtigen Einfluss, den die Kohlen-Saurespannung des Blutes auf dessen Sauerstoffbingung ubt. *Skand. Arch. Physiol.* 16:402, 1904.

Bond, A. N. An Analysis of the Response of Salamander Gills to Changes in the Oxygen Concentration of the Medium. *Develop. Biol.* 2:1–20, 1960.

Brett, J. R. The Metabolic Demand for Oxygen in Fish, Particularly Salmonids, and a Comparison with Other Vertebrates. *Resp. Physiol.* 14:151–170, 1972.

Bretz, W. L., and Schmidt-Nielsen, K. Patterns of Air Flow in the Duck Lung. *Fed. Proc.* 29:662, 1970.

Dawson, W. R., and Poulson, I. Oxygen Capacity of Lizard Blood. *Am. Midl. Nat.* 68:154–164, 1962.

Duncker, H. R. Structure of Avian Lungs. *Resp. Physiol.* 14:44–63, 1972.

Gleysteem, J., and Stroud, R. Respiration Frequency, Tidal Volume and Minute Volume: Vertebrates. *Respiration and Circulation.* P. Altman and D. Dittmer, eds. Bethesda, Md.: Federation of American Societies for Experimental Biology, 1970.

Holeton, G. F. Gas Exchange in Fish with and without Hemoglobin. *Resp. Physiol.* 14:142–150, 1972.

Jackson, D. C. Metabolic Depression and Oxygen Depletion in the Diving Turtle. *J. Appl. Physiol.* 24:503–509, 1968.

Johansen, K. Air Breathing in Fishes. *Fish Physiology.* Vol. 4. W. S. Hoar and D. J. Randall, eds. New York: Academic Press, 1970.

———. Heart and Circulation in Gill, Skin and Lung Breathing. *Resp. Physiol.* 14:193–210, 1972.

Johansen, K., and Lenfant, C. A Comparative Approach to the Adaptability of O_2-Hb Affinity. *Oxygen Affinity of Hemoglobin and Red Cell Acid Base Status.* M. Rorth and P. Adstrup, eds. New York: Academic Press, 1972.

Krogh, A. The Rate of Diffusion of Gases through Animal Tissues, with Some Remarks on the Coefficient of Invasion. *J. Physiol.* 52:291–308, 1919.

Krogh, A., and Krogh, M. On the Tensions of Gases in the Arterial Blood. *Skand. Arch. Physiol.* 23:179–192, 1910.

Larimer, J. L., and Schmidt-Nielsen, K. A Comparison of Blood Carbonic Anhydrase of Various Mammals. *Comp. Biochem. Physiol.* 1:19–23, 1960.

Olsen, C. R., Elsner, R., and Hale, F. C. "Blow" of the Pilot Whale. *Nature* 163:953–955, 1969.

Parer, J. T., and Metcalfe, J. Oxygen Transport by Blood in Relation to Body Size. *Nature* 215:653–654, 1967.

Rahn, H. Aquatic Gas Exchange: Theory. *Resp. Physiol.* 1:1–12, 1966.

Rahn, H., and Fenn, W. O. *A Graphical Analysis of the Respiratory Gas Exchange: The O_2-CO_2 Diagram.* Bethesda, Md.: American Physiological Society, 1955.

Randall, D. J. Gas Exchange in Fish. *Fish Physiology.* Vol. 4. W. S. Hoar and D. J. Randall, eds. New York: Academic Press, 1970.

Romer, A. S. Skin Breathing—Primary or Secondary? *Resp. Physiol.* 14:183–192, 1972.

Satchell, G. H. *Circulation in Fishes.* New York: Cambridge, 1971.

———. The Reflex Coordination of the Heartbeat with Respiration in the Dogfish. *J. Exp. Biol.* 37:719–731, 1960.

Scheid, P., and Piiper, J. Direkte Messung der Stromgerichtung der Atemluft in der Entenlung. *Pflueger Arch.* 319:R59, 1970.

Schmidt-Nielsen, K. *How Animals Work.* New York: Cambridge, 1972.

Scholander, P. F., Hargens, A. R., and Miller, S. L. Negative Pressure in the Interstitial Fluid of Animals. *Science* 161:321–328, 1968.

Schumann, D., and Piiper, J. Der Sauerstoffbedarf der Atmung Bei Fischen nach Messungen an der narkotisierten Schleie (*Tinca tinca*). *Pflueger Arch.* 288:15–26, 1966.

Weathers, W. W., and White, F. N. Hematological Observations on Populations of the Lizard *Sceloporus occidentalis* from Sea Level and Altitude. *Herpetologica* 28:172–175, 1972.

Wood, S. C., and Johansen, K. Adaptation to Hypoxia by Increased HbO_2 Affinity and Decreased Red Cell ATP Concentration. *Nature, New Biology* 237:278–279, 1972.

6

OSMOREGULATION AND EXCRETION

Leon Goldstein

The functions of the osmoregulatory and excretory systems are to control the osmotic pressure, ionic composition, and volume of the body fluids. These functions bring the osmoregulatory and excretory systems into direct contact with most other systems in the body. The physicochemical processes of all cells are affected by the osmotic pressure of their environment. Krogh pointed out that the development of higher mental processes is incompatible with an extracellular osmolarity equivalent to more than 1 percent NaCl. Similarly, if the volume of body fluids was not regulated, dilution or concentration of the extracellular fluids would lead to adverse effects on intracellular processes (such as metabolism) and to disruption of physical dynamic forces operating between extracellular compartments, such as the blood capillaries and the interstitial fluid.

The normal operation of nerve and muscle is dependent on control not only of the total osmolarity of their fluids but also of the ionic composition of both their intracellular and extracellular media. Thus the concentrations of ions such as Na^+, K^+, Ca^{2+}, and Mg^{2+} (as well as ratios of one ion to another) in these fluids are critical for neurotransmission and muscular contraction. Perturbations in the concentrations or ratios of these ions in the extracellular fluid are corrected by the body's osmoregulatory systems.

Regulation of acid-base balance in body fluids is shared between the osmoregulatory-excretory systems and the respiratory system. In higher animals there is a remarkable degree of coordination between these two systems, which results in very close control of the pH of the body fluids.

The disposal of most metabolic end products, with the exception of CO_2, falls to a major degree upon the kidneys. This link between metabolism and excretion is so close that the ability to eliminate end products is used as an index of the function of the excretory system.

In higher animals the operation of the osmoregulatory and excretory systems is regulated by the endocrine system. In this chapter and in the chapter on endocrines (Chapter 9) several examples of links between the two systems are discussed.

1 • GENERAL CONSIDERATIONS

1.1 Regulation of Salt and Water Metabolism

Osmoregulation is the ability of an organism to hold relatively constant both the total electrolyte content and volume (water content) of its cells. The regulation of volume and electrolyte content is functionally inseparable since the amount of water in an organism (or cell) is dependent in large part on the solute content of the organism. In metazoan animals, cells are surrounded by a fluid known as the extracellular fluid. This is the "internal environment" (*milieu interieur*) that, as the great French physiologist Claude Bernard pointed out, must remain relatively constant in composition for the organism to function properly. As stated by L. J. Henderson:

> Stability may sometimes be afforded by the natural environment, as in
> seawater. In other cases, an integument may sufficiently temper the external changes. But by far the most interesting protection is afforded, as in
> man and the higher animals, by the circulating liquids of the organism,
> the blood plasma and the lymph, or as Claude Bernard called them, the
> milieu interieur. In his opinion, which I see no reason to dispute, the existence and constancy of the physicochemical properties of these fluids is
> a necessary condition for the evolution of free and independent life.[1]

The ability of a species to stabilize the composition and volume of its internal environment (extracellular fluid) is critical to the survival and success of the species in its external environment. In higher vertebrates the composition of the extracellular fluid is closely controlled (Table 6.1).

Many invertebrates and some vertebrates can tolerate relatively high electrolyte concentrations in their body fluids. However, August Krogh pointed out a number of years ago that the evolution of higher vertebrates was highly dependent on the regulation of the internal electrolyte concentration at about 1 percent or less. Higher concentrations than this seem to be incompatible with the functioning of the various cells in the body, especially those in the central nervous system. Even more critical than the electrolyte concentration are the tolerance limits for the volume of

[1] From L. J. Henderson, *Blood: A Study in General Physiology*. New Haven: Yale University Press, 1928, p. 20.

TABLE 6.1 Concentrations of Selected Solutes in Normal Human Blood Serum

Solute	Mean	Range
Calcium (meq/liter)	5.2	4.5–5.8
Glucose (mg/100 ml blood)	90	65–110
Magnesium (meq/liter)	0.8	0.4–1.1
Potassium (meq/liter)	4.6	3.6–5.6
Sodium (meq/liter)	140	131–154
Chloride (meq/liter)	105	98–108

Source: From *Hawk's Physiological Chemistry*, 14th ed., by B. L. Oser, ed. Copyright © 1965 by McGraw-Hill Book Company. Used with permission of McGraw-Hill Book Company.

circulating body fluid. This is due to the fact that the various body tissues are dependent on an efficient circulation for the delivery of O_2 and metabolic substrates and the removal of metabolic products. The brain again seems to be the most sensitive area, since it is almost absolutely dependent on an adequate supply of oxygen, and CNS function ceases rapidly during circulatory insufficiency. However, animals differ markedly in their tolerance to changes in circulating volume, and it is difficult to set a lower limit on the circulating fluid volume that is compatible with life.

Feedback loops operate to regulate both the volume and concentration of body fluids of vertebrates (Figure 6.1). If the volume of fluids in the body becomes too large, changes are brought about that lead to increase in the excretion of water and in some cases solute. The opposite occurs during volume depletion. Similarly, if the concentration of solutes in the body becomes excessive, there is an increase in the excretion of solutes and vice versa during solute depletion. Most of the feedback systems regulating salt and water balance are under hormonal control. This subject is discussed in detail in Chapter 9.

1.2 Osmoticity and Tonicity

As discussed in Chapter 2, osmosis is the net movement of water across a semipermeable membrane due to a difference in solute concentration on the two sides of the membrane. The solution on the side of the membrane containing the lower concentration is said to be *hypoosmotic* to the solution on the more concentrated side. The latter solution is termed *hyperosmotic* to the solution on the dilute side. If the solutions on the two sides of the membrane come into osmotic equilibrium, they are said to be *isosmotic*; the osmotic concentrations of the two solutions are equal. When cells (or tissues and even organisms) are placed in solutions that are more concentrated or dilute than they are, water tends to move across the cells,

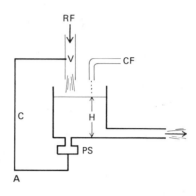

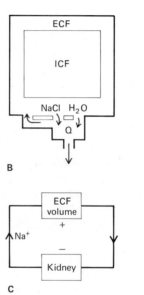

FIGURE 6.1 Operation of feedback loops in the regulation of extracellular fluid volume. (**A**) Model system illustrating control of fluid volume by negative feedback. C—circuit to valve (V) regulating flow of fluid through pipe (RF); CF—pipe with constant flow. H—height of fluid in tank. Pressure sensor (PS) responds to changes in height. (**B**) Regulation of extracellular fluid volume (ECF) by filtration and reabsorption or excretion of salt and water in kidneys (Q). ICF—intracellular fluid volume. (**C**) Operation of feedback loop in the regulation of ECF volume by kidney. Volume of extracellular fluid related to amount of Na^+ retained by kidney.

and they either shrink (*crenation*) or swell (*lysis*). Solutions that cause cells to shrink are said to be *hypertonic* and those in which cells swell are termed *hypotonic*. If the cells undergo no volume change when placed in a solution, the solution is said to be *isotonic*.

Thus there is a clear distinction between *osmoticity* and *tonicity*. The former term refers to the relative concentration of a solution, whereas the latter indicates whether or not it will cause a volume change in cells that it is bathing. There are various combinations possible for the use of these terms. For example, an isosmotic solution containing a solute to which the cell membrane is impermeable is also isotonic; however, an isosmotic solution of solute capable of penetrating the cell membrane will cause the cell to increase in volume and is therefore hypotonic.

1.3 Volume and Composition of Body Fluids

The body fluids may be divided into two main compartments: the intracellular and extracellular fluids (Figure 6.2). The extracellular fluids can be further subdivided into interstitial fluid, that is, the fluid between cells; plasma, the fluid circulating in the blood vessels; and special fluids known as the transcellular fluids. The latter fluids include the digestive, cerebrospinal, ocular, pleural, peritoneal, and synovial fluids. Unlike the extracellular fluid the intracellular fluid cannot be conveniently divided into separate compartments. However, this does not mean that the intracellular fluid is a homogeneous mixture. On the contrary, one look at a microscopic section of almost any cell in the body reveals the heterogeneous nature of the intracellular fluid. In addition, there is variation in the composition of the intracellular fluid from cell to cell in different tissues and even within the same tissue.

In man and other mammals total body water comprises about 50 to 70 percent of total body weight; approximately 20 percent is extracellular and 30 to 40 percent is intracellular. Total body water in submammalian species is generally greater than that in mammals, ranging from 70 to 80

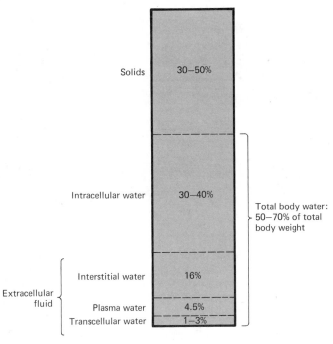

FIGURE 6.2 Body fluid compartments. The percentages shown are representative of mammalian species.

TABLE 6.2   Major Osmotic Constituents in Plasma and Muscle of Several Vertebrates

	Plasma, mM							Muscle, mmole/kg H₂O						
	Na	K	Cl	Mg	Ca	Other	Osmolarity	Na	K	Cl	Mg	Ca	Other	Osmolarity
Seawater	470	10	550	50	10		1070							
Providence tap water	2.3 ×10⁻³	2.5 ×10⁻⁴	3.7 ×10⁻³	2.9 ×10⁻⁴	2.7 ×10⁻³									
Fishes Hagfish (*Myxine glutinosa*)	550	10	565	20	5		1150	130	145	105	10	3	AA 70 TMAO 210	990
Flounder (*Platichthys flesus*)	In SW: 195 / In FW: 155	5 / 5	165 / 115				365 / 305	15 / 10	160 / 155	40 / 30			AA 70 TMAO 30 / AA 45 TMAO 15	
Dogfish (*Squalus acanthias*)	265	4	250	3	7	Urea 350 TMAO 70	1010	30	190				Urea 350 TMAO 270	
Coelacanth (*Latimeria chalumnae*)	180		200	15	4	Urea 355 TMAO 110	1180	30	75	35	15	2	Urea 420 TMAO 290	
Tetrapods Frog (*Rana sp.*)	110	3	90	3	3		230	15	140	1	15	5	AA 10 Urea 2	
Turtle (*Chrysemys picta*)	135	3	85	5	6									
Duck (*Anas platyrhyncus*)	140	3	105	2	2	Urea 5 AA 3								
Rat	150	6	120	2	3	Urea 5	325	15	150	5	2		AA 30 Urea 5	350
Man	150	5	105	3	5		290							

TMAO—trimethylamine oxide; AA—amino acids; SW—seawater; FW—fresh water.

Sources: D. Bellamy and I. Chester Jones, *Comp. Biochem. Physiol.* 3:175, 1961; R. Lange and K. Fugelli, *Comp. Biochem. Physiol.* 15:283, 1965; J. W. Burger, *Sharks, Skates and Rays*, P. W. Gilbert, R. F. Mathewson and D. P. Rall, eds. Baltimore: Johns Hopkins Press, 1967; L. Goldstein, S. C. Hartman, and R. P. Forster, *Comp. Biochem. Physiol.* 21:719, 1967; H. W. Smith, *Biol. Rev.* 11:49, 1936; P. Lutz and J. D. Robertson, *Biol. Bull.* 141:553, 1971; W. T. W. Potts and G. Parry, *Osmotic and Ionic Regulation in Animals.* New York: Macmillan, 1964; D. S. Dittmer, ed., *Blood and Other Body Fluids.* Washington, D.C.: Federation of American Societies of Experimental Biology, 1961; B. L. Oser, ed., *Hawk's Physiological Chemistry*, 14th ed. New York: McGraw-Hill, 1965.

percent of the total body weight. This difference is due mainly to an increase in intracellular water in lower vertebrates. Extracellular water is similar or somewhat lower than in mammals. There does not appear to be any clear-cut relation between the water content of the environment and the volume of the body fluids in related groups of vertebrates living in different environments.

The major solutes in the extracellular fluid of vertebrates are sodium chloride and, in some species, urea and trimethylamine oxide (Table 6.2 and Figure 6.3). Potassium and, in some species, urea and trimethylamine oxide are the main solutes of intracellular fluid. There are a variety of other solutes, such as Ca^{2+}, Mg^{2+}, and amino acids, found in the body fluids of vertebrates. Although they account for a minor part of the total osmolarity of the body fluids, they are important in the functioning of various physiological processes, and their concentrations are generally closely regulated. For example, Ca^{2+} plays an important role in the action of nerve and muscle, and the concentration of this ion in the extracellular fluids of these tissues is critical for their proper function.

The osmolarity of the body fluids of vertebrates ranges from about 220 to 1000 mOsm. With the exception of the hagfish, the vertebrates fall into two groups with regard to osmolarity of the body fluids. The majority of vertebrates have a body fluid osmolarity of less than 350 mOsm. On the other hand, there is a fairly sizable but nevertheless minor group made up

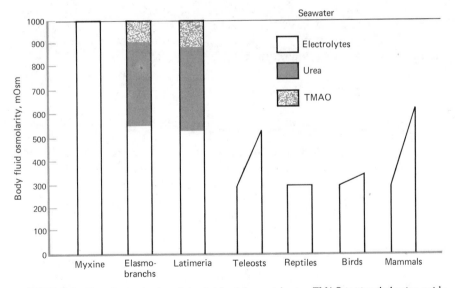

FIGURE 6.3 Osmolarity of extracellular fluids of the vertebrates. TMAO is trimethylamine oxide. (Modified from A. P. M. Lockwood, *Animal Body Fluids and Their Regulation.* Cambridge: Harvard University Press, 1966.)

of the chondrichthyes and the coelacanth (*Latimeria chalumnae*), in which body fluid osmolarity is 1000 mOsm or higher. The elevated osmolarity of this group is due to the presence of relatively high concentrations of organic solutes such as the nitrogenous compounds urea and trimethyl-amine oxide.

1.4 Environmental Problems

The types of osmoregulatory problems that animals face are directly related to the habitat in which they live. Most of the earth is covered with an aquatic environment; approximately two-thirds to three-fourths of the earth's surface is covered with water. Animals living in the sea have plenty of the wrong kind of water ("Water, water everywhere, nor any drop to drink . . .", S. T. Coleridge, *Ancient Mariner*); the osmolarity of seawater is higher than that of most vertebrate body fluids, and seawater contains high concentrations of electrolytes such as Mg^{2+}, Ca^{2+}, and SO_4^{2-}, which tend to upset the ionic balance of the internal body fluids. In contrast animals living in fresh water are continually threatened with being swamped and diluted by their environment. Finally, terrestrial animals are faced with a more or less desiccating environment in which they often have to travel great distances to obtain water.

Although the oceans differ somewhat in different regions of the earth, more marked variations are found in inland bodies of water. The latter can vary from almost "pure" water (less than 1 mOsm/liter) to concentrations several times greater than seawater. Seawater is composed mainly of sodium chloride, magnesium, and sulfate. On the other hand, the major ions in fresh water are calcium and carbonate. The salt lakes in the United States and Canada are similar to the oceans, with Na, Cl, Mg, and SO$_4$ being the chief ions.

1.5 Methods of Osmoregulation

There are five major mechanisms of osmoregulation in vertebrates (Figure 6.4): (1) regulation of water entry by changes in permeability of external surfaces (such as skin and gills) or drinking behavior; (2) regulation of water loss by changes in renal function or permeability of external surfaces; (3) regulation of the rate of renal loss of electrolytes; (4) cutaneous, oral, or branchial acquisition of ions or extrarenal excretion of ions; and (5) variation of the osmotic pressure of body fluids by changes in the concentrations of physiologically inert solutes. Each of these mechanisms will be discussed in this chapter, not as separate entities but rather as they participate in osmoregulation in different vertebrates.

Animals may be classified into two major categories with respect to their osmoregulatory processes (Figure 6.5). *Osmoconformers* are animals

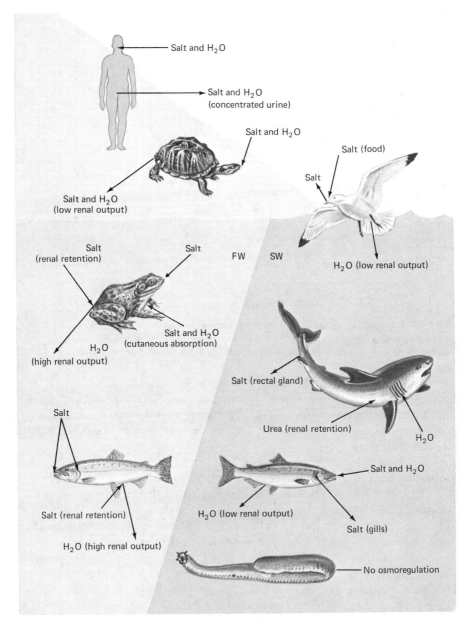

FIGURE 6.4 Mechanisms of osmoregulation in representative vertebrates. FW—freshwater forms; SW—marine forms.

whose body fluid osmotic concentration is similar to that of the environment. *Osmoregulators* are those whose internal osmotic concentration re-

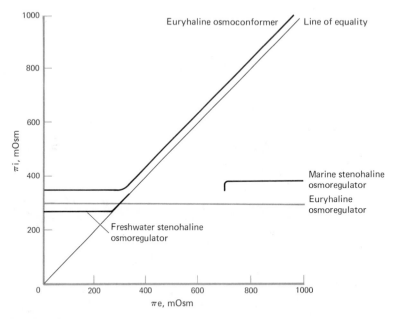

FIGURE 6.5 Patterns of osmoregulation in aquatic organisms. Line of equality illustrates the situation of perfect conformation between osmolarity of internal and external environment. πi—osmolarity of internal body fluids; πe—osmolarity of environment.

mains independent of external osmotic concentration within certain limits. Many invertebrates, especially marine invertebrates, are osmoconformers. The marine hagfish, *Myxine*, is one of the only osmoconforming vertebrates known.

Many fishes can tolerate and live in waters of varying salinity. For example, the eel and salmon spend much of their life in either fresh water or salt water but migrate to a salt-water or freshwater environment to spawn. These fishes are called *euryhaline*. However, there are other fishes that cannot survive more than a modest change in the salinity of their environment on either side. These are referred to as *stenohaline*.

The terms euryhaline, stenohaline, osmoconformer, and osmoregulator are only relative and not meant to imply absolute differences between groups. For example, stenohaline species can tolerate small changes in environmental salinity, and there are limits of salinity changes that euryhaline organisms can withstand. Similarly, the osmoregulatory mechanisms that enable an osmoregulator to maintain salt-water balance in a given salinity range may break down outside that range. Thus when the osmoregulator enters an environment outside of a certain salinity range, the osmolarity of its body fluids changes and it becomes an osmoconformer.

1.6 Osmoregulation in an Isosmotic Environment

Most marine invertebrates, the *Ascidia*, and the vertebrate *Myxine* are isosmotic to their external environment; that is, they have the same osmotic concentration as the seawater in which they live. Although the hagfish has the same total electrolyte concentration as seawater, there are marked differences in the concentrations of individual ions within its body fluids. The concentrations of magnesium, calcium, and sulfate are lower in the blood of *Myxine* than in seawater, and the concentrations of sodium and chloride are somewhat higher. The regulation of divalent ion concentrations, which is a function of the kidney, is probably necessary for proper performance of the central nervous system and muscle. Both systems are very sensitive to the concentrations of these ions in the body fluids (see Chapters 3 and 4).

The elasmobranchs—sharks, skates, rays, and the coelacanth, *Latimeria chalumnae*—are also isosmotic with their environment, but they maintain isosmosity by retention of organic solutes such as urea and trimethylamine oxide (TMAO), in contrast to the invertebrates and hagfish, which are isosmotic with the sea by virtue of having the same electrolyte concentration. Urea and TMAO are both relatively inert, nontoxic, readily diffusible compounds and can be retained in relatively high concentration (up to 0.5 M) by vertebrates with no disturbance of normal physiological processes. Urea and, to some extent, TMAO have the further advantage of being in ready supply to many vertebrates due to the presence of the requisite biochemical pathways for their biosynthesis in the livers of many species.

Although elasmobranchs do not drink seawater, some sodium chloride does enter these fishes via the gills. The excess salt is excreted by either the kidneys or a special gland present at the end of the digestive tract—the *rectal gland*. This unusual gland is found only in elasmobranch fishes and the coelacanths, but glands with similar functions are found in the head region of marine birds and reptiles.

1.7 Osmoregulation in a Hypoosmotic Environment

Invertebrates, fishes, amphibia, and reptiles living in fresh water are hyperosmotic to their environment; that is, the osmotic concentration of their body fluids greatly exceeds that of fresh water. They are all faced with two major osmotic problems. First, since they are more concentrated than their environment, water tends to rush across any exposed body surface and dilutes their body fluids. Second, fresh water is usually quite low in electrolyte content, especially sodium and chloride, and the animals are threatened with electrolyte losses. Three general solutions to these problems have evolved in different animals: (1) a decrease in water permeability, (2) an increase in water excretion, and (3) an increase in salt uptake. One way to cut down on water loss is to decrease the water permeability of cell membranes in tissues exposed to the environment. Indeed, the cell membranes of freshwater protozoans and fish eggs are relatively impermeable to water. Similarly, the skin of aquatic amphibia is less permeable to water than that of terrestrial amphibia; the permeability of the isolated skin of the aquatic frog *Rana septentrionalis* is approximately one-third to one-fourth that of the terrestrial toad, *Bufo americanus*. The skins of two semiaquatic species, *Rana pipiens* and *Hyla versicolor*, have intermediate levels of permeability.

Many animals employ water pumps to eliminate excess water. For example, freshwater protozoans contain contractile vacuoles, which continually pump water out of the cytoplasm. Water pumping in vertebrates is done by the kidney. In freshwater vertebrates water and salts are filtered in the glomerulus of the kidney. The salts are reabsorbed as the urine passes down the renal tubules, and the water is excreted in the form of a dilute urine. There is a good correlation between the rate of water filtration and excretion by the kidney and availability of fresh water in the environment (Figure 6.6). For example, if one examines a closely related group of animals, such as the fishes, it is apparent that those living in fresh water have significantly higher rates of filtration and excretion of water than their counterparts living in the sea. Furthermore, in euryhaline fishes glomerular filtration rate and urine flow change dramatically when the fishes migrate from fresh water to seawater and vice versa. The exact nature of the mechanisms bringing about these changes in kidney function is unknown. However, it is likely that circulatory and hormonal factors are both involved.

Salt pumps have also evolved to make up for ions lost from the body to a dilute environment. Freshwater fishes and the larval forms of many amphibia have gills that can pump sodium and chloride against steep chemical and electrical gradients. In many amphibia the skin functions in a similar manner. The ability of freshwater organisms to absorb salt from the dilute environment in which they live is remarkable. For example, freshwater fishes can absorb Na^+ and Cl^- from water containing less than 1 mM NaCl. This process takes place against steep chemical gradients, since the concentration of Na^+ and Cl^- in the extracellular fluid of these fishes is

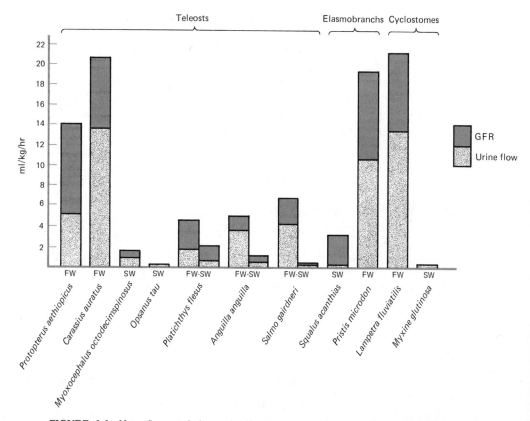

FIGURE 6.6 Urine flows and glomerular filtration rates (GFR) in representative freshwater and marine fishes. FW—fish in fresh water; SW—fish in seawater; FW-SW—fish in either fresh water or seawater. (See Bentley, 1971.)

greater than 100 mM. The work involved in this process requires a significant input of the fish's metabolic energy.

1.8 Osmoregulation in a Hyperosmotic Environment

Fishes, reptiles, mammals, birds, and the few amphibia living in the sea are hypoosmotic to their environment; they have a lower electrolyte concentration than that of the sea. These animals face three major problems. First, they are threatened with desiccation due to the tendency for water to move from them into the more concentrated environment. Second, they are threatened with ionic imbalance since magnesium, sulfate, and calcium, which are more concentrated in the environment than in their body fluids, are continually entering the body and tending to upset the ionic balance of the internal environment. Third, excess salt is continually entering their bodies either by diffusion across exposed epithelial membranes (for example, gills) or by ingestion in the diet. Different solutions have evolved in the various groups of animals living in the sea to combat these problems. The teleost fishes maintain water balance by drinking seawater, absorbing it through the gut, and desalinizing the seawater in the gills. Fishes living in fresh water drink little or no water (Figure 6.7).

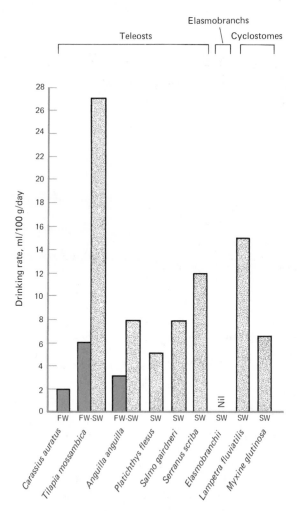

FIGURE 6.7   Drinking rates in representative freshwater and marine fishes. Abbreviations are similar to those used in Figure 6.6. (See Bentley, 1971.)

However, marine teleosts continually swallow water, which is absorbed in the gut. The studies of Homer Smith and those of more recent investigators have shown that approximately 70 to 80 percent of the water imbibed by these fishes is absorbed by the gut. Most of the sodium chloride present in the ingested seawater is also absorbed. The excess sodium chloride is removed from the bloodstream by the gills, leaving "osmotically free" water behind. Most of the divalent ions present in the ingested seawater are not absorbed, and the small amounts that do pass across the intestinal wall are excreted by specific transport systems in the kidney.

The elasmobranchs have evolved a different solution to the problem of living in the sea. They drink little or no seawater but instead retain innocuous nitrogenous end products such as urea and TMAO in their body fluids. The concentrations of these substances are sufficient to raise the osmotic pressure of the fish's body fluids to a point equal to, or slightly higher than, that of the surrounding seawater. Thus these fishes are not threatened with water loss and have no need to drink seawater. Other vertebrates have used this same trick to solve their osmotic problems. The coelacanth (*Latimeria chalumnae*), once thought to be extinct and a relative of the crossopterygian fish that supposedly gave rise to the tetrapods, also retains urea and TMAO to maintain osmotic balance in the sea. Similarly, the crab-eating frog (*Rana cancrivora*) of Southeast Asia retains large amounts of urea in its body fluids when it enters a saline environment.

In addition to gills, other organs have evolved for elimination of salt from the body fluids of marine vertebrates. The elasmobranchs, as previously discussed, have a salt-secreting gland located in the wall of the large intestine (rectal gland) that secretes a concentrated sodium chloride solution. Marine birds and reptiles have nasal or orbital glands that function in a similar manner to eliminate excess sodium chloride from the body. The secretions of all these glands are hyperosmotic to seawater.

One way of preventing water loss is to cut down on the avenues of water excretion. A main route of water loss in vertebrates is the urine. As indicated previously, if one compares urine flow rates in marine fishes to these same rates in freshwater fishes, it is apparent that the flow rate in marine forms is a great deal less. The same phenomenon may be seen in euryhaline fishes transferred from a freshwater to a saline environment. A marked reduction in urine flow ensues when the fishes are transferred from a hypo- to a hyperosmotic environment (Figure 6.6).

Mammals living in the sea—for example, whales, seals, and dolphins—do not have a special gland for eliminating salts from the body to help them maintain solute-water balance in the sea. Instead they obtain free water from the metabolism of food and the production of a concentrated urine by the kidney. For example, ingestion of 1 kg of herring, which has a total body water of about 75 percent and body fluid osmolarity of approximately 350 mOsm, will yield 600 ml of free water to the seal when the seal's kidneys concentrate these fluids approximately five-fold during the production of a concentrated urine. Additional free water will be generated during metabolism of the fat, protein, and carbohydrate in the herring.

1.9   Osmoregulation in a Terrestrial Environment

Representative species from most of the major phyla inhabit various terrestrial parts of the earth. The osmoregulatory problem that most of the animals living on land face is water loss. Water availability on land ranges from excessive in such places as rain forests to almost complete absence in the deserts. Water loss from body surfaces is related directly to the temperature and inversely to the humidity in which the animal is living. Various solutions to the problem of water loss have evolved. Decreased permeability of body surfaces is one of the major ways in which water loss is prevented. The waxy cuticle of insects and thick skin of toads are two examples of this method of combating water loss. Another means of reducing water loss is reduction in urine flow. For example, urine flow rates in desert-dwelling species are considerably lower than they are in related species living in areas of greater water availability. Mammals living in the desert are able to reduce urine flow rate by the production of a highly concentrated urine. For example, many of the desert rodents can produce urine that is 15 to 20 times more concentrated than plasma. In contrast the ordinary rat can concentrate its urine only 5 to 6 times greater than plasma osmolarity.

In addition to physiological adaptations reducing water loss, desert-dwelling animals have evolved behavioral adaptations, such as burrowing and nocturnal activity, that allow them to avoid the heat and aridity of the desert. Food selection may also be used to advantage. For example, selection of plant seeds with a high fat and low protein content is advantageous since the oxidation of 1 g of fat yields 1.07 g of water, whereas 1 g of carbohydrate yields only 0.6 g of water, and 1 g of protein yields only 0.4 g of water.

2 • SPECIAL OSMOREGULATORY SYSTEMS

2.1   Salt Secretion by Glands and Gills

Animals living in an environment of restricted freshwater availability—for example, the sea and desert—must possess mechanisms for excreting excess solutes and conserving water in order to survive. Some of the more interesting adaptations of this type are found in marine animals. Marine teleost fishes regulate salt-water balance by branchial (gill) secretion and marine elasmobranchs by a salt-secreting rectal gland. Marine birds and reptiles use salt-secreting glands located in the head region to excrete excess solute. Some terrestrial birds and reptiles also possess salt-secreting glands, but their function is not clear. Marine mammals do not have these glands but are able to excrete salt in a highly concentrated urine.

Salt Gland of Marine Birds. The salt gland of marine birds is located above the orbit of the eye and opens into the nasal cavity (Figure 6.8). The gland is composed of several lobes containing central canals that are encircled by branching secretory tubules having a rich blood supply. The cells of these tubules have deep basal infoldings lined with mitochondria—a histological picture similar to that found in the cells of many other structures involved in the transepithelial transport of sodium chloride. The nasal gland is capable of secreting relatively high concentrations of sodium chloride: 1000 to 2000 mOsm. The osmolarity of seawater is around 1000 mOsm and that of the birds' plasma about 280 mOsm. Thus these birds can generate "fresh water" by desalinization of fluids taken in

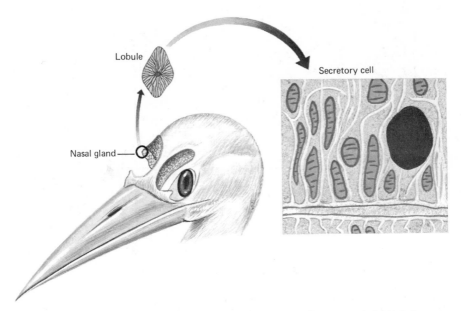

FIGURE 6.8 "Nasal" gland (salt-secreting) of marine birds. Gland composed of lobules containing epithelial cells that produce a secretion hyperosmotic to seawater. (Modified from K. Schmidt-Nielsen, *Circulation* 21:955, 1960. By permission of the American Heart Association, Inc.)

their diet and can maintain salt-water balance without drinking fresh water.

Fluid production by the gland is not continuous; rather, it is responsive to the salt intake of the bird. After ingestion of saline or seawater, fluid output rises sharply (Figure 6.9). The concentration of sodium chloride remains about the same in nasal secretion at varying flow rates, and salt excretion is related to rate of fluid production, not the concentration of salt in the fluid. The nasal gland is much more effective in excreting sodium chloride than the kidneys. Schmidt-Nielsen found that a 1.4-kg seagull given 134 ml of seawater containing 54 mmoles of NaCl excreted approximately 44 mmoles of NaCl via the nasal gland and 4 mmoles of NaCl via the kidneys during the following 3 hr. Thus the nasal glands eliminated over 80 percent of the ingested NaCl, whereas the kidneys excreted less than 10 percent.

The precise manner in which fluid production is controlled is still uncertain. Fluid secretion can be provoked by administration of either sodium chloride or sucrose to the birds so that a rise in osmolarity rather than salt concentration of the birds' body fluids appears to be the effective stimulus. The nasal gland is innervated by postganglionic fibers from the parasympathetic nervous system. Electrical stimulation of these nerves or injection of the parasympathetic neurotransmitter, acetylcholine, into

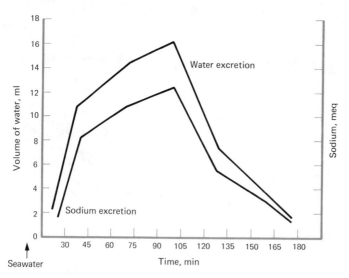

FIGURE 6.9 Nasal excretion of sodium and water by a black-backed gull following ingestion of seawater. Arrow indicates time of administration of 134 ml seawater to a 1.4-kg gull. (Data from K. Schmidt-Nielsen, *Circulation* 21:955, 1960. By permission of the American Heart Association, Inc.)

vessels going to the gland both cause secretion. However, it is not clear whether the neuronal pathway is directly involved in the response of the gland to salt loading. The adrenal glands seem to be involved somehow in the salt response since hemiadrenalectomy diminishes the response and total adrenalectomy obliterates it.

The postnatal development of the nasal gland is dramatically affected by the level of salt in the bird's diet. In an interesting study on the development of salt glands in newly hatched ducks the level of salt in the birds' drinking water was shown to have a significant effect on both the size and the differentiation of the salt gland. The glands of ducklings that drank water to which salt had been added were approximately three times larger than the glands of birds that drank fresh water. The increase in size of the gland was due to increase in both size and number of secretory cells in the glands of the salt-treated birds. The size of the gland may also be increased in adult ducks by adding salt to their drinking water. Within a few days after transferring adult ducks from fresh water to saline, the size of the nasal gland increased about threefold. The birds given saline to drink were able to secrete sodium chloride, in response to a test dose, much faster than birds given fresh water (Figure 6.10). The enhanced rate of

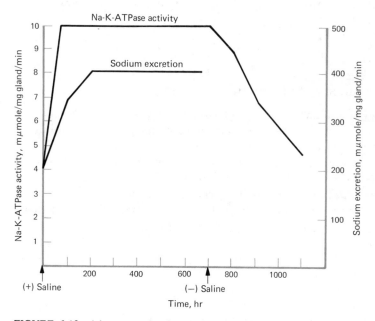

FIGURE 6.10 Adaptation of sodium-potassium ATPase activity and sodium excretion in nasal glands of ducks maintained on saline. (+) saline and (−) saline indicate timepoints at which NaCl was added and removed from drinking water. (Data from G. L. Fletcher, I. M. Stainer, and W. N. Holmes, *J. Exp. Biol.* 47:375–391, 1967.)

sodium chloride excretion may be related to the concomitant increase in the activity of Na^+, K^+-activated transport ATPase (see Chapter 2, Sections 3 and 12) in the glands of salt-treated birds.

Salt Gland of Elasmobranch Fishes.   Although elasmobranchs do not drink seawater, some salt does enter these fishes either by diffusion across the gills or via the diet. This excess salt is excreted via the kidneys and the rectal gland. The latter organ is a fingerlike (digitiform) outgrowth at the end of the large intestine. It is most highly developed in the sharks, intermediate in the skates, and rather small in rays.

The secretion of the rectal gland is isosmotic with the blood (approximately 1000 mOsm) but consists almost solely of sodium chloride (450 mM). Small quantities of other ions (K^+, Ca^{2+}, Mg^{2+}) and urea are also secreted, but these solutes make up no more than 10 percent or so of the total osmolarity.

In contrast to urine formation, which is more or less continuous, fluid production by the rectal gland is sporadic. The reason for the intermittent nature of this fluid flow is not understood. Since rectal gland secretion can be provoked by experimental maneuvers that expand the size of the extracellular compartment, the gland may be involved in the regulation of extracellular volume and respond to changes in the size (and perhaps osmolarity) of this compartment.

2.2   Salt Secretion by Branchial Epithelium

As mentioned previously, marine teleosts maintain osmotic balance by drinking seawater and excreting sodium chloride (the main solute in seawater) via the gills. Other ions, such as Mg^{2+}, Ca^{2+} and SO_4^{2-}, which may enter the body via the gut, are excreted by the kidneys. The gills, therefore, play a major role in maintaining salt-water balance in these fishes. The teleost gill is composed of a series of filaments branching off a central supporting cartilaginous bar (Figure 6.11). The filaments are made up of several different cell types: supporting ("pillar") cells, mucous cells, and respiratory epithelial cells. Keys and Willmer (1932) described large granular-staining cells in the branchial epithelium of eels adapted to seawater, which they labeled "chloride" cells and assigned the function of secreting sodium chloride. Although the existence of these cells has been debated over the years, recent studies have confirmed the original hypothesis of Keys and Willmer that these mitochondria-rich cells function in the transport of sodium chloride across the gill filament. One of the more convincing studies has been done by Philpott, who employed the technique of electron diffraction to prove the existence of high concentrations of chloride in the putative chloride-secreting cells of the euryhaline

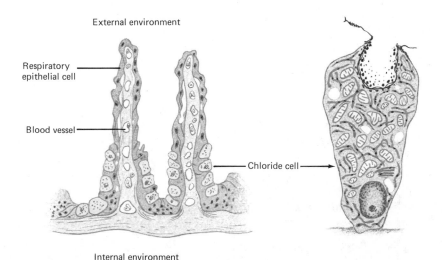

External environment

Respiratory epithelial cell

Blood vessel

Chloride cell

Internal environment

FIGURE 6.11 Gill leaflet of marine teleost containing the "chloride cells." Note apical "pit" of chloride cell (thought to be site of salt secretion). (Modified from A. Keys and E. N. Willmer, *J. Physiol.* (*London*) 76:368, 1932. By permission of Cambridge University Press.)

fish *Fundulus*. Gill filaments from seawater-adapted *Fundulus* were fixed in silver acetate–osmium tetroxide to localize chloride. The chloride cells readily took up the silver stain; the apical portion of these cells was the part stained most intensely, and the stain seemed to surround a cavity that opened onto the surface of the cell. The electron diffraction pattern of the granules stained with silver was identical to that of pure silver chloride.

Many physiological studies have been carried out on salt secretion by the gills of marine and euryhaline teleosts. Recent studies have shown that up to 50 percent of injected radioactive labeled sodium is excreted by gills of marine teleosts in 1 hr (Figure 6.12). This amazingly high rate of sodium "turnover" is due in large part to an exchange of internal sodium with that of the fish's environment. This indicates that the gill is highly permeable to sodium in both directions and that salt balance is achieved by adjusting the rates of salt influx (entry) and outflux (exit). When euryhaline fishes are transferred from seawater to fresh water, there is an immediate reduction in sodium outflux followed by a slower further decline. The result is a large reduction in total exchange of sodium as well as net secretion. Thus the immediate response of transfer to fresh water appears to be a decrease in permeability of gills to sodium, preventing loss of internal sodium. The secondary response is a slower decline in secretion of sodium by the gills.

In contrast to the situation with sodium chloride, the gills of marine teleosts are less permeable to water than those of freshwater teleosts. Simi-

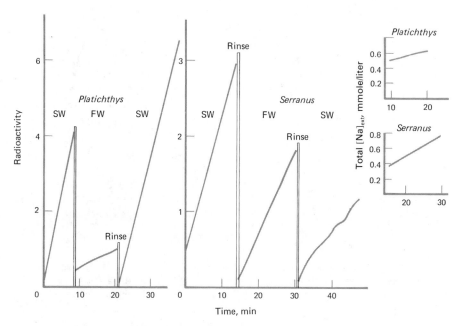

FIGURE 6.12 Relative appearance rates of ^{24}Na in seawater-adapted *Platichthys* and *Serranus* transferred from seawater (SW) to fresh water (FW) and back to seawater. Inserts show appearance of total external sodium concentration while fish were in fresh water. (From J. Maetz, *Phil. Trans. Roy. Soc.* 262:234, 1971. With permission of The Royal Society.)

larly, a reduction in osmotic water flow across the gills occurs when euryhaline teleosts are transferred from fresh water to seawater, despite the fact that the osmotic gradient between the fish and its environment is greater in seawater (approximately 600 mOsm) than in fresh water (approximately 250 mOsm). The mechanism by which branchial permeability is reduced in seawater is not understood. However, this process may be under endocrine control, since the presence or absence of adrenal cortical hormones and prolactin are known to have marked effects on both water and salt permeability of the teleost gill.

It has recently been shown that external potassium is essential for maintaining sodium balance in marine teleosts. Potassium in seawater is thought to be exchanged for a portion of the sodium transported from blood to seawater, and it is this portion of the sodium efflux that may be responsible for maintaining the concentration gradient of sodium across the gill. The Na$^+$-K$^+$ exchange suggests that Na$^+$,K$^+$-activated adenosine triphosphatase (transport ATPase) may play a role in gill ion exchange. Indeed, when the killifish, *Fundulus*, is transferred from fresh water to seawater, there is a significant increase in branchial Na$^+$,K$^+$-ATPase. It is not known whether the enzyme activity increases in the same cells as those present in fresh water, or whether new cells develop in seawater that have a higher level of Na^+,K^+-ATPase and therefore transport capacity. The possibility of the latter is suggested by experiments on salmon in which there was a stimulation in turnover of epithelial cells of the basal region of gill lamellae in fishes transferred from fresh water to seawater. Thus new cells with increased transport ability may appear in the gill when euryhaline fishes migrate from fresh water to seawater or vice versa. Although most of the work on salt transport by the fish gill has concentrated on the mechanism of Na^+ transport by the branchial epithelium, recent studies indicate that Cl^- is transported actively across the gill of marine teleost. Electrical measurements have shown that the body fluids of marine teleosts are positive with respect to the external environment. This evidence, taken together with the fact that Cl^- concentration in seawater is about two times that of the fish's extracellular fluid, indicates that Cl^- is being pumped against an electrochemical gradient during transport from blood to seawater across the gills. Much more work needs to be done on the mechanism of salt transport by the fish gill before we can begin to understand the exact mechanisms by which sodium and chloride are handled by this organ.

2.3 Salt Transport across Amphibian Skin and Bladder

As mentioned previously, animals living in fresh water face the problem of continuous expansion and dilution of their body fluids by their environment. Most amphibia live or return to breed in fresh water and must possess mechanisms for maintaining their internal electrolyte concentration well above that of a hypoosmotic environment. The kidneys of amphibia can produce copious quantities of very dilute urine, but some salt loss is unavoidable, and the salt must be replaced. Other than dietary sources, the major means of salt uptake in amphibia is via the skin. In 1935 Huf showed that an isolated sac of frog skin bathed with Ringer's saline solution on both sides could transport Cl^- from outside to inside against a concentration gradient. The process was inhibited by cyanide and stimulated by lactate and pyruvate, showing its dependence on metabolism —that is, active transport. Later Krogh showed that intact frogs deprived of salt by being kept in distilled water could absorb salt (actually Cl^-) from very dilute solutions. Although sodium and chloride are absorbed together, their absorption is not absolutely linked since Cl^- can be absorbed from NH_4Cl, KCl, and $CaCl_2$ solutions, and sodium can be absorbed from $NaHCO_3$ and Na_2SO_4 solutions. It seems that independent systems are responsible for the transport of Na^+ and Cl^- across the skin.

Chloride absorption in the absence of sodium takes place by a process that exchanges HCO_3^- on the inside of the skin for chloride on the outside. Similarly, sodium absorption in the absence of chloride is thought to occur via an exchange of external sodium for internal ammonium or hydrogen ions.

Tadpoles are unable to transport salt across the skin before metamorphosis. They depend on uptake of salt by the gill and gut (diet) for their salt needs, and in this sense they resemble freshwater fish. During metamorphosis, however, when the gills are lost, the skin acquires the ability to transport sodium against a concentration gradient.

The ease with which the frog skin may be removed and maintained in vitro has made this tissue a popular tool for the study of transepithelial transport. The Danish physiologist Hans Ussing pioneered the use of radioisotopes in the study of the transepithelial processes operating to transport sodium chloride across the frog skin. The kinetics of transport can be studied with the aid of radioactive isotopes of sodium, ^{22}Na and ^{24}Na. In these experiments the isolated skin is mounted and sealed between two half-chambers containing amphibian Ringer's solution (Figure 6.13). Electrodes are placed in each chamber so that the electric potential between the two chambers can be recorded. Movement of sodium in

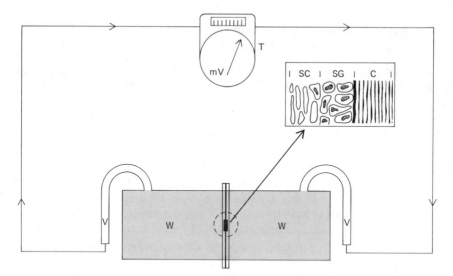

FIGURE 6.13 Apparatus used to measure ion transport and electric potential across frog skin. W—chambers containing Ringer's solution on each side of skin; V—electrodes leading to millivoltmeter (T) used to measure potential difference across skin. Frogskin (shown in insert) is composed of three layers: an outer layer, stratum corneum (SC); a middle layer, stratum germinativum (SG); and an inner layer, corium (C). (Modified from H. H. Ussing and K. Zerahn, *Acta Physiol. Scand.* 23:110–127, 1951.)

each direction across the skin, called *unidirectional fluxes,* can be followed by measuring the rate of movement of radioactive sodium across the skin. This technique can be used to determine whether the movement of electrolyte across the skin is due to passive forces (down the electrochemical gradient) or is the result of active transport (against an electrochemical gradient). (See Chapter 2 for a discussion of active and passive transport.) Using this technique Ussing found that the movement of chloride across the Cu^{2+}-treated frog skin was passive and followed the electrochemical gradient across the skin. However, the transport of sodium from the outside to inside the skin was against the electrochemical gradient and could not have been due to passive forces, that is, diffusion. The fact that this sodium flux is inhibited by metabolic poisons such as dinitrophenol suggests that there is an active transport system responsible for the movement of sodium from outside to inside the skin.

The origin of the potential difference across the frog skin is thought to be related to the permeability properties of the two sides of the skin to sodium and potassium. The outside surface of the skin behaves like a sodium battery; the potential difference across the skin is directly related to the concentration difference of Na^+ across the outside surface. The inside surface behaves like a potassium battery; the potential difference across the skin is directly related to the concentration difference of K^+ across the inside surface. Thus the skin is thought to behave like two batteries, one a sodium and the other a potassium, in series.

A number of years ago Ussing and Koefoed-Johnsen proposed a theoretical model to account for the movement of NaCl across the frog skin. In this model the outside surface of the skin was assumed to be selectively and passively permeable to Na^+ and permeable to Cl^- in a nonselective way. The inside surface of the skin was assumed to be permeable to K^+ but not to Na^+. Sodium entering the skin from the outside surface down an electrochemical gradient was transported across the inside surface in exchange for potassium by a "pump" located at the inside surface of the skin. The pump was depicted as a Na^+-K^+ exchange pump that did not contribute to the potential difference across the skin. Rather, the potential was thought to be due to the net movement of sodium ions from the solution bathing the outside to that bathing the inside of the skin. Net movement of chloride across the skin could then be explained by diffusion of the anion down the electrical gradient created by the movement of sodium ions.

The nature of the "sodium pump" in frog skin is still not clear. Three possible mechanisms have been suggested (Figure 6.14): (1) a Na^+-K^+ exchange pump, (2) an electrogenic (nonneutral or charge-producing) sodium pump, (3) separate but closely linked pumps for actively transporting sodium and chloride. Evidence in support of the Na^+-K^+

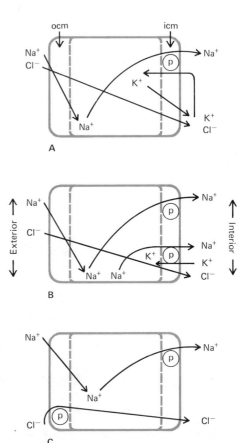

FIGURE 6.14 Hypothetical models for the transport of sodium chloride across a frogskin. Three possible mechanisms for transporting sodium chloride across the skin are: (**A**) a Na⁺-K⁺ exchange pump; (**B**) an exchange pump plus an electrogenic (charge-producing) Na⁺ pump; (**C**) separate but closely linked pumps for actively transporting sodium and chloride. Models are simplified by assuming that frog skin behaves as a single epithelial layer. ocm and icm—outer and inner cell membrane; p—pump involved in active transport of ions.

pump is the dependence of sodium transport on the presence of K⁺ on the inside of the skin and inhibition of sodium transport by ouabain, a specific inhibitor of Na⁺,K⁺-activated ATPase and sodium transport. However, the ouabain effect is not specific since Cl⁻ and H₂O movement across the skin is also inhibited, and the dependence of Na⁺ transport on the presence of K⁺ may be due to the effect of the latter on the permeability of the skin to Na⁺ rather than an exchange process. Thus the second possibility, that of an electrogenic pump, must be considered; that sodium is pumped across the inside of the skin without exchange for K⁺ moving in the opposite direction. It is also possible that there is more Na⁺ pumped out of the cell than there is K⁺ pumped in—a loosely coupled Na⁺-K⁺ exchange pump.

The third possibility, that both sodium and chloride are pumped together by independent processes, is supported by the experiments of

Krogh (described earlier) showing that both Na^+ and Cl^- could be absorbed independent of each other across the frog skin in vivo. These earlier experiments were extended by Barker-Jorgensen, who measured sodium and chloride fluxes and potential differences across the frog skin in vivo. In contrast to the results obtained with the isolated frog skin, unequivocal evidence was obtained showing that Cl^- as well as Na^+ was actively transported across the frog skin in vivo. It is possible that all three pumps exist and operate in vivo. Some sodium may be pumped in exchange for potassium while an additional fraction is transported without exchange. Similarly, a part of the chloride crossing the skin may move passively down an electrochemical gradient while the rest is actively pumped. Indeed, it has recently been shown that Cl^- may be actively transported across the isolated frog skin if the Cl^- concentration on the outside of the skin is kept relatively low compared to that on the inside of the skin. However, the net flux of the Cl^- is low compared to passive flux of Cl^- observed when Cl^- concentration on the outside is high and equal to that inside the skin.

The frog skin has been a very useful model for study of the transport of ions across epithelial cells. The information obtained from studies on the skin has been applied to other organs, such as the kidney, which are less amenable to in vitro study of electrophysiological processes. Thus many of the models and experiments of transepithelial transport of electrolytes across the renal tubules are based on results obtained previously on the frog skin.

The urinary bladders of amphibia are reservoirs for the retention of salt and water by these animals. Urine formed in the kidneys passes to the bladder, where it is processed before being eliminated from the body. In aquatic amphibia (such as frogs) the bladder aids in osmoregulation by reabsorbing most of the sodium chloride that has escaped reabsorption in the renal tubules, thus rendering the final urine markedly hypoosmotic. In terrestrial amphibia (such as toads) the bladder serves as a water reservoir as well as a salt-conserving organ. The urinary bladders of many terrestrial amphibia can hold water equivalent to 30 to 50 percent of the animal's body weight.

The reabsorption of both salt and water from the bladder is under hormonal control. When the animals are well hydrated, mainly salt is reabsorbed, accompanied by little or no water reabsorption. Salt reabsorption by the bladder is regulated by the adrenal glands. Hormones secreted by the adrenals facilitate the uptake of sodium chloride by the bladder.

The urinary bladder of amphibia is a highly distensible sac, consisting of a single layer of epithelial cells along with some loose connective tissue and smooth muscle (Figure 6.15). This simple structure and amenability of the bladder to in vitro experimentation has given the physiologist

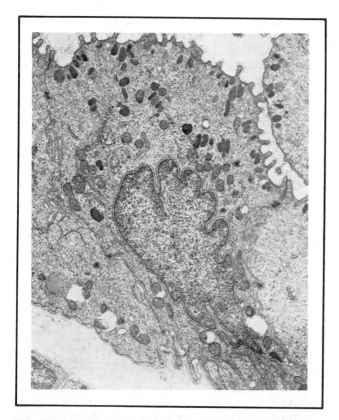

FIGURE 6.15   Electron micrograph of toad bladder. The large cell in the center of micrograph is a granular cell, the most predominant cell type in the bladder mucosal epithelium. It extends from the urinary space (at the top) to the basement membrane. (Courtesy of J. W. Mills)

an excellent opportunity to study the processes involved in the transport of salt and water by this organ.

The toad bladder, like the frog skin, is capable of transporting Na^+ against an electrochemical gradient. Studies on the isolated toad bladder bathed in amphibian Ringer's solution have shown that a potential difference of 50 mV exists across the bladder wall; the mucosal surface is negative with respect to the serosal surface. Na^+ is readily transported from the mucosal to serosal surface against this electrical gradient, indicating that the transport process is active. The current (applied from an external source) needed to reduce the potential difference across the bladder to zero ("short-circuit current") is equal to the electrical equivalents of sodium ion transported across the bladder under these conditions, indicating that the potential difference across the bladder wall is due to the transport of Na^+.

If an osmotic gradient is established across the bladder wall by placing a dilute amphibian Ringer's solution on the mucosal side and normal amphibian Ringer's solution on the serosal side, sodium chloride is reabsorbed from the bladder urine with little or no water movement. If antidiuretic hormone is added to the serosal side, both osmotic movement of water and sodium transport are greatly accelerated.

3 • VERTEBRATE KIDNEY

3.1 Evolutionary Considerations

The vertebrate kidney has had an interesting evolutionary history (Figure 6.16). Primitive vertebrates probably evolved in the sea, possessed little ability to osmoregulate, and may have resembled the present-day myxinoid fishes—for example, the hagfish. The hagfish has a primitive-type kidney with a series of large, oval glomeruli running down each side of the kidney and emptying into collecting ducts (*archinephric ducts*) that drain directly to the outside surface. The main function of this kidney is probably the excretion of divalent ions (Ca^{2+}, Mg^{2+}, and SO_4^{2-}) and some metabolic end products. With the evolution of the freshwater fishes there was a need to filter and excrete large quantities of water. We see in present-day freshwater fishes large, well-developed glomeruli, which filter the blood plasma at relatively high rates, and the presence of the two nephron segments, the proximal and distal tubules, between the glomerulus and the collecting ducts. These two segments are responsible for reabsorption of the bulk of the salt, as well as glucose, amino acids, and other nutrients filtered at the glomerulus. Marine teleosts, which do not have the same need to filter water and reabsorb salt to the same degree as their freshwater counterparts, have glomeruli greatly reduced in size (and even absent in some forms) and do not have distal tubules. The main function of the kidney of marine teleosts is the excretion of divalent ions. The kidneys of elasmobranchs are quite elaborate, with well-developed glomeruli and nephrons differentiated into several segments. The complex organization of the elasmobranch kidney may be related to the large quantities of urea and TMAO that these fishes filter and reabsorb from the glomerular filtrate. The amphibian kidney is basically similar to that of the freshwater teleost, which is consistent with the basic similarity in function that the two kidneys perform in these freshwater organisms.

When the terrestrial animals evolved from the ancestral amphibians and left their aquatic environment, the kidney had to be modified to re-

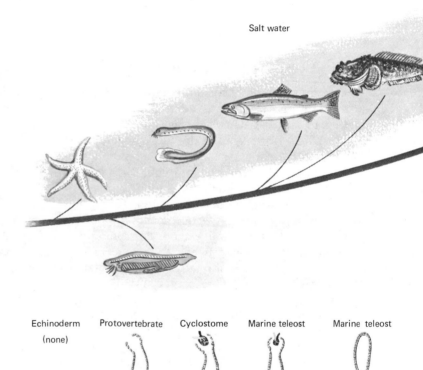

FIGURE 6.16 Diagrammatic representation of evolutionary history of the vertebrate kidney. Nephrons of representative vertebrates (as well as that of the hypothetical protovertebrate) are shown on bottom line below corresponding species. Note progressive increase in segmentation of nephron during evolution. Shaded areas indicate marine environment. (Modified from H. W. Smith, *From Fish to Philosopher,* by permission of Little, Brown and Co. Copyright © 1953 by Homer W. Smith.)

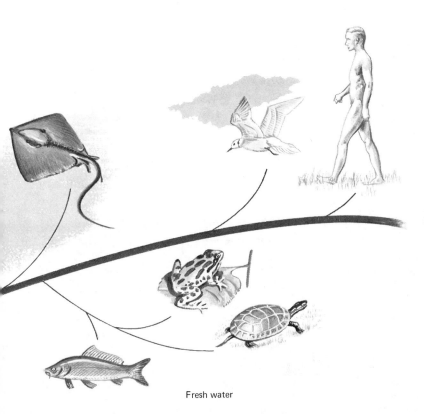

Fresh water

asmabranch   Freshwater teleost   Amphibian   Reptile   Bird   Mammal

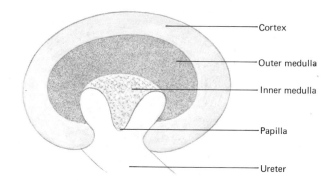

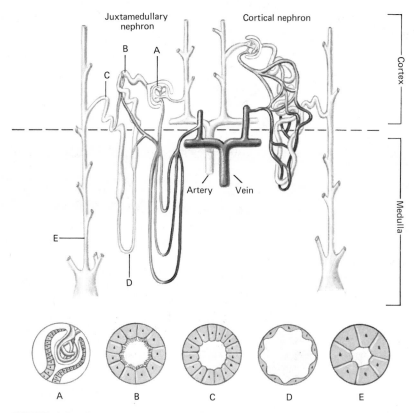

FIGURE 6.17 Diagrammatic representation of a mammalian kidney. Top section of diagram shows a cross section of rabbit kidney. The middle section illustrates the nephron segments found in the cortex and medulla. The bottom section depicts the histological appearance of the different nephron segments. Preglomerular arteriolar, postglomerular arteriolar, and venous blood are shown in white, gray, and black, respectively. (Modified from H. W. Smith, *The Kidney.* New York: Oxford University Press, 1951.)

duce water loss through the urine. However, in animals on land the kidney had to assume almost sole responsibility for the excretion of nitrogenous wastes, which were previously able to diffuse across the gills and skin of animals living in an aquatic environment. For birds and reptiles the solution to this problem was to reduce the size of the glomeruli and excrete nitrogen waste in the form of uric acid by tubular secretion. The mammals, however, retained the amphibian traits both of urea excretion and relatively high glomerular filtration rates. A unique segment of the kidney, the loop of Henle (Figure 6.17), evolved in mammals and enabled them to produce a urine more concentrated than plasma and thereby avoid water loss.

3.2 Processes of Urine Formation

Three major processes are involved in urine formation by the vertebrate kidney: glomerular filtration, tubular reabsorption, and tubular secretion. The glomerulus is a capillary bed specialized for high filtration rates. The blood pressure in the glomerular capillaries is higher than in any other capillary bed in the body. Furthermore, the filtration area per gram of tissue in the glomerular capillaries is about 50 times that in other tissues, such as skeletal muscle. The high blood pressure and large filtration area ensure a rapid rate of filtration of plasma in the glomerulus. The elegant micropuncture studies done by A. N. Richards and his colleagues showed that the fluid filtered at the glomerulus is an ultrafiltrate of plasma. In these studies small samples of fluid (less than 1 μliter) were withdrawn from Bowman's capsule by use of micropipettes (1 μ or less in diameter) and analyzed by microanalytical methods for electrolytes, organic solutes, and protein. The average concentration of all solutes, with the exception of protein (which was present in negligible concentrations), was the same in the filtrate as in plasma. Thus urine formation begins with the mechanical separation of a portion of the water and nonprotein solutes from the plasma circulating through the glomeruli.

Since urine formation begins with filtration of plasma in the glomerulus, it is desirable to have a method for measuring glomerular filtration rate for quantitative studies of renal function. Smith and his colleagues devised a technique for measuring glomerular filtration rate in intact animals, including man. This technique takes advantage of the fact that certain high-molecular-weight substances such as the polyfructoside, inulin, are filtered in the glomerulus and neither reabsorbed, secreted, nor metabolized in the renal tubules. Thus any inulin excreted in the urine could have gained access only by glomerular filtration, and the total amount of inulin excreted in the urine is equal to the glomerular filtration rate times the plasma concentration. Since all the inulin present in the urine is derived from plasma filtered in the glomeruli, a simple equation can be used to calculate glomerular filtration rate if one knows the concentration of inulin in the urine (U_{In}), the urine flow rate ($\dot{V}$) and the plasma inulin concentration (P_{In}):

$$\text{GFR} = \frac{U_{In}\dot{V}}{P_{In}}$$

For example, if 1 ml of plasma contains 1.0 mg of inulin and if 120 ml of plasma are filtered through the glomeruli each minute, then 120 mg of inulin will be delivered to the tubules and excreted in the urine each minute. Since most of the water filtered at the glomerulus is reabsorbed as the urine flows down the tubules, the 120 mg of inulin will be excreted in about 1.0 ml of urine, and the urinary concentration of inulin will be 120 mg/ml. From these data and the previous equation we can calculate that the glomerular filtration rate in this case is 120 ml/min; that is,

$$\text{GFR} = \frac{U_{In}\dot{V}}{P_{In}} = \frac{120 \text{ mg/ml} \times 1.0 \text{ ml/min}}{1.0 \text{ mg/ml}} = 120 \text{ ml/min}$$

This means of measuring glomerular filtration rate is called the *inulin clearance technique* since it measures the amount of plasma "cleared" of inulin per unit time. This technique can also be used to determine the fate of various other endogenous and exogenous substances in the renal tubules. Substances both filtered in the glomeruli and secreted by the renal tubules are removed from plasma more efficiently than inulin and have plasma clearance rates greater than that of inulin. In contrast those substances filtered and reabsorbed back into the circulation by the renal tubules have smaller clearances than inulin. Thus one can determine whether a substance is reabsorbed or secreted by the renal tubules by comparing the clearance of the substance to that of inulin (Figure 6.18). For example, the clearance of glucose by the kidney is less than 1 percent of that of inulin, indicating that this metabolite is extensively reabsorbed by the renal tubules. On the other hand, many foreign substances have clearances much greater than that of inulin, indicating that they are transported from blood to urine by the renal tubules (tubular excretion or secretion).

3.3   Reabsorption of NaCl in the Renal Tubule

The major work of the vertebrate kidney is involved in the reabsorption of NaCl from the glomerular filtrate (Figure 6.19). Although many other substances are reabsorbed by the renal tubules, the fact that NaCl concentration is at least 10 times higher than the concentration of any other solute

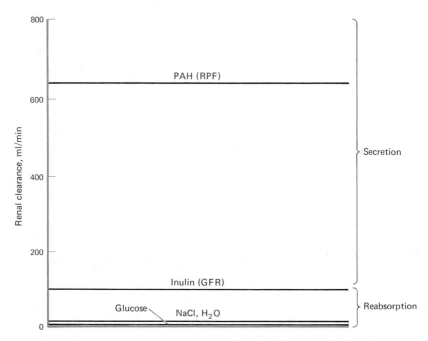

FIGURE 6.18 Renal clearance values for man. The renal clearance value for inulin is equal to the glomerular filtration rate (GFR). Therefore substances whose clearances are less than inulin are reabsorbed by renal tubules; those that are greater than inulin are secreted. Clearance of p-aminohippurate (PAH) is equal to renal plasma flow (RPF).

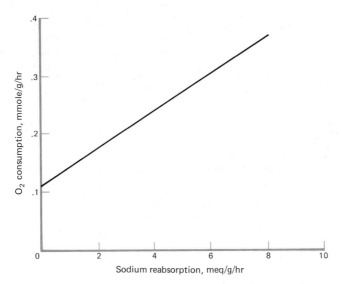

FIGURE 6.19 Relation of renal oxygen consumption to sodium reabsorption. The direct relation between oxygen consumption and sodium reabsorption indicates that the major work of the kidney is the reabsorption of sodium chloride filtered in the glomeruli. (Modified from G. Torelli, E. Milla, A. Faelli, and S. Costantini, *Am. J. Physiol.* 211:576—580, 1966.)

means that most of the energy expended by the kidney goes towards the reabsorption of this salt. This can be shown by measuring changes in renal oxygen consumption when the amount of NaCl filtered at the glomerulus is altered experimentally. Under these conditions we can demonstrate a direct relation between the change in NaCl filtered (and therefore reabsorbed) and the change in oxygen consumption; increases in filtered NaCl produce proportional increases in oxygen consumption.

The mechanism of salt reabsorption in the renal tubule has received a great deal of attention. The nature of sodium reabsorption in the different segments of the renal tubule has been studied by the micropuncture technique (Figure 6.20). These studies have shown that two-thirds or more of the NaCl filtered in the glomerulus is reabsorbed by the proximal segment of the tubule. A similar fraction of the filtered water is also reab-

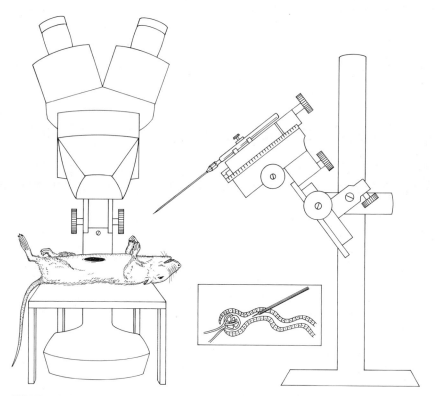

FIGURE 6.20   Diagram of apparatus used in renal micropuncture technique. Micromanipulator (at right) is used to position micropipette into precise location in the nephron. Small (μliter or less) quantities of fluid are drawn into pipette and analyzed by microchemical or radioisotopic tracer techniques. Insert diagrams the appearance of micropipette in lumen of renal tubule.

sorbed in the proximal tubule, and therefore the reabsorbate is isosmotic to plasma. Thus the concentration of sodium chloride at the end of the proximal tubule is essentially the same as in plasma. Although no concentration gradients are present for sodium chloride across the proximal tubule, the reabsorption of salt is not passive; rather, studies have shown that sodium chloride can be reabsorbed against a concentration gradient artificially produced across the proximal tubule, and that metabolic poisons inhibit the reabsorption of NaCl in the proximal tubule.

The classical concept for the reabsorption of NaCl in the proximal tubule is that sodium ions enter the cytoplasm from the lumen of the tubule, diffuse across the cell, and are "pumped" across the antilumenal surface into the extracellular fluid by an active transport system. In this model Cl^- follows the movement of Na^+ passively; it diffuses down the electrical gradient created by active transport of Na^+. This mechanism is similar to the one proposed for the active transport of sodium across the frog skin. However, there is an increasing body of evidence to indicate that when NaCl enters the cell, instead of diffusing to the basal surface it is pumped into the lateral spaces between cells (Figure 6.21). At the apical surface of the cell these spaces are closed off from the lumen by tight junctions, but at the basal surface these spaces are open. Thus NaCl transported along the lateral surface of cells enters the spaces, leading to the production of an osmotic gradient with the highest concentration of solute being present at the apical (lumenal) ends of the cells. This solute gradient pulls water from the cytoplasm (and thus lumen) into the spaces, expanding the spaces and leading to a bulk flow of fluid into the extracellular fluid and then into capillary circulation surrounding the tubules. This model for sodium chloride transport across epithelial membranes was first proposed by Diamond and supported by experimental evidence obtained in a variety of epithelial tissues. In favor of the operation of this model in the proximal tubules of the vertebrate kidney, Bodil Schmidt-Nielsen and her colleagues have found that when lizards are filtering, and therefore reabsorbing fluid from the lumen of the proximal tubule, the lateral spaces between the cells of the proximal tubule are dilated, whereas in lizards not filtering or reabsorbing fluid, these spaces are closed. It has been suggested that the Starling forces (hydrostatic pressure and protein oncotic pressure; see Chapter 5, Section 4.2) that govern the movement of fluid across capillary walls may also operate to regulate fluid reabsorption in the proximal tubule. In this model, fluid movement from the lateral spaces into the peritubular capillaries is governed by the difference in hydrostatic pressure and oncotic pressure between the lateral spaces and capillary lumen.

Most of the 30 percent or so of the NaCl escaping reabsorption in the proximal tubule is reabsorbed by the distal tubule and collecting ducts

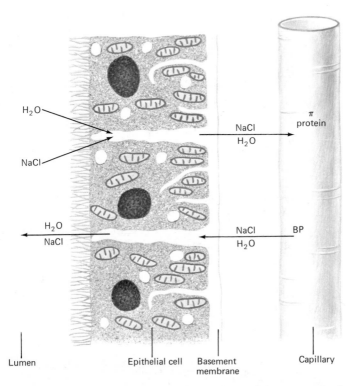

H_2O

$NaCl$

H_2O
$NaCl$

NaCl
H_2O

NaCl
H_2O

π
protein

BP

Lumen   Epithelial cell   Basement   Capillary
                                  membrane

FIGURE 6.21   Diagrammatic representation of intercellular movement of
NaCl and water across the renal tubule. π protein—plasma oncotic pressure;
BP—peritubular capillary blood pressure. π protein favors reabsorptive
movement of NaCl and water, whereas BP opposes it.

and in birds and mammals by the loop of Henle. The distal reabsorption of
sodium chloride is characterized by the development of steep concentration gradients of this salt between the lumen of the distal tubule and the
surrounding circulation. Concentration gradients of 100/1 (blood/lumen)
are common in the distal tubule of freshwater vertebrates. The presence of
these gradients and the fact that the electropotential difference across the
tubule (approximately +50 mV) opposes the passive movement of sodium
ion from lumen to blood indicate that the reabsorption of sodium in the
distal tubule is by active transport.

There is a good correlation between the cytological differentiation
of the cells in the distal tubule and water availability in the environment.
Among lizards the gecko, which lives in a relatively moist environment
and produces a dilute urine, has distal tubular cells in which there are
deep basal infoldings lined with mitochondria, indicating that these cells
have the capacity for pumping NaCl against concentration gradients (Fig-

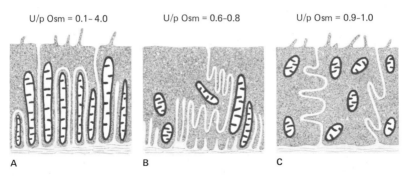

U/p Osm = 0.1–4.0   U/p Osm = 0.6–0.8   U/p Osm = 0.9–1.0

A   B   C

FIGURE 6.22 Diagrammatic representation of ultrastructure of distal tubular cells of mammals and lizards: (**A**) mammals, (**B**) gecko, and (**C**) horned toad. The gecko lives in a moist environment, whereas the horned toad lives in arid environments. U/p Osm— urine-to-plasma osmolarity ratios. (Based on J. S. Roberts and B. Schmidt-Nielsen, *Am. J. Physiol.* 211:485, 1966.)

ure 6.22). On the other hand, in the horned toad and Galapagos lizard, which live in relatively arid environments and do not excrete dilute urine, the distal tubular cells are poorly differentiated; they do not have well-developed basal infoldings and contain very few mitochondria.

In all mammals and some birds the nephrons have a segment not found in the kidneys of other animals: the loop of Henle. This segment of the nephron is involved in a highly specialized function—that is, the production of a concentrated urine. The loop is thought to work as a countercurrent multiplier system (Figure 6.23). Urine entering the descending limb of the loop from the proximal tubule is isosmotic with plasma. The urine becomes hyperosmotic to plasma in the descending limb by the passive entry of salts from, and loss of water to, the interstitium. The urine is maximally concentrated at the tip of the loop in the papilla of the kidney. After turning the bend of the loop, the urine then becomes hypoosmotic to plasma by the active extrusion of salts, without water, from the ascending limb. The result of these gyrations is the net loss of salt from the loop of Henle as it traverses the medulla and an increase in osmolarity of the medullary interstitium. This by itself does not result in the production of a concentrated urine. However, the urine leaving the loop of Henle loses water in the distal tubule and collecting ducts of the renal cortex to become isosmotic to plasma and reduced in volume. This urine then returns to the medulla via the collecting ducts and becomes concentrated by equilibrating with the hyperosmotic environment of the medullary interstitium.

The efficiency of the countercurrent multiplier system of the renal medulla theoretically should increase in a direct relation to the length of the loop of Henle. Schmidt-Nielsen and her colleagues have shown that

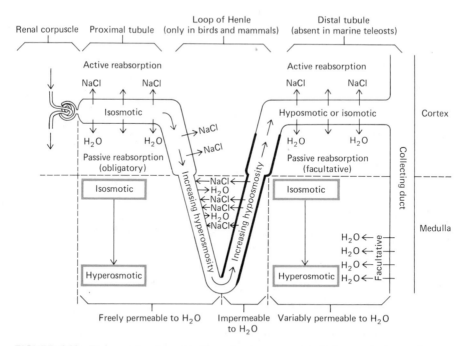

FIGURE 6.23 Diagram of sodium chloride and water movements in the renal tubules of verte-brates. [Modified from William S. Hoar, *General and Comparative Physiology,* 2d ed., © 1975, p. 310. (After Pitts, 1959.) Reprinted by permission of Prentice-Hall, Inc., Englewood Cliffs, N.J.]

there is a good correlation between the medullary thickness (taken as an index of loop length) and urine-concentrating ability among a variety of mammals (Figure 6.24). The mammals that have the longest loops are those that produce the most concentrated urines: the desert rodents. The mammals with the shortest loops are those that live in moist habitats and excrete dilute urines: the beaver and mountain beaver.

3.4 Hormonal Control of Salt and Water Reabsorption

In mammals salt reabsorption is regulated in part by the action of aldos-terone, a steroid secreted by the adrenal cortex (Chapter 9, Section 10). This hormone affects the reabsorption of sodium chloride in the distal seg-ment of the nephron-distal convoluted tubule and collecting ducts. In the absence of aldosterone the body tends to lose sodium and retain potas-sium. These effects can be immediately reversed by administration of the steroid. The control of salt reabsorption by aldosterone is found in mammals, birds, reptiles, and amphibia. Aldosterone is not present in

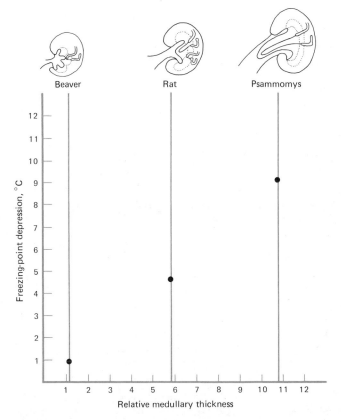

FIGURE 6.24 Relation of renal medullary thickness to urinary concentrating ability in three representative mammals. (From B. Schmidt-Nielson and R. O'Dell, *Am. J. Physiol.* 200:1119, 1961.)

fishes. The principal adrenal steroid found in teleost fishes is corticosterone, and this hormone acts to regulate salt transport in the gills rather than in the kidneys of these organisms.

The principal water balance hormone in mammals is the antidiuretic hormone (Chapter 9, Section 10). This hormone is produced in the hypothalamus but is released from the posterior lobe of the pituitary. The secretion of the hormone is controlled by cells in the supraoptic nucleus of the hypothalamus, which in turn are activated by osmoreceptors also located in the hypothalamus. As Verney showed in a classic experiment on the control of water reabsorption in the dog kidney, an increase in the osmotic pressure of the blood produces a prompt increase in secretion of antidiuretic hormone by the posterior pituitary. The system is so sensitive that a rise in osmotic pressure of only 1 percent can produce a 90 percent reduction in urine flow. The antidiuretic hormone is released by the posterior pituitary into the circulation and carried to the kidney. The hormone is thought to promote water reabsorption by increasing the permeability of the cells lining the distal segments of the nephron to water.

The antidiuretic hormone of mammals is a polypeptide called *arginine vasopressin* (or *lysine vasopressin* in pigs). However, a slightly different antidiuretic hormone is present in birds, reptiles, amphibians, and fishes. This latter hormone, called *arginine vasotocin,* promotes the reabsorption of water in the renal tubules of birds, reptiles, and amphibia as well as in the skin and bladder of many amphibia. However, in fishes the hormone not only fails to increase water reabsorption but may actually have a diuretic effect when injected into these organisms. The significance of this latter observation is not at all clear.

A complete discussion of the hormonal control of salt and water balance in vertebrates is found in Chapter 9.

3.5 Tubular Secretion (Excretion)

Many substances eliminated by the kidney are excreted simply by glomerular filtration. However, a variety of other exogenous and endogenous compounds are removed by renal tubular secretion in addition to filtration. Tubular secretion refers to a process whereby substances are actively transported from the peritubular capillaries to the tubular urine. There are specific active transport processes for the secretion of each group of chemical compounds. Substances secreted by the renal tubules are often poorly filterable at the glomerulus; in many instances these compounds are bound to plasma proteins and cannot cross the glomerular membranes. In lower vertebrates tubular secretion supplements the low rate of glomerular filtration in the elimination of electrolytes and metabolic products by the kidneys.

As mentioned previously, glomerular filtration rates are low, and in some cases absent, in marine teleosts. In these fishes divalent ions (such as Ca^{2+}, Mg^{2+}, and SO_4^{2-}) and some metabolic products (creatine, creatinine, and TMAO) are excreted mainly by tubular secretion. Many of these secretory systems have been retained during the evolution of the vertebrates. For example, uric acid is secreted by the kidneys of many higher vertebrates including man, and the chief method of eliminating nitrogenous waste in birds and reptiles is tubular secretion of this nitrogenous end product.

All vertebrates examined have a system in their renal tubules for the secretion of organic acids. This system, which is relatively nonspecific and transports a wide variety of foreign as well as natural compounds, has provided the comparative physiologist with an important experimental tool for the study of tubular secretion. The proof of renal tubular secretion, which was doubted to exist up until the late 1920s, was obtained by E. K. Marshall in an experiment demonstrating the renal excretion of the organic dye, phenol red, in the aglomerular goosefish, *Lophius americanus.* Since this fish has no functional glomeruli in its kidneys, the only way in which the dye could have been excreted by the kidneys was through tubular secretion. Studies by R. P. Forster and his colleagues on renal tubules isolated from the kidneys of fishes and amphibia and maintained in saline media in vitro have shown that the organic acid secretory system is localized in the proximal segment of the nephron. This system shows all the properties commonly associated with active transport processes: competition for secretion between different organic acids, dependence on oxidative metabolism, inhibition by metabolic poisons, and temperature dependency.

Quantitative studies on renal tubular secretion may be done in intact animals by the renal clearance technique using the equation

$$T_{OA} = U_{OA}\dot{V} - (\text{GFR} \times P_{OA})$$

where T_{OA} is rate of tubular secretion, U_{OA} the urinary concentration of organic acid, $\dot{V}$ the urine flow rate, GFR the glomerular filtration rate, and P_{OA} the plasma concentration of organic acid. At low plasma concentration T_{OA} will increase as P_{OA} concentration increases (Figure 6.25). However, a point will be reached where the transport process is saturated with OA and a maximal rate of tubular secretion is reached. This is referred to as Tm_{OA}, the maximal rate of tubular secretion. This value is relatively constant within a specific species and is a measure of both functional proximal tubular mass and the relative activity of the transport system within that species.

Removal of some organic acids such as p-aminohippurate (PAH) from the peritubular capillaries by the renal tubules is so efficient that its concentration in blood leaving the kidney (renal venous blood) is very low. This allows a practical application of the Fick principle for measuring renal blood flow in man and other animals.

This principle, which was originally applied to the measurement of cardiac output, states that the amount of blood flowing through an organ can be calculated by determining the quantity of substance X removed by the organ and the arteriovenous concentration difference of X according to the following equation:

$$\text{blood flow} = \frac{\text{total } X \text{ removed}}{\text{arterial } X - \text{venous } X}$$

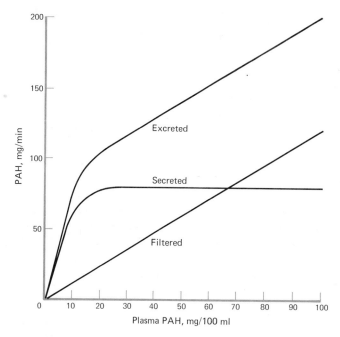

FIGURE 6.25 Relation of filtration, secretion, and excretion rates of p-aminohippurate (PAH) to plasma concentrations in man.

For calculating renal plasma flow this equation becomes

$$RPF = \frac{\dot{V}U_X}{RA_X - RV_X}$$

where RPF is the renal plasma flow, V the urine flow rate, U_X the concentration of X in the urine, RA_X the renal arterial plasma concentration of X, and RV_X the renal venous concentration of X. Any substance removed from the renal circulation may be used to calculate renal plasma flow using this equation. However, if one uses a compound that is completely removed from the renal circulation, then the concentration of X in renal venous plasma reduces to 0, and the equation becomes

$$RPF = \frac{\dot{V}U_X}{A_X}$$

which is the clearance equation described above. Since many organic acids such as p-aminohippurate are nearly completely removed from blood circulating through the kidney, the clearance of these substances is equal to the renal plasma flow:

$$RPF = \frac{\dot{V}U_{OA}}{A_{OA}}$$

Renal blood flow is calculated by dividing RPF by 1 minus the hematocrit:

$$RBF = \frac{RPF}{1 - Hct}$$

In fishes, amphibia, and reptiles there is a dual blood supply to the kidney (Figure 6.26); an arterial supply, similar to that in the mammalian kidney, originates in the renal artery (or arteries) and terminates in the peritubular capillaries. In addition, however, these species also have a portal circulation that originates in the renal portal vein and terminates in peritubular capillaries. Thus arterial and venous portal blood both supply the peritubular circulation. This latter type of circulation is partially retained in birds but is not found in the mammalian kidney.

Recent experiments by A. C. Barger have shown that blood flow in the mammalian kidney is composed of at least four components. There is a rapid blood flow supplying the outer cortex, a slower flow in the inner cortex and outer medulla, and a very slow flow in the inner medulla. The

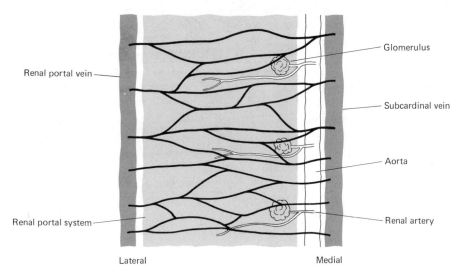

FIGURE 6.26 Diagrammatic representation of the dual blood supply of the amphibian kidney. (Redrawn from J. P. Hayslett, *J. Clin. Invest.* 52:1314–1319, 1973.)

slowest flow rate is found in the circulation of the perirenal fat. The regulation of these different flow rates is critical for the proper operation of the mammalian kidney.

3.6   Hydrogen Ion Excretion in the Renal Tubules

One of the most important aspects of tubular secretion, especially in higher vertebrates, is the excretion of H^+ by the renal tubules. This process is of paramount importance in maintaining the proper pH of the body fluids and the acid-base balance of the individual. Animals that eat an alkaline-ash diet (herbivores) produce an excess of base during digestion (see Chapter 8, Section 2.1). This extra base is excreted mainly by the kidneys in the form of HCO_3^-. On the other hand, animals on an acid-ash diet (carnivores and omnivores) produce an excess of H^+ during digestion. The H^+ is excreted mainly in a buffered form by the renal tubules.

The process of H^+ excretion by the renal tubules has been studied most extensively in man, dog, and rat. These studies have shown that very little of the H^+ secreted by the renal tubules is eliminated as free hydrogen ion but rather as $H_2PO_4^-$ and NH_4^+ salts (Figure 6.27). The basic process underlying H^+ secretion by the renal tubule is thought to be a Na^+-H^+ exchange in which H^+ produced in the cell is exchanged at the lumenal membrane for Na^+ in the tubular lumen. In this process the secretion of H^+ by the tubular cells is coupled to the transport of a fraction of the sodium reabsorbed by the renal tubule. Hydrogen ion secreted by the renal tubule is thought to be produced by the hydration of CO_2 (that is, $CO_2 + H_2O \leftrightarrows H_2CO_3 \leftrightarrows H^+ + HCO_3^-$) in the renal cells, a process catalyzed by a highly active enzyme known as carbonic anhydrase. Once the H^+ ion enters the renal tubule, it combines with buffer salt (mainly Na_2HPO_4), which enters the tubular lumen by glomerular filtration, or with NH_3, which diffuses into the lumen from the tubular cells. Elimination of H^+ in the form of $H_2PO_4^-$ and NH_4^+ allows the body to excrete significant amounts of acid (approximately 100 meq/day in man) without lowering the pH of the urine to levels that would be injurious to the tubular cells; urine pH rarely falls below 4.0.

The HPO_4^{2-} that buffers H^+ in the tubular lumen arises during the cellular metabolism of phospholipids and phosphoproteins in the body. In contrast ammonia excreted in the urine is produced mainly in the cells surrounding tubular lumen and enters the lumen directly via nonionic diffusion.

In a series of elegant studies Pitts and his colleagues have shown that in the dog approximately two-thirds of the ammonia produced by the renal cells arises by the deamination of glutamine. This amino acid is ex-

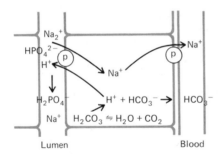

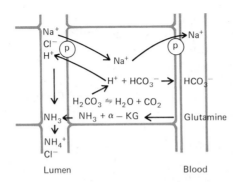

FIGURE 6.27 Sodium-hydrogen ion exchange mechanisms in renal tubule. p refers to Na^+-H exchange pump or Na^+ pump. (Based on R. F. Pitts, *Fed. Proc.* 7:418, 1948.)

tracted by the renal cells from the blood passing through the kidney and is hydrolyzed in the cells, under the action of an enzyme known as glutaminase, to ammonia plus glutamic acid. Glutamic acid may in turn be oxidatively deaminated in the renal cells (by another enzyme known as glutamic acid dehydrogenase) to ammonia plus α-ketoglutaric acid.

The ammonia-producing system is adaptive in mammals. Prolonged (days) systemic acidosis leads to a progressive increase in the excretion and production of ammonia by the mammalian kidney. The increased production is due to an acceleration in the rates of extraction and deamidation of glutamine by the renal tubular cells. Although in some species (rat) the acid-induced rise in ammonia production is accompanied by an increase in the level of glutaminase in the kidney, in other species (dog) ammonia production rises in the absence of any change in renal glutaminase concentration. Numerous studies have been done on this subject, but the exact mechanism by which systemic acidosis induces a rise in renal ammonia production is still not known.

Little is known about H^+ excretion in lower vertebrates. The few studies that have been done suggest that the processes of H^+ excretion by the renal tubules of lower vertebrates are similar to those operating in mammals but more limited. Marine teleosts and elasmobranchs, which eat an acid-ash diet, excrete an acid urine. Much of the H^+ secreted by the renal tubules of these fishes is eliminated in the form of $H_2PO_4^-$. Although the rate of H^+ excretion is relatively constant in fishes under normal conditions, this rate can be artificially elevated by administration of HPO_4^{2-}, which combines with H^+ in the tubular lumen and is eliminated in urine as $H_2PO_4^-$.

Acid excretion in the frog is similar to that in mammals but slower to respond to a stimulus. In mammals excretion of ammonia and $H_2PO_4^-$ rise within a couple of hours following administration of acid. In *Rana catesbiana* acid excretion following a single infusion of HCl increases two days after treatment and remains elevated for another two days. The difference in time constants for the response of acid excretion between mammals and amphibia may reflect differences in metabolic rate between the two groups, due perhaps to the difference in body temperature.

Acid excretion in birds is quite similar to that in mammals; acid is excreted in the form of NH_4^+ and buffer salt. However, the buffer salt, in contrast to the situation in mammals, is mainly urate. Excretion of NaH_2PO_4 constitutes only a minor fraction of the acid excreted. The reason for this difference is the fact that glomerular filtration rate may be relatively low in birds (compared to mammals), resulting in a low rate of filtration of Na_2HPO_4 into the tubular lumen. In contrast uric acid, the major end product of nitrogen metabolism in birds, is secreted by the cells lining the renal tubule into the lumen, thus providing a much greater supply of urate to neutralize H^+ than phosphate. The time constant for the response of acid excretion in the chicken is similar to that in mammals, the rise in acid excretion commencing within 1 to 2 hr following administration of acid to the bird.

The source of ammonia excreted by the kidneys of birds and submammalian species is unknown. Glutamine is present in the plasma of these species, but renal extraction studies comparable to those done in mammals, showing the role of this amino acid in renal ammonia production, have not yet been performed in submammalian species. In many of these latter species alanine and glycine are the chief amino acids in plasma. It is entirely possible, therefore, that these amino acids are the sources of urinary ammonia in lower vertebrates.

4 • NITROGEN METABOLISM AND OSMOTIC REGULATION

There is a close link between the form of nitrogen metabolism and the osmoregulatory needs of an organism. Environmental factors, especially water availability, have a marked effect on the nature of the end products formed during nitrogen metabolism. There are three main end products of nitrogen metabolism in animals: ammonia, urea, and uric acid. The sources of these end products are the proteins and nucleic acids in the animal's own cells or those ingested in the diet. Ammonia is produced in many of the deaminating reactions that occur during the metabolism of both proteins (see Chapter 7, Section 3.3) and nucleic acids, but the main pathway of ammonia formation in vertebrates is probably that of transamination of amino acids to glutamic acid and subsequent deamination of the latter amino acid to ammonia, catalyzed by the ubiquitous enzyme glutamic acid dehydrogenase (Chapter 7, Section 3.3); that is,

$$
\underset{\text{Amino acid}}{R-\overset{\overset{\displaystyle NH_2}{|}}{\underset{\underset{\displaystyle H}{|}}{C}}-COOH} + \underset{\alpha\text{-ketoglutaric acid}}{HOOC-\overset{\overset{\displaystyle H}{|}}{\underset{\underset{\displaystyle H}{|}}{C}}-\overset{\overset{\displaystyle H}{|}}{\underset{\underset{\displaystyle H}{|}}{C}}-\overset{\overset{\displaystyle O}{\|}}{C}-COOH} \xrightarrow{\text{Transaminase}}
$$

$$
\underset{\text{Keto acid}}{R-\overset{\overset{\displaystyle O}{\|}}{C}-COOH} + \underset{\text{Glutamic acid}}{HOOC-\overset{\overset{\displaystyle H}{|}}{\underset{\underset{\displaystyle H}{|}}{C}}-\overset{\overset{\displaystyle H}{|}}{\underset{\underset{\displaystyle H}{|}}{C}}-\overset{\overset{\displaystyle NH_2}{|}}{\underset{\underset{\displaystyle H}{|}}{C}}-COOH}
$$

and

$$
\underset{}{HOOC-\overset{\overset{\displaystyle H}{|}}{\underset{\underset{\displaystyle H}{|}}{C}}-\overset{\overset{\displaystyle H}{|}}{\underset{\underset{\displaystyle H}{|}}{C}}-\overset{\overset{\displaystyle NH_2}{|}}{\underset{\underset{\displaystyle H}{|}}{C}}-COOH} \xrightarrow[\text{dehydrogenase}]{\text{Glutamic acid}}
$$

$$
\underset{\alpha\text{-ketoglutaric acid}}{HOOC-\overset{\overset{\displaystyle H}{|}}{\underset{\underset{\displaystyle H}{|}}{C}}-\overset{\overset{\displaystyle H}{|}}{\underset{\underset{\displaystyle H}{|}}{C}}-\overset{\overset{\displaystyle O}{\|}}{C}-COOH} + NH_3
$$

Urea is formed by two biosynthetic pathways in animals. One, which is found in many invertebrates and in the teleost fishes, involves the degradation of uric acid produced in nucleic acid metabolism (the "uricolytic" pathway, Figure 6.28). In the other pathway urea is formed via the ornithine-urea cycle (Figure 6.29). In this cycle nitrogen enters both as ammonia and the α-amino group of aspartic acid and leaves as the two nitrogen atoms of urea. This cycle is localized mainly in the liver and is the

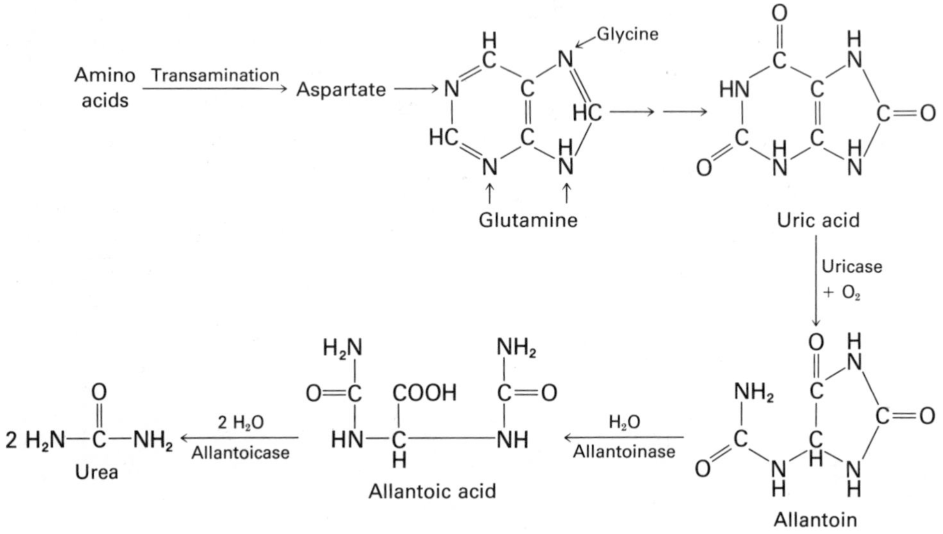

FIGURE 6.28 Uricolytic pathway for urea formation.

major route of urea formation in all vertebrates examined, with the exception of the teleost fishes.

Uric acid is the major end product of nitrogen metabolism in insects, birds, and nonchelonian reptiles (snakes, lizards). In these animals the normal biosynthetic pathway involved in the formation of purines is adapted to handle the relatively large amount of nitrogenous end products formed during protein catabolism (breakdown). The sources of nitrogen atoms incorporated into urate are the amino nitrogen of glycine, the α-amino nitrogen of asparate, and the amide nitrogen of glutamine. Animals forming uric acid as a major nitrogenous end product lack the uricolytic enzymes found in many other vertebrates.

Aquatic organisms are generally ammonotelic; they form and excrete ammonia as the major end product of nitrogen metabolism. Ammonia has the advantage of requiring no expenditure of energy during its formation; in fact energy is released in many of the biochemical reactions producing ammonia. In addition, it requires no special organ for excretion; it readily diffuses across body surfaces such as gills and skin. However, ammonia is toxic to all organisms, and it cannot be stored as such in the body. Thus animals that are strictly ammonotelic require a constant supply of water into which ammonia can be excreted. Animals that live out of the water, or aquatic organisms that leave the water periodically, require some mechanism for detoxifying ammonia. The method most com-

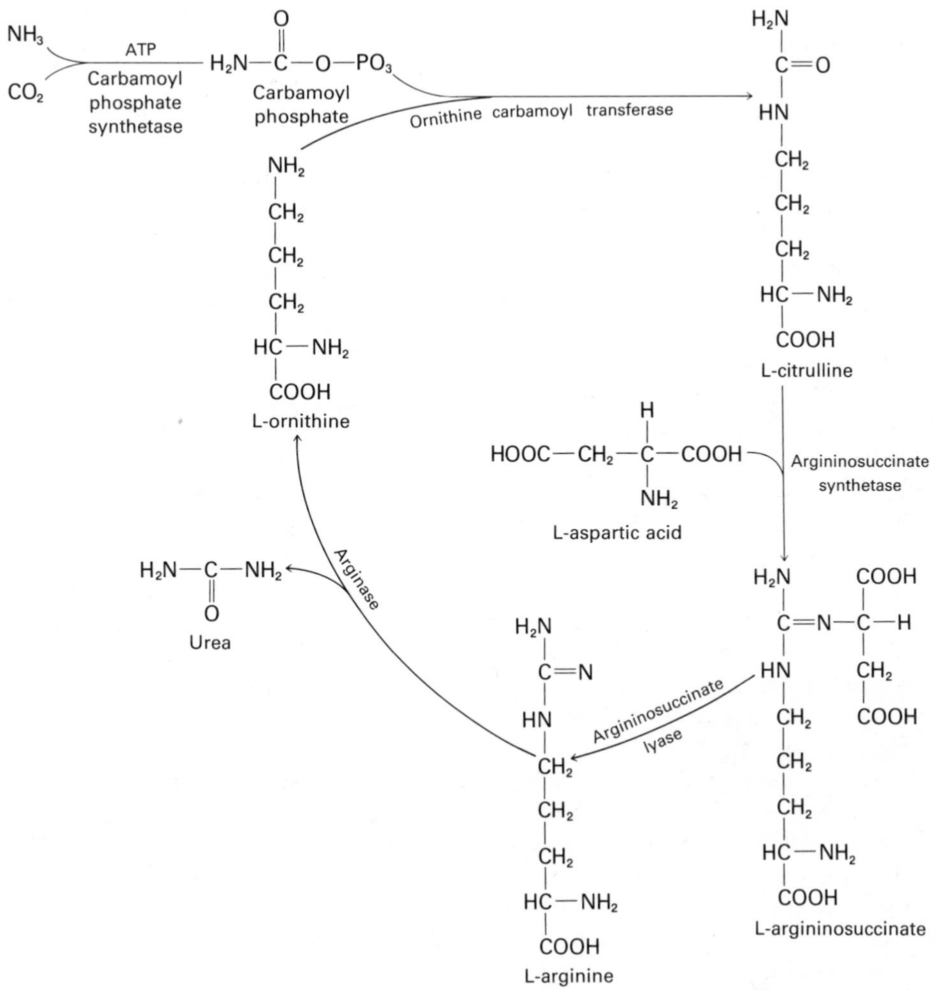

FIGURE 6.29 Ornithine-urea cycle pathway of urea formation.

monly used by vertebrates for the detoxification of ammonia is the incorporation of this product into urea via the Krebs ornithine-urea cycle. A dramatic example of this method of detoxifying ammonia is seen in the African lungfish, *Protopterus,* which undergoes periodic estivation (a state of dormancy brought about by lack of water) when the lakes in which it lives dry out. In water the fish is ammonotelic. However, in the dry seasons it burrows into the mud and surrounds itself with a water-impervious cocoon. During this period heart and circulation rates fall to very low levels and excretion virtually ceases. If the fish were to continue producing ammonia as the main end product of nitrogen metabolism, this toxic base would accumulate to dangerously high levels. Instead, however, the ammonia is detoxified by incorporation into the nontoxic product urea, and this latter compound accumulates to levels as high as 1 to 2 percent of the fish's body weight during extended periods of estivation. The accumulation of urea not only serves to avoid ammonia intoxication but also has the theoretical advantage of decreasing water loss from the body fluids. The increased concentration of urea in the body fluids tends to lower the vapor pressure of these body fluids and helps to prevent evaporative water loss.

There are two surviving families of lungfishes, *Lepidosirenidae* and *Ceratodontidae*, and these two groups differ markedly in their ecological behavior. The African lungfish, *Protoperterus*, of the former family estivates out of water during the dry season, whereas the Australian lungfish, *Neoceratodus forsteri*, of the latter family uses its lung only as an accessory breathing organ and never leaves the water. The environmental status of the South American lungfish, *Lepidosiren paradoxa*, of the family *Lepidosirenidae* falls between that of *Protopterus* and the Ceratodontid, *Neoceratodus*. In between *Protopterus*, which forms a dry, water-impervious cocoon during its estivation, and *Neoceratodus*, which never estivates, *Lepidosiren* estivates in damp or wet burrows. These differences in ecological behavior place different osmotic stresses on each type of fish. The threat of desiccation and ammonia intoxication during estivation is greatest in *Protopterus*, less in *Lepidosiren*, and nonexistent in *Neoceratodus*. In an investigation on the relative activities of the ornithine-urea cycle in the livers of the three lungfishes, it was found that the highest level occurred in *Protopterus*, an intermediate level was found in *Lepidosiren*, and the lowest level was present in *Neoceratodus* (Figure 6.30). Thus there is a close correlation between the environmental stresses and the activity of the ornithine-urea cycle in the three types of lungfishes.

Retention of organic solutes such as urea and TMAO for purposes of osmoregulation is employed by the Chondrichthyes (sharks, rays, skates, and chimera), the Coelacanth, *Latimeria chalumnae*, and euryhaline amphibia such as the crab-eating frog of Southeast Asia, *Rana cancrivora*. These organic solutes are synthesized endogenously in the organisms and are retained at relatively high concentrations by virtue of their low rates of excretion. The gills of fishes are relatively impermeable to these solutes, and the kidneys of fishes retaining urea and TMAO have mechanisms for reabsorbing over 90 percent of the solutes filtered in the glomerulus. In contrast other vertebrates such as man excrete 50 percent or more of the filtered urea. Euryhaline amphibia living in hyperosmotic environments

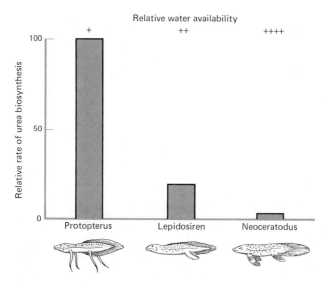

FIGURE 6.30   Relative rates of urea biosynthesis in three modern lungfishes. Number of plus signs indicates relative water availability in fish's environment. (See D. Funkhouser, L. Goldstein, and R. P. Forster, *Comp. Biochem. Physiol.* 41A:439—443, 1972.)

maintain high concentrations of urea by decreasing glomerular filtration rate and thus reducing the amount of urea excreted in the urine. Plasma osmolarity of *Rana cancrivora* living in fresh water is 290 mOsm, and plasma urea concentration is 40 mM. In 80 percent seawater, plasma osmolarity rises to 830 mOsm and urea concentration to 350 mM. Sodium concentration increases from 125 meq/liter in fresh water to 250 meq/liter in 80 percent seawater. Thus most of the increase in plasma osmolarity is due to the rise in plasma urea concentration; electrolyte concentration, as reflected by plasma sodium concentration, changes little compared to urea.

   The rise in plasma urea concentration that occurs when the frog moves from fresh water to seawater may be due to either an increase in the rate of biosynthesis of urea or a decrease in the rate of excretion, or both. Thus plasma urea concentration is determined by the ratio of the rate of urea biosynthesis to the rate of clearance (removal) of urea from the body fluids according to the equation:

$$\text{plasma urea (mM)} = \frac{\text{urea synthesis (mmole/unit time)}}{\text{urea clearance (liters/unit time)}}$$

In adult frogs urea is cleared from the body fluids mainly by renal excretion. When *Rana cancrivora* is in 80 percent seawater, urine flow is approximately 1 percent of the flow rate in fresh water, and urea clearance is reduced markedly. The reduction in urea clearance leads to a significant rise in plasma urea concentration, as observed in Figure 6.31. However, the contribution of urea biosynthesis to alterations in plasma urea concentration under these conditions is unknown, since there is no information available on the effect of changes in environmental salinity on rates of urea biosynthesis in this species.

The South African toad, *Xenopus laevis,* is another amphibian that can tolerate a mild degree of salinity. Although the toad does not normally encounter salt water, it can adapt to a hyperosmotic environment in the laboratory. When placed in saline solution (300 mOsm), body fluid urea concentration rises markedly until the osmolarity of the internal fluids is slightly higher than that of the medium. Experiments have shown that the rise in urea concentration is due both to an increase in the rate of urea biosynthesis in the liver and a decrease in clearance and excretion of urea in the kidney. Both mechanisms are necessary for the full rise in urea accumulation to occur.

All marine elasmobranchs retain urea and trimethylamine oxide to maintain osmotic balance (or slight superiority) to the marine environment. Urea constitutes 35 to 45 percent of the osmotically active solutes in the body fluids and TMAO about 10 percent. The high concentrations of

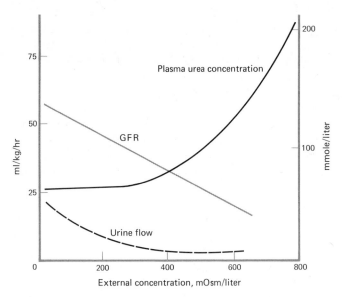

FIGURE 6.31   Renal function in the crab-eating frog (*Rana cancrivora*) as a function of environmental salinity. (Data from K. Schmidt-Nielsen and P. Lee, *J. Exp. Biol.* 39:167, 1962.)

urea are achieved by production of the end product in the liver and retention by the gills and kidneys. Although all elasmobranchs maintain relatively high concentrations of urea in their body fluids, several of these fishes that enter brackish and even fresh water reduce the concentration of urea in their body fluids in order to decrease the difference in osmotic pressure between themselves and their environment; otherwise these fishes would be burdened with a tremendous influx of water from the dilute environment. Even so, body fluid osmolarity in these fishes remains significantly above that of the environment. The reduction in internal urea concentration is accomplished by different mechanisms in different elasmobranchs. In the lemon shark, *Negaprion brevirostris*, urea content is reduced by an increased rate of excretion of the end product. However, in the skate, *Raja erinacea*, a reduction in total body urea concentration is brought about by a decrease in the rate of urea biosynthesis when the fish enters a dilute environment.

The kidneys also play an important role in the adaptation of euryhaline elasmobranchs to different salinities. In seawater the kidneys help to maintain a relatively high internal osmotic pressure by reabsorbing over 90 percent of the urea and TMAO filtered at the glomerulus. However, when the fishes enter dilute seawater, there is an inhibition of the reabsorption of these two end products in the renal tubules, and as little as 50 percent of the filtered urea is reabsorbed by the renal tubules under these conditions (Figure 6.32). TMAO reabsorption is also reduced, but less drastically than that of urea. Recent studies have shown that the reabsorption of urea in the renal tubules of the elasmobranch kidney is regulated by the operation of the Starling forces (capillary hydrostatic and oncotic pressure) in the peritubular capillaries surrounding the renal tubules. Inhibition of urea reabsorption, similar to that seen during environmental dilution, was observed following experimental maneuvers that would be expected to increase capillary hydrostatic pressure and decrease protein oncotic pressure. It is quite possible, therefore, that the changes produced in the internal environment when the fishes enter brackish or fresh water (for example, expansion of the extracellular fluid volume) lead to similar alterations in the Starling forces of the peritubular capillaries and bring about alterations in renal reabsorption of urea in this manner.

Recent studies have shown that the coelacanth, *Latimeria chalumnae*, osmoregulates in a manner similar to that of the elasmobranchs; that is, it maintains high concentrations of urea and trimethylamine oxide in its body fluids. This fish, which until recently was thought to have been extinct for the past 400 million years, is closely related to the crossopterygian fish that gave rise to the tetrapods: amphibia, reptiles, birds, and mammals. Thus the production and retention of organic solutes for osmoregulatory purposes is an old trick which has been passed on and re-

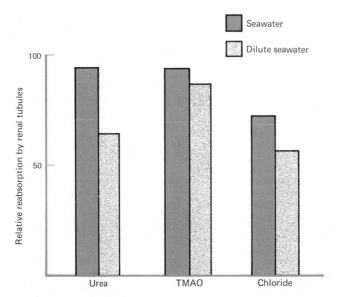

FIGURE 6.32 Effect of environmental dilution on reabsorption of urea and other solutes by renal tubules of the little skate, *Raja erinacea.* (Data from L. Goldstein and R. P. Forster, *Am. J. Physiol.* 220:742, 1971.)

tained during vertebrate evolution. In fact urea retention is used in the mammalian kidney to aid in the production of an osmotic gradient by the countercurrent multiplier system of the renal medulla.

REFERENCES

Barker-Jorgensen, C., Levi, H., and Zerahn, K. On Active Uptake of Sodium and Chloride Ions in Anurans. *Acta Physiol. Scand.* 30:178–190, 1954.

Bentley, P. J. *Endocrines and Osmoregulation.* New York: Springer-Verlag, 1971.

———. The Physiology of the Urinary Bladder of Amphibia. *Biol. Rev.* 41:275–316, 1966.

Bernard, C. *An Introduction to the Study of Experimental Medicine.* New York: Dover, 1957.

Burger, J. W. Roles of the Rectal Gland and the Kidneys in Salt and Water Excretion in the Spiny Dogfish. *Physiol. Zool.* 38:191–196, 1965.

Conte, F. P., and Lin, D. H. Y. Kinetics of Cellular Morphogenesis in Gill Epithelium during Sea Water Adaptation of *Oncorhynchus* (Walbaum). *Comp. Biochem. Physiol.* 23:945–957, 1967.

Ellis, R. A., Goertemiller, C. C., Jr., Delellis, R. A., and Kablotsky, Y. H. The Effect of Salt Water Regimen on the Development of the Salt Glands of Domestic Ducklings. *Develop. Biol.* 8:286–308, 1963.

Epstein, F. H., Katz, A. I., and Pickford, G. E. Sodium and Potassium-Activated Adenosine Triphosphatase of Gills: Role in Adaptation of Teleosts to Salt Water. *Science* 156:1245–1247, 1967.

Forster, R. P., Goldstein, L., and Rosen, J. K. Intrarenal Control of Urea Reabsorption by Renal Tubules of the Marine Elasmobranch, *Squalus acanthias. Comp. Biochem. Physiol.* 42A:3–12, 1972.

Goldstein, L., and Campbell, J. W., eds. *Nitrogen Metabolism and the Environment.* New York: Academic Press, 1972.

Griffith, R. W., Umminger, B. L., Grant, B. F., Pang, P. K. T., and Pickford, G. E. Serum Composition of the Coelacanth, *Latimeria chalumnae Smith. J. Exp. Zool.* 187:87–102, 1974.

Henderson, L. J. *Blood: A Study in General Physiology.* New Haven: Yale University Press, 1928.

Hodler, J., Heinemann, H. O., Fishman, A. P., and Smith, H. W. Urine pH and Carbonic Anhydrase Activity in the Marine Dogfish. *Am. J. Physiol.* 183:155–162, 1955.

Kerstetter, T. H., and Kirschner, L. B. Active Chloride Transport by the Gills of Rainbow Trout (*Salmo gairdneri*). *J. Exp. Biol.* 56:263–272, 1972.

Krogh, A. *Osmotic Regulation in Aquatic Animals.* New York: Cambridge University Press, 1939.

Lewy, J. E., and Windhager, E. E. Peritubular Control of Proximal Tubular Fluid Reabsorption in the Rat Kidney. *Am. J. Physiol.* 214:943–945, 1968.

Marshall, E. K. The Aglomerular Kidney of the Toadfish (*Opsanus tau*). *Bull. Johns Hopkins Hosp.* 45:95–102, 1929.

Motais, R., and Garcia-Romeu, F. Transport Mechanisms in the Teleostean Gill and Amphibian Skin. *Ann. Rev. Physiol.* 34:141–176, 1972.

Phillips, J. G., Holmes, W. N., and Butler, D. G. The Effect of Total and Sub-Total Adrenalectomy on the Renal and Extra-Renal Response of the Domestic Duck (*Anas platyrhynchos*) to Saline Loading. *Endocrinology* 69:958–969, 1961.

Philpott, C. W. Halide Localization in the Teleost Chloride Cell and Its Identification by Selected Area Electron Diffraction. *Protoplasma* 60:7–23, 1965.

Pitts, R. F. Renal Production and Excretion of Ammonia. *Am. J. Med.* 36:720–742, 1964.

Potts, W. T. W., and Parry, G. *Osmotic and Ionic Regulation in Animals.* New York: Macmillan, 1964.

Richards, A. N. Processes of Urine Formation. *Proc. Roy. Soc. (London)* B126:398–432, 1939.

Riegel, J. A. *Comparative Physiology of Renal Excretion.* New York: Hafner, 1972.

Schmidt-Nielsen, B., and Davis, L. E. Fluid Transport and Tubular Intercellular Spaces in Reptilian Kidneys. *Science* 159:1105–1108, 1968.

Shannon, J. A., and Smith, H. W. The Excretion of Inulin, Xylose and Urea by Normal and Phlorizinized Man. *J. Clin. Invest.* 14:393–401, 1935.

Ussing, H. H. Transport of Electrolytes and Water across Epithelia. *The Harvey Lectures.* Series 59. New York: Academic Press, 1965.

Wolbach, R. A. Renal Regulation of Acid-Base Balance in the Chicken. *Am. J. Physiol.* 181:149–156, 1955.

Yoshimura, H., Yata, M., Yuasa, M., and Wolbach, R. A. Renal Regulation of Acid-Base Balance in the Bullfrog. *Am. J. Physiol.* 201:980–986, 1961.

METABOLISM

Anthony R. Leech

Nutrition and metabolism are closely related subjects, and although considered in separate chapters, they are logically introduced together. Metabolism is the integrated chemical changes that take place within an organism. The input into metabolism is the diet of the animal, and Chapter 7 begins with a chemical description of the variety of foodstuffs ingested by animals. Before entering the tissues of the animal, the foodstuff molecules are subjected to modification both by digestion processes and by the action of microorganisms, as described in Chapter 8. Studies of the food requirements of an animal together with preabsorptive processing constitute the science of nutrition. Metabolism serves to link these input molecules with two outward manifestations of life: first, with the structure of the organism in the process of growth and, second, with the ability of the organism to expend energy—for example, in movement. As these demands will differ in detail from species to species and from tissue to tissue, so too will the pattern of metabolism vary, as discussed in Chapter 7.

Although the processes of nutrition, digestion, and metabolism are frequently described in textbooks of biochemistry, these areas developed historically as branches of physiology. Only after an explosive growth period did the study of metabolism break away from physiology to provide the foundations for the new science of biochemistry. Since the study of metabolism overlaps both biochemistry and physiology, it serves as a link between these subjects. This is particularly apparent in the field of endocrinology; however, in other areas, such as muscle contraction and ion transport, it is equally impossible to divorce metabolism from the physiology of the process.

Interest in nutrition has been accelerating in recent years. The main reasons for this are environmental and medical in nature. With the steadily increasing demand on the world's food supply, a great deal of attention has been paid to the nutritional requirements of man and his domesticated animals. To determine the effects of diet on growth and health requires not only feeding experiments but a full understanding of metabolism. The second reason for the upsurge in interest is the increasing awareness of links between diet and the chronic diseases of man. Comparative studies serve to concentrate attention on relevant features and can also help by providing models for nutrition-linked diseases, such as arteriosclerosis, which can be induced and studied in a variety of experimental animals.

1 • INTRODUCTION

What an animal eats turns into that animal. In this chapter some of the processes involved in this conversion are described at a biochemical level. Animals, though diverse in their feeding biology, are all heterotrophic; that is, they are nutritionally dependent on a supply of organic molecules. These molecules have been synthesized by other organisms, so at the bottom of every food chain there must be an autotrophic organism capable of synthesizing organic molecules from inorganic substances. Quantitatively the most important autotrophs are green plants, which utilize light energy to achieve the synthesis of organic molecules from carbon dioxide.

An animal's food must provide it not only with material for growth but also with a source of the energy necessary to drive synthetic reactions. Energy is also needed for locomotion and for maintaining the uneven distribution of cellular constituents characteristic of living organisms. There is not always a clear-cut distinction between the kinds of ingested molecules used to generate energy and those providing materials for the structure of the organism. In many cases the biochemical conversions that occur, collectively known as metabolism, will vary according to physiological circumstances. Even when an animal is no longer growing, its tissues are undergoing continuous replacement, with the rate of breakdown of any one constituent exactly equaling its rate of formation. Isotopic tracer studies have shown that some 400 g of protein are broken down and resynthesized by an adult man each day.

An early finding in the study of comparative biochemistry was that the major metabolic processes were extraordinarily similar in all organisms. For example, glucose oxidation in insect flight muscle proceeds by the same chemical steps as glucose oxidation in human heart muscle. This realization had the effect of diminishing interest in comparative biochemistry, and attention was concentrated on convenient laboratory animals such as the white rat. Greater diversity is, naturally, seen in the pathways leading to end products such as toxins, pheromones, and pigments, the occurrence of which is limited to certain species only. However, even major metabolic pathways exhibit quantitative differences between species and may, in some cases, be absent. The relative importance of metabolic pathways may also vary from tissue to tissue in the same organism. One of the aims of metabolic biochemistry is to relate these variations to the biology of the animal—for example, to its diet, to its ability to go without food for extended periods, or to the availability of oxygen.

2 • CHEMISTRY OF FOOD

One way of considering metabolism is to start with the input materials, that is, with the molecules available to an animal in its food. These will be of the same chemical type as the body substances of the animal itself. The quantitatively important molecules in the diet fall into four classes, namely, carbohydrates, lipids, proteins, and nucleic acids. This leaves a large heterogeneous group of other molecules, each of which is likely to be present in relatively small amounts. Some of these, the vitamins and minerals, are of particular importance since they cannot be synthesized from other substances.

2.1   Carbohydrates

Carbohydrates are polyhydroxy aldehydes and ketones, of which glucose is the best-known example. The structural formulas of some common carbohydrates are given in Figure 7.1. Most naturally occurring simple carbohydrates (monosaccharides) have between three and seven carbon atoms. In the larger monosaccharides a cyclic structure is favored in which the carbonyl group and a hydroxyl group link to form an internal hemiacetal. Many monosaccharides occur both in the free state and combined into larger molecules. This combination usually involves a glycoside linkage between the carbonyl carbon atom of a monosaccharide and a hydroxyl (or amino) group of a second molecule. If this second molecule is a monosaccharide, a disaccharide is formed. Polymerization can continue if the disaccharide has a free carbonyl group until a polysaccharide is ultimately formed. Since an aldohexose, such as glucose, has five hydroxyl groups, any of which can link to a second monosaccharide, considerable diversity of polysaccharide structure is possible. This is further increased by the possibility of branching and by variations in the sequence and nature of the monosaccharide units. One further degree of variation stems from the fact that in cyclic monosaccharides the carbonyl carbon atom is asymmetric, so two optical isomers can exist, differing in the orientation of the hydroxyl group available for the formation of the glycoside linkage. The two

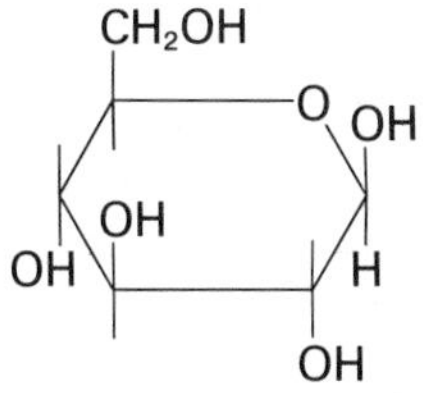

β-D-glucose, an aldohexose
monosaccharide

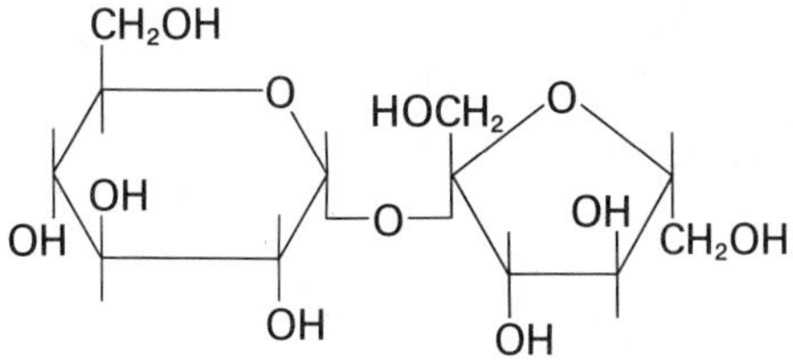

Sucrose, a disaccharide

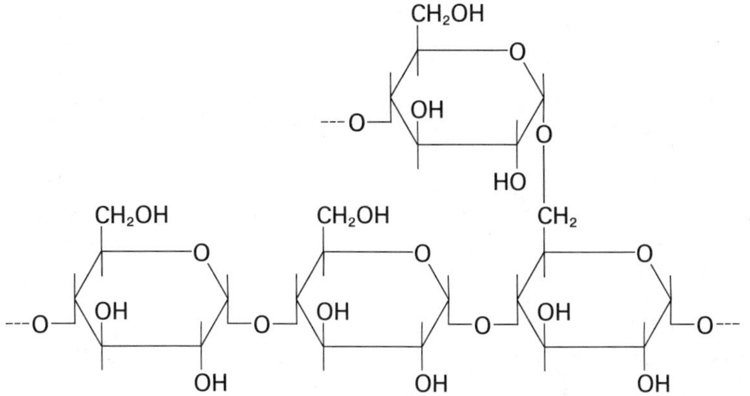

Part of a molecule of the polysaccharide glycogen
showing α(1→4) links and an α(1→6) branch point

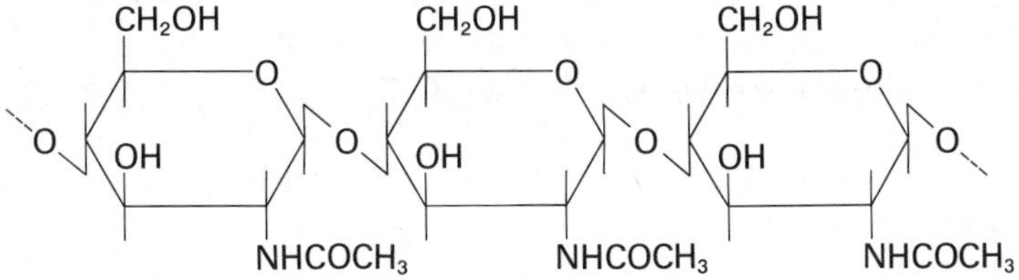

Part of a molecule of the polysaccharide chitin showing
N-acetyl-glucosamine units linked β(1→4)

FIGURE 7.1 Structural formulas of typical carbohydrates occurring in the diets of animals.

possibilities are termed α and β, and this seemingly trivial stereochemical detail can lead to striking differences in the overall properties of the polysaccharides. Thus amylose (a component of starch) is a linear polymer of α-glucose units linked between carbon atoms 1 and 4. It is freely soluble in water in marked contrast to cellulose, which differs structurally from amylose only in that the glycoside linkages in cellulose are β. A further conse-

TABLE 7.1   Dietary Carbohydrates

Carbohydrate	Type	Monosaccharides Present	Food Source
Starch	Polysaccharide	Glucose	Grains, tubers, pulses
Cellulose	Polysaccharide	Glucose	Plants
Glycogen	Polysaccharide	Glucose	Animals
Chitin	Polysaccharide	N-acetylglucosamine	Arthropod exoskeleton
Pentosans	Polysaccharide	Pentoses	Fruit, gums
Sucrose	Disaccharide	Glucose, fructose	Cane and beet sugar
Lactose	Disaccharide	Glucose, galactose	Milk
Maltose	Disaccharide	Glucose	Soybeans
Trehalose	Disaccharide	Glucose	Insects, fungi
Glucose	Monosaccharide	—	Fruits, honey
Fructose	Monosaccharide	—	Fruits, honey

quence of this difference is that separate enzymes are required for the hydrolysis of the two polysaccharides.

Polysaccharides have two major roles in organisms. Some provide a biologically convenient means of storing an energy-yielding fuel in the cell, whereas others have a structural function. Foremost among the energy-storing polysaccharides in animals is glycogen, an $\alpha(1 \rightarrow 4)$-linked polymer of glucose with $\alpha(1 \rightarrow 6)$ links at branch points. Examples of polysaccharides with a structural function are cellulose in plants and chitin in arthropod cuticle. Some of the carbohydrates present in the diet of animals are given in Table 7.1.

2.2   Lipids

Lipids are a less chemically homogeneous group than are carbohydrates and are most usefully defined as derivatives of fatty acids (that is, medium- to long-chain monocarboxylic acids). The group therefore includes fats, oils, waxes, and phospholipids, as well as more complex derivatives. The structural formulas of some typical lipids are given in Figure 7.2. The higher fatty acids occur unesterified only in small amounts, but their esters with the trihydric alcohol, glycerol, are widely distributed. These neutral triglycerides (fats and oils) serve as intracellular energy stores and typically contain unbranched fatty acids, each with an even number (between 12 and 24) of carbon atoms and with up to four double bonds. The fatty acid composition of the triglycerides is frequently charac-

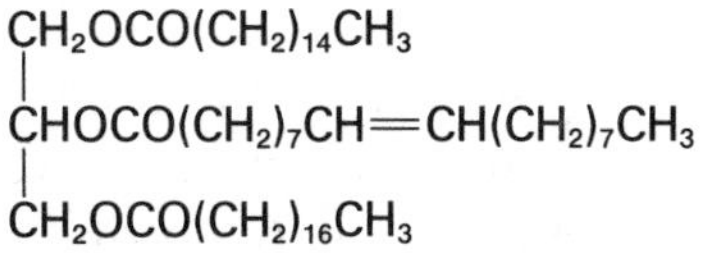

$CH_2OCO(CH_2)_{14}CH_3$

$CHOCO(CH_2)_7CH=CH(CH_2)_7CH_3$

$CH_2OCO(CH_2)_{16}CH_3$

A neutral triglyceride containing a palmityl, an oleyl,
and a stearyl group

$CH_3(CH_2)_{24}CH_2OCO(CH_2)_{14}CH_3$

A wax (hexacosanyl palmitate) occurring in beeswax

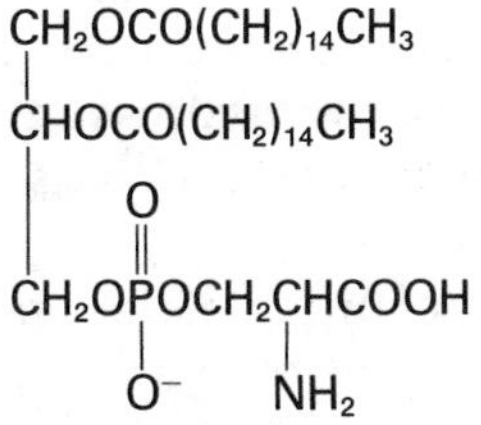

$CH_2OCO(CH_2)_{14}CH_3$

$CHOCO(CH_2)_{14}CH_3$

$$\underset{\underset{O^-}{|}}{\overset{\overset{O}{\|}}{CH_2OPOCH_2CHCOOH}}\quad \underset{NH_2}{|}$$

A phospholipid
(phosphatidyl serine)
containing two
palmityl groups

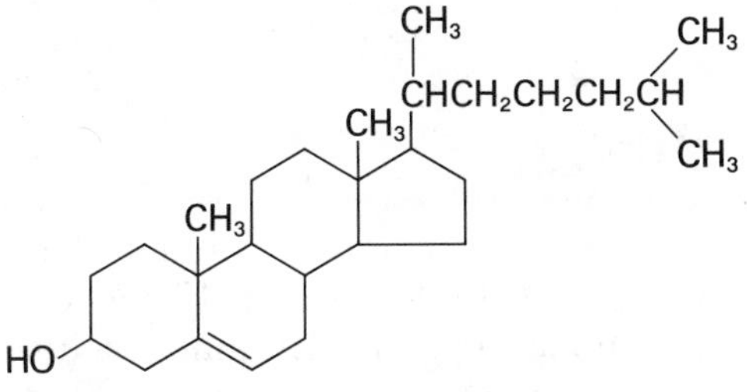

A steroid (cholesterol)

FIGURE 7.2 Structural formulas of representatives of the major classes of lipid molecules (depicted in their un-ionized forms).

teristic of a species. Triglycerides with a high proportion of unsaturated fatty acids (that is, those containing double bonds) have a relatively low melting point and are more abundant in cold-blooded animals since it appears that the lipid store must be fluid at body temperature (see Chapter 10). Certain unsaturated fatty acids, such as linoleic, linolenic, and arachidonic acids, cannot be synthesized by some animals but are required for optimal health and must therefore be present in the diet.

Most waxes are esters of long-chain fatty acids, not with glycerol, but with long-chain monohydric alcohols. They too function as energy stores in some organisms, particularly in deep-sea animals, since some waxes have a low melting point. Waxes of slightly different composition perform structural functions as in beeswax and in the water-repellant wax layer on insect cuticle.

A third major group of lipids, the complex lipids, are found mostly in membranes, where they are an important structural component. They are well suited to membrane formation since they contain hydrophilic groups as well as hydrophobic ones (see Chapter 2). In phospholipids the hydrophilic group is an alcohol (for example, serine, inositol, choline) esterified with phosphoric acid, which forms a second ester link with the terminal hydroxyl group of glycerol. The remaining two hydroxyl groups of the glycerol are esterified to fatty acids. In yet more complex lipids, glycerol is replaced by the long-chain alcohol, sphingosine.

The steroids are often included within the definition of lipids, more on account of their solubility properties than on the basis of chemical structure. Steroids are distinguished by the presence of the polycyclic cyclopentanophenanthrene structure (Figure 7.2). Only those steroids serving structural functions, for example, cholesterol in membranes, are likely to be ingested in any quantity. Many other steroids have specialized functions (such as sex hormones) and are present in small amounts only (see Chapter 9).

2.3   Proteins

Most carbohydrates and lipids lack nitrogen, and it is protein that forms the major dietary source of this element. Proteins are polymers of amino acids, of which twenty kinds are commonly found. The amino acids found in proteins have, with the exception of proline, an amino ($-NH_2$) and a carboxyl ($-COOH$) group attached to the same carbon atom. This carbon atom also bears the side group characteristic of that amino acid. The structural formulas of some representative amino acids are shown in Figure 7.3. The side groups are seen to be chemically diverse, varying in complexity from a single hydrogen atom (glycine) to an indole nucleus (tryptophan). In proline the amino group is incorporated into a five-membered ring with the side chain.

About half of the amino acid types that are required for protein synthesis can be manufactured by higher animals from nonamino acid precursors. The remainder cannot be synthesized and are known as the essential amino acids. These must be present in the diet if normal growth is to occur. The list of essential amino acids varies from one species to another, but man requires leucine, isoleucine, lysine, phenylalanine, methi-

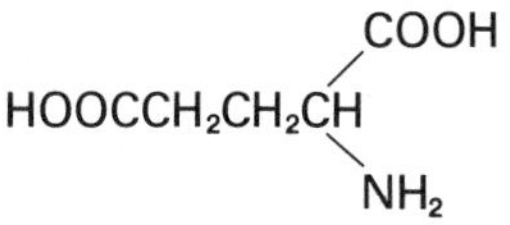

Glutamic acid

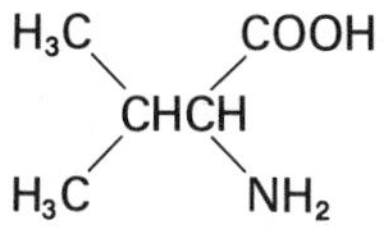

Valine

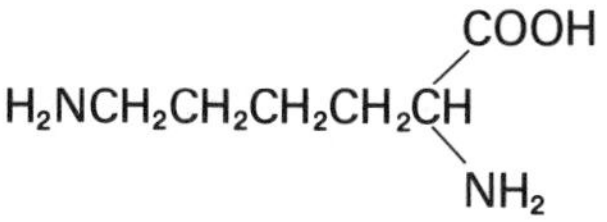

Lysine

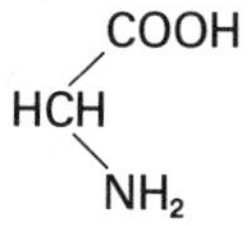

Glycine

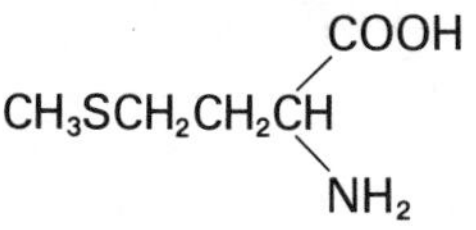

Methionine

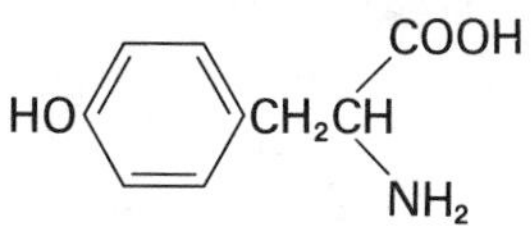

Tyrosine

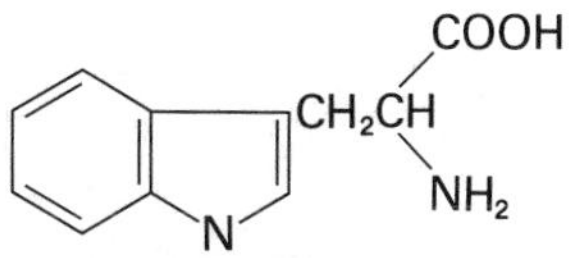

Tryptophan

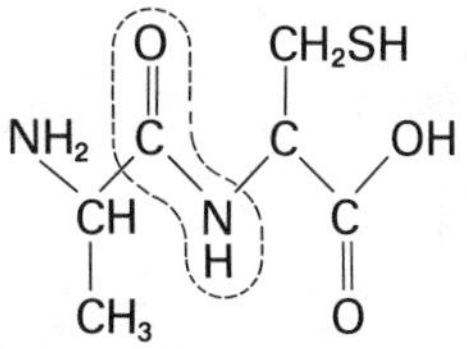

Alanylcysteine, a dipeptide, showing the formation
of a peptide bond

FIGURE 7.3 The structural formulas of some representative amino
acids and a dipeptide.

onine, threonine, tryptophan, and valine. Two others, tyrosine and cysteine, can be synthesized only from members of the group of essential amino acids, and a third, histidine, cannot be synthesized at an adequate rate by infants. Young rats require a dietary supply of arginine; young chickens, a supply of glycine. Amino acids are precursors in the synthesis of many other nitrogenous compounds—for example, histamine (from histidine), pyrimidines (from aspartic acid), creatine (from glycine and arginine), and catecholamines (from tyrosine).

Proteins are formed by the condensation of the carboxyl group of one amino acid with the amino group of a second. This process is repeated to form linear polymers of defined sequence. Natural proteins typically contain chains of up to several hundred amino acids and can be composed of more than one such chain. The enzymes responsible for catalyzing the chemical conversions described later in this chapter are all proteins. Their specificity and catalytic activity are determined by the nature and distribution of groups (for example, —OH, —SH, —COOH, and —NH$_2$) on the amino acid side chains. Other proteins have the property of ordered self-association and aggregate to form such structures as the myofibrils of muscle (actin and myosin) and the microfibrils of hair (keratin).

2.4   Nucleic Acids

Although they are quantitatively less important as a nitrogen source than proteins, the nucleic acids play a vital role in transmitting and expressing

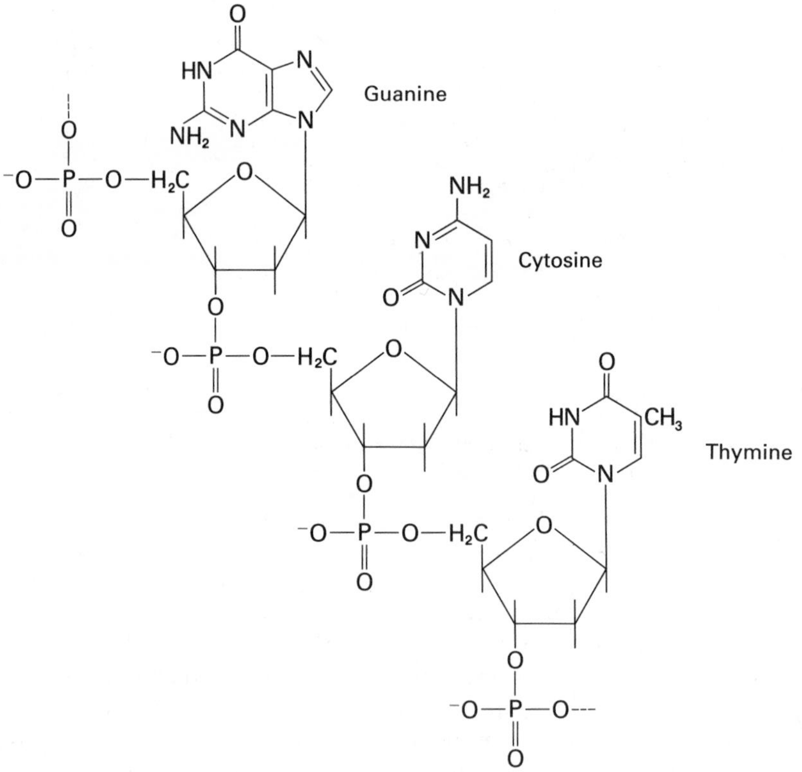

FIGURE 7.4   Structural formula of part of a DNA molecule.

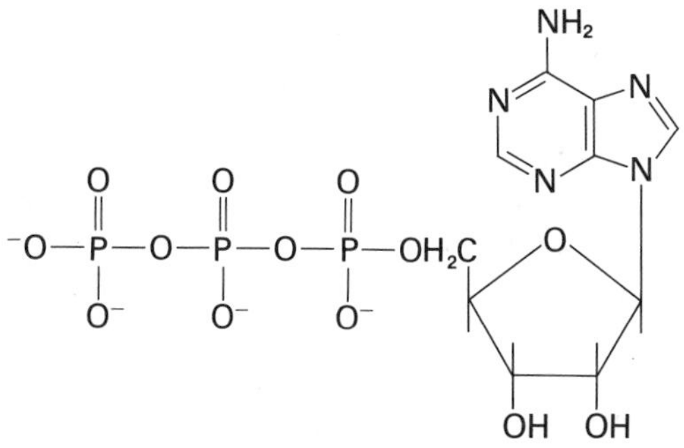

FIGURE 7.5 Structural formula of adenosine 5'-triphosphate (ATP).

genetic information. In the process of cell division the deoxyribonucleic acid (DNA) of the nucleus is copied precisely, so each somatic cell in an organism contains exactly the same genetic information. DNA is a double-stranded molecule, each strand of which is an immensely long, linear polymer of nucleotide units. Each of the nucleotides consists of a nitrogenous base (adenine, guanine, cytosine, or thymine) attached to a monosaccharide (2-deoxyribose) by an N-glycoside link (Figure 7.4). Also attached to the deoxyribose, by an ester link, is a phosphate group through which polymerization is achieved, since the phosphate group can form a second ester link with the deoxyribose portion of the adjacent nucleotide. The two strands of DNA are linked by specific hydrogen bonding between bases. The sequence of bases along the DNA determines protein synthesis.

Outside the nucleus is found a variety of similar molecules, the ribonucleic acids (RNA). They differ from DNA in the nature of the sugar (ribose) and in the fact that the base thymine is replaced by uracil. Ribonucleic acids are single-stranded and show a greater variety in function than does DNA. All are concerned with the translation of genetic information into protein structures. As well as being precursors of nucleic acids, several mononucleotides, such as adenosine triphosphate (ATP), function as intermediates linking energy-producing reactions to energy-yielding reactions (Figure 7.5).

2.5 Vitamins

A large number of compounds, vital to the existence of an organism, do not fall into any of the major classes of molecules described and are present in tissues in relatively small amounts. Most of these can be synthesized in adequate amounts by animals from the major components of the diet, but a small number cannot. These are the vitamins and must be present in the diet. The vitamins are chemically diverse, and they have been classified on the basis of function and solubility properties. Many of the vitamins are required for the synthesis of coenzymes and prosthetic groups (regions of enzyme molecules that are not composed of amino acids). It should be noted that some compounds that do belong to the major chemical classes described (such as certain amino acids and fatty acids) cannot be synthesized and are, like vitamins, required specifically in the diet. Animals vary somewhat in their synthetic capabilities, so vitamins for one species may not be vitamins for another. In some cases an-

TABLE 7.2 Vitamins of Human Nutrition and Their Roles

Vitamin	Major Role	Effect of Deficiency
Retinol (A)	Precursor of retinal, a component of visual purple	Night blindness: keratomalacia
Calciferol (D)	Absorption of calcium and bone deposition	Rickets
Tocopherol (E)	Antioxidant	Sterility (in rats)
Vitamin K	Prothrombin synthesis	Prolonged bleeding
Ascorbic acid (C)	Hydroxylation reactions	Scurvy
Thiamin (B_1)	Prosthetic group in some enzymes acting on keto compounds	Beriberi
Riboflavin (B_2)	Precursor of FAD and FMN (prosthetic groups in some enzymes catalyzing oxidations)	Degeneration of skin and mucous membranes
Nicotinic acid (niacin)	Precursor of NAD and NADP (coenzymes in some oxidations)	Pellagra
Pyridoxine (B_6)	Precursor of prosthetic group in enzymes of amino acid metabolism	Skin inflammation
Pantothenic acid	Precursor of coenzyme A involved in lipid metabolism	—
Biotin	Prosthetic group involved in carboxylations	—
Folic acid (pteroyl glutamic acid)	Transfer of formyl groups	Anemia
Cyanocobalamin (B_{12})	Transfer of methyl groups	Pernicious anemia

Note: Some vitamins exist in more than one interconvertible form.

imals obtain vitamins from the bacteria living symbiotically in their alimentary canal. The vitamin requirements for higher animals are fairly similar, and those essential for human nutrition are listed in Table 7.2.

2.6 Minerals

A number of inorganic substances are required for adequate growth—some in substantial amounts and others only in trace amounts. Many metal ions play an essential part in enzymic catalysis and may be bound either strongly or weakly to enzymes. Magnesium, iron, manganese, zinc, and copper ions most usually fulfill such roles. Calcium is also required, and it serves as an important constituent of skeletal materials: as the phosphate in the endoskeleton of vertebrates and as the carbonate in the exoskeleton of some invertebrates. Calcium ions also have a role in regulating certain cellular processes, particularly muscle contraction. The nonmetals—sulfur, phosphorus, and iodine—can be incorporated into essential metabolites from their ions. Other ions—sodium, potassium, and chloride—are required in larger amounts and are transported to adjust osmotic concentrations and maintain cell volume (see Chapter 6). The transport of sodium and potassium ions is particularly important since differences in their concentrations across membranes give rise to electrical potential differences that form the basis of the conduction of impulses by nerves and muscles (see Chapter 3).

There are three other simple inorganic molecules that can be taken up by animals and incorporated into body constituents, namely, water, oxygen, and carbon dioxide. The absence of nitrogen from this list is significant and determines the dependence of all animals on a source of organic nitrogen compounds.

3 • INTERMEDIARY METABOLISM

Each of the multitude of chemical reactions occurring in living organisms is catalyzed by an enzyme that accelerates, often very greatly, the rate at which the reaction occurs. More than a thousand enzymes have been described, and at first sight there appears to be a bewildering variety of reactions occurring in living organisms. However, the basic types of reaction involved (such as methylation, oxidation of alcohols, and hydrolysis of esters) are relatively few. The diversity of enzymes results from their specificity not only for the chemical groups undergoing reaction but in many enzymes for the whole reactant (substrate) molecule. This specificity enables individual reactions to be regulated by the activity of the appropriate enzyme.

It is a general observation that biological changes are gradual rather than sudden, and this applies equally to biochemical conversions, which proceed by series of relatively simple enzyme-catalyzed steps. To make sense out of this array of reactions, biochemists dissect out major routes of conversion and designate them as metabolic pathways. In relatively few instances are these pathways physically organized into enzyme complexes, so in most cases intermediates are substrates for more than one enzyme. Metabolism is in reality a multibranched web of reactions. Some restriction is imposed, however, on the possible reactions of intermediates due to the existence of membrane-bounded compartments (for example, nucleus and mitochondria) within the cell. Enzyme activities may be different in different compartments, and the membranes impose barriers to the free diffusion of many metabolic intermediates.

Although all chemical reactions are in principle reversible, not all metabolic reactions are at equilibrium under physiological conditions. If, overall, pathways were at equilibrium, there would be no net flux through them. In general the requirement that the pathway be out of equilibrium is achieved by having some reactions very far from equilibrium and other reactions close to equilibrium. This means that some, but not all, reactions in a pathway can proceed in either direction. The reactions that are far from equilibrium are prevented from attaining equilibrium by the relatively low activity of the enzyme involved. In some reactions the attainment of a close-to-equilibrium state would mean that the concentrations of substrate or products would be impractically high or low.

Since no organism can disobey the laws of thermodynamics, there must be an input of energy into the organism that balances the energy expended in locomotion and in driving synthetic reactions. This input is in the form of chemical energy in ingested foodstuffs. Energy-producing and -utilizing reactions are coupled together by the involvement of common chemical intermediates, of which reduced coenzymes and nucleoside triphosphates are particularly important. Many of the degradative reactions employed by organisms involve the oxidation of organic substances, with the concomitant reduction of a coenzyme, which is most frequently nicotinamide adenine dinucleotide (NAD^+) (see Figure 7.6). The reduced form of the coenzyme (NADH) can be used as a reducing agent in synthetic reactions, thus coupling breakdown and synthesis. However, not all energy-yielding reactions are oxidations, and not all energy-requiring reactions are reductions. A more general coupling intermediate, adenosine triphosphate (ATP), has evolved for this purpose (see Figure 7.5). NADH can generate ATP as a result of oxidation of the reduced coenzyme in the electron transport pathway and the associated phosphorylation of ADP. Some ATP is also formed in reactions not associated with the electron transport pathway as, for example, in the substrate level phospho-

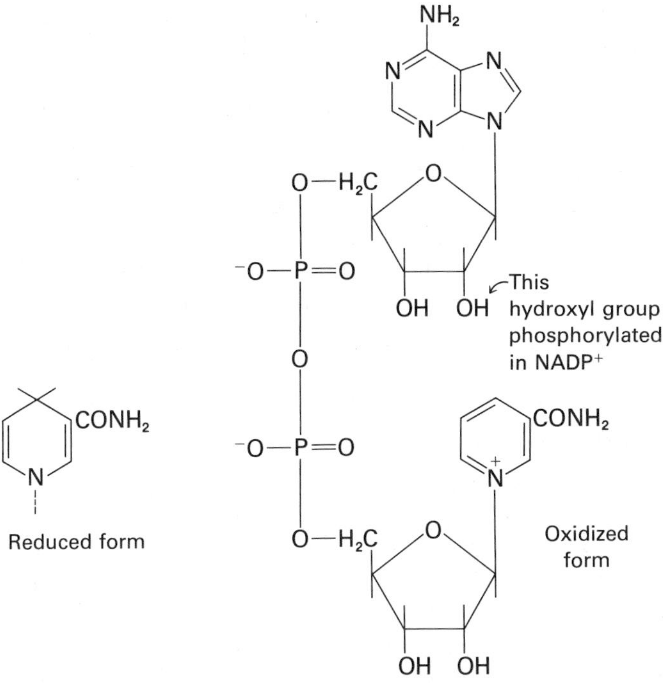

FIGURE 7.6 Structural formulas of oxidized and reduced forms of nicotin-amide adenine dinucleotide (NAD$^+$).

rylation of glycolysis (see Section 3.1). ATP is used not only in synthetic reactions but as the energy source for muscle contraction, active transport, and light production (in luminescent organisms). For this reason ATP is termed a "high-energy compound," thereby recognizing that reactions in which it is the phosphorylating agent will be more favorable than equivalent reactions in which inorganic phosphate is involved. This can be seen if the equilibrium constant[1] for the phosphorylation of glucose by ATP is compared with that for the phosphorylation by phosphate:

glucose + ATP $\rightleftharpoons$ glucose-6-phosphate + ADP $K_{eq} = 4 \times 10^3$
glucose + H$_3$PO$_4$ $\rightleftharpoons$ glucose-6-phosphate + H$_2$O $K_{eq} = 1.2 \times 10^{-3}$

The ability of ATP to function in this way is a consequence of its chemical structure, so the products of the reaction are at a lower energy

---

[1] The equilibrium constant (K_{eq}) of a reaction is the ratio of the product of the concentration of the substances formed in the reaction to the product of the concentration of the reactants (as the equation is written) when the reaction is at equilibrium.

level than ATP itself. In an analogous manner organic chemists find that acetyl chloride is a better acetylating agent than is acetic acid. Other "high-energy" phosphates are present in cells, but all can be synthesized from ATP. A further significant property is that the simple hydrolysis of ATP, though energetically very favorable, does not occur at appreciable rates under physiological conditions for kinetic reasons.

With these general principles in mind it is possible to look in more detail at the reactions that are quantitatively most important in animal metabolism and to see how this basic pattern has been modified in different species. The processes by which food substances are degraded in the alimentary canal and absorbed by the animal are described in Chapter 8.

3.1   ATP Synthesis from Carbohydrates

Perhaps the environmental feature having the greatest effect on the course of metabolism is the availability of oxygen. Vertebrates and many invertebrates are dependent on a continuous supply of oxygen for their survival, but other invertebrates can tolerate the temporary or permanent absence of oxygen. Some, indeed, are poisoned by oxygen. Even vertebrates possess some tissues which can function in the absence of oxygen, including certain muscles, kidney medulla, and erythrocytes. The pathways of ATP synthesis from carbohydrate differ according to whether oxygen is available or not; the aerobic pathway will be described first.

The pathway of aerobic carbohydrate metabolism can be divided into three stages for ease of description: first, the oxidation of glucose to pyruvate (glycolysis);[2] second, the further oxidation of pyruvate to carbon dioxide by the tricarboxylic acid cycle; and, third, the synthesis of ATP from reduced coenzymes by electron transport and oxidative phosphorylation. These pathways are common to all aerobic tissues and very only in detail from species to species.

The glycolytic pathway is summarized in Figure 7.7. In the first reactions two ATP-requiring phosphorylations convert glucose to fructose-1,6-diphosphate, which is then split to give two triose phosphate molecules. In a complex reaction of central importance in glycolysis, these triose phosphates (as glyceraldehyde-3-phosphate) are oxidized by NAD^+ and phosphorylated by inorganic phosphate. The remaining reactions of glycolysis are involved in the transfer of phosphate groups onto ADP to form ATP. A total of four molecules of ATP are synthesized for each glucose molecule degraded, but since two molecules of ATP were used in the

---

[2] Since organic acids are completely ionized at physiological hydrogen ion concentrations, biochemists are accustomed to referring to such acids as their anions.

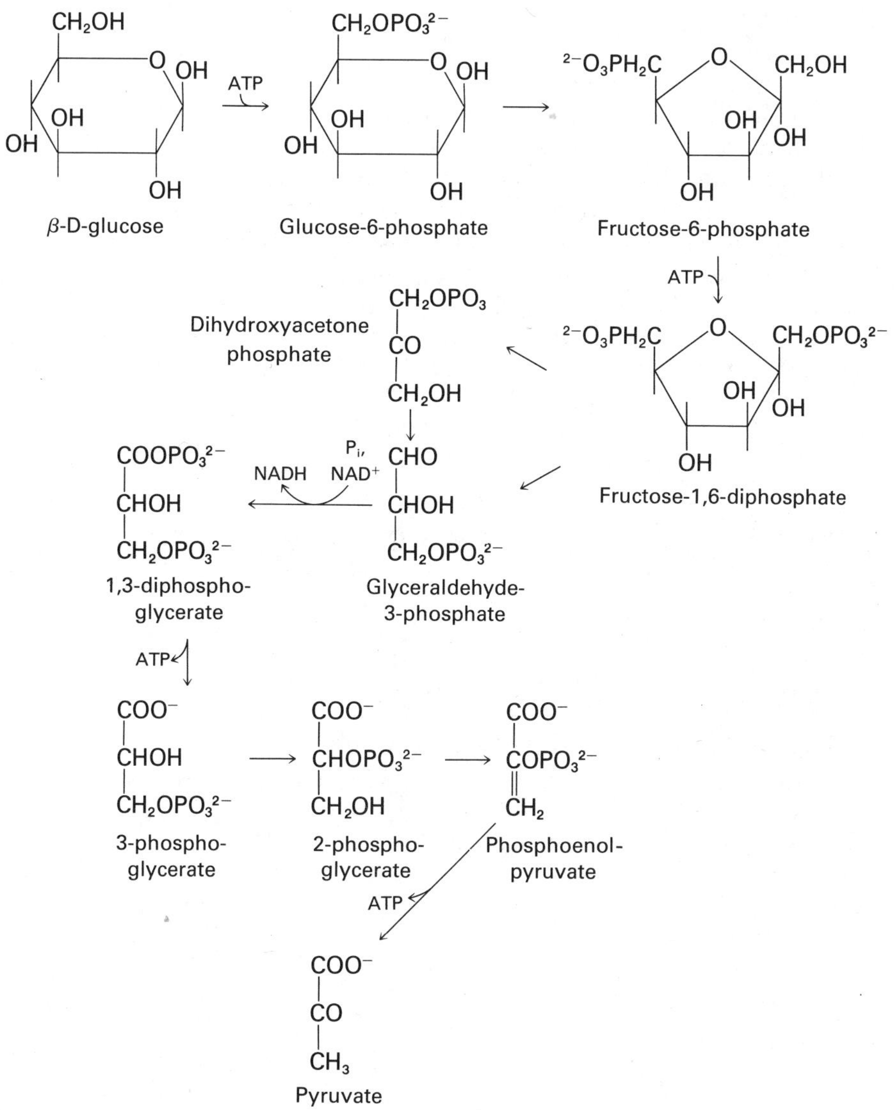

FIGURE 7.7 The pathway of glycolysis from glucose to pyruvate.

initial phosphorylations, the net yield is two ATP molecules per glucose molecule. In many animals glucose is stored as its polymer, glycogen, which enters glycolysis by reaction with inorganic phosphate to form glucose-1-phosphate. This isomerizes to glucose-6-phosphate, so for each glucose unit of glycogen degraded three molecules of ATP are formed. In this way some of the energy required to synthesize glycogen from glucose is regained.

Under aerobic conditions the pyruvate produced in glycolysis enters the mitochondrion. Here its oxidative decarboxylation is coupled to the transfer of the resulting acetyl group to coenzyme A (CoA). The acetyl-CoA enters the tricarboxylic acid cycle (known also as Krebs' cycle or the citric acid cycle) by acetylating oxaloacetate to form citrate. This six-carbon compound is oxidized in a series of steps (Figure 7.8) with associated decarboxylations to form oxaloacetate once more. This molecule can now accept a further acetyl group, thus perpetuating the cycle.

As can be seen from Figure 7.7 and 7.8, for each molecule of glucose oxidized ten molecules of NAD^+ and two of protein-bound FAD have

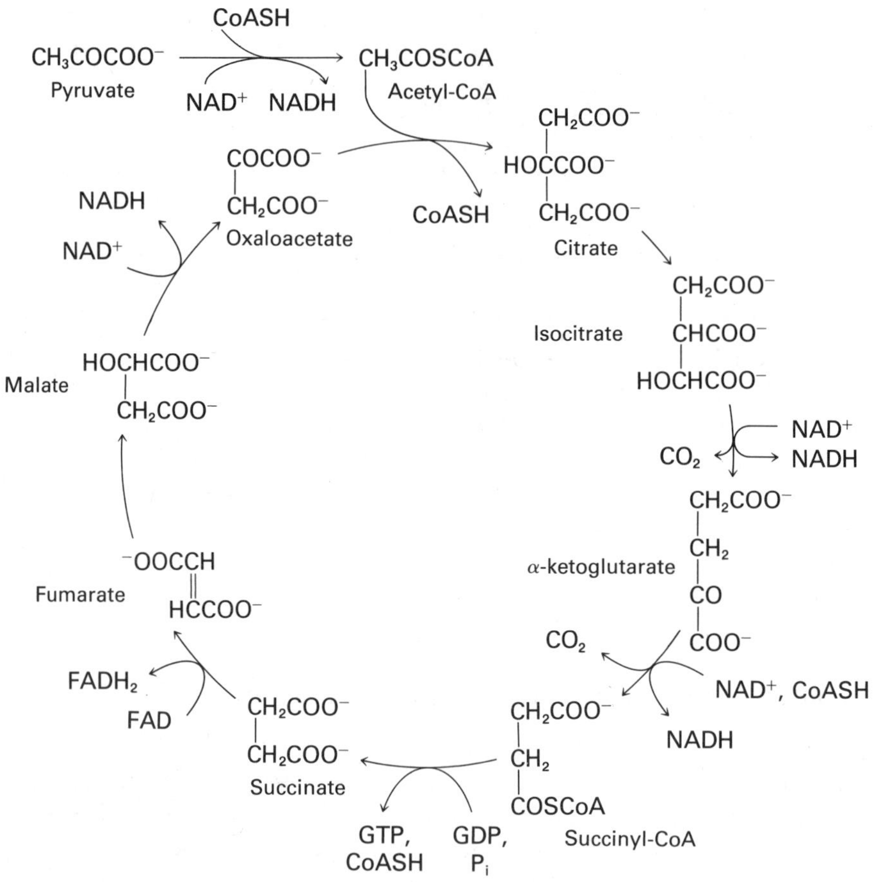

FIGURE 7.8 The tricarboxylic acid cycle.

been reduced. It is during the subsequent oxidation of these coenzymes that the largest part of the ATP yield is produced. Oxidation of these reduced coenzymes takes place not in a single reaction but via a series of enzyme-bound reducible intermediates forming the electron transport pathway (Figure 7.9). Through this pathway electrons are transferred from the reduced coenzymes, via flavoprotein, coenzyme Q, and cytochromes, to molecular oxygen which is reduced to water. The transport of each pair of electrons down this pathway is coupled to the synthesis of three molecules of ATP (from ADP) in a manner not yet fully elucidated. Electron pairs entering at the flavoprotein level yield two molecules of ATP. The coupling of this electron transport to oxidative phosphorylation is normally "tight," and thus the inhibition of phosphorylation (for example, due to the absence of ADP) totally prevents electron transport. Conversely, when oxygen is absent, electron transport cannot occur, so oxidative phosphorylation ceases.

Two of the ten molecules of NADH have been generated in the cytoplasm, but the reactions of electron transport and oxidative phosphoryl-

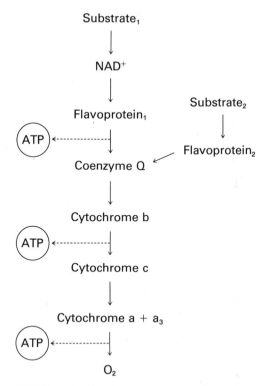

FIGURE 7.9 The electron transport pathway showing sites of ATP synthesis.

ation occur (like those of the tricarboxylic acid cycle) inside mitochondria. Therefore, to realize their ATP-generating potential, these NADH molecules formed in glycolysis must cross the inner mitochondrial membrane. However, this barrier is not permeable to NADH, and different species have devised different means for getting this reducing power into the mitochondrion. In insect flight muscle, which among invertebrate tissues has the greatest need for high rates of oxidation due to the high energy cost of flight, the glycerol phosphate shuttle (Figure 7.10) performs this function. Because the mitochondrial membrane is permeable to both dihydroxyacetone phosphate and glycerol phosphate, one revolution of the

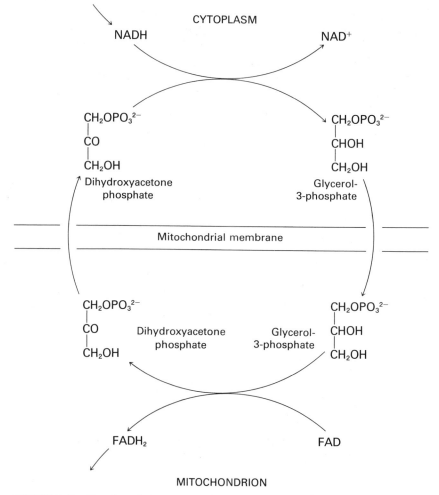

FIGURE 7.10 The glycerol phosphate shuttle for the transport of reducing equivalents from the cytoplasm into the mitochondrion.

shuttle results in the reduction of one molecule of flavoprotein inside the mitochondrion at the expense of one molecule of NADH in the cytoplasm. Two molecules of ATP can then be synthesized from each molecule of flavoprotein reduced.

This mechanism probably does not operate in vertebrates since the activity of the mitochondrial glycerol phosphate dehydrogenase is too low in most tissues. However, another shuttle, similar in principle but different in detail, has been proposed for vertebrate muscle (Figure 7.11). The NADH in the cytoplasm reduces oxaloacetate to malate in a reaction catalyzed by malate dehydrogenase. This malate can cross the mitochondrial membrane into the mitochondrion, where the same enzyme catalyzes the reduction of mitochondrial NAD⁺ by the malate. The shuttle is completed by the transport of oxaloacetate back into the cytoplasm, which itself requires another shuttle involving transamination reactions. Although three

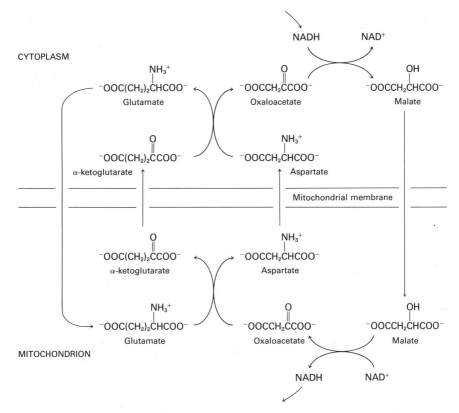

FIGURE 7.11 The malate-oxaloacetate shuttle for the transport of reducing equivalents from the cytoplasm into the mitochondrion.

TABLE 7.3   Summary of ATP Molecules Produced by the Oxidation of
             One Molecule of Glucose

Process	Reduced Coenzyme	ATP Yield
Glycolysis	2 NADH	6
	—	2
Pyruvate dehydrogenase	2 NADH	6
Tricarboxylic acid cycle	6 NADH	18
	2 FADH$_2$	4
	—	2 (as GTP)
		Total 38

ATP molecules can be synthesized from the NADH transported into the mitochondrion, part of this energy must be utilized in the active transport of components of the shuttle. This is necessary to ensure that the shuttle remains out of equilibrium and hence has a net direction. A summary of the ATP molecules produced at different stages from one glucose molecule is presented in Table 7.3.

In the absence of oxygen no oxidative phosphorylation can occur, so the tissue becomes entirely dependent on the substrate level phosphorylations of glycolysis. However, this pathway produces NADH and, unless it can be reoxidized, the NAD$^+$ in the cell will rapidly be used up, and glycolysis will stop. The problem of the reoxidation of glycolytic NADH under anaerobic conditions has not been solved in the same way by all anaerobic organisms. All have, however, employed variations on the same theme, that is, using the pyruvate produced in glycolysis or a derivative thereof as an oxidizing agent. It is in vertebrates that the simplest variation of this ploy is seen. Anaerobic tissues of vertebrates possess high activities of the enzyme lactate dehydrogenase, which catalyzes the reaction:

$$CH_3CO \cdot COOH + NADH + H^+ \longrightarrow CH_3CH(OH) \cdot COOH + NAD^+$$

Since vertebrates are only locally anaerobic, the lactate passes into the blood and is transported to the liver, where it can be used for the resynthesis of glucose. Although yeast is not an animal, its anaerobic metabolism has long attracted the attention of man. This organism oxidizes its NADH, not by the use of pyruvate but by acetaldehyde formed by the decarboxylation of pyruvate. The end products of this anaerobic metabolism are therefore carbon dioxide and ethanol, both of which are excreted by the yeast cell.

It will be appreciated that the ATP yield from anaerobic pathways is very low compared with that from aerobic pathways, but some invertebrates have managed to make use of another ATP-producing reaction in the absence of oxygen. This involves the reduction of fumarate to succinate by a flavin-linked dehydrogenase, which at the same time reoxidizes the glycolytic NADH. This reaction (fumarate reductase) was first discovered in the pig roundworm (*Ascaris lumbricoides*) and has since been reported from other helminth parasites. It is, in principle, a reversal of the succinate dehydrogenase reaction occurring in the tricarboxylic acid cycle, although a different enzyme is involved. ATP synthesis probably occurs during electron transport from NADH to reduce the flavoprotein. Consideration must be given to the pathway by which fumarate is synthesized (Figure 7.12). *Ascaris* lacks pyruvate kinase, and the phosphoenolpyruvate produced in glycolysis is carboxylated to oxaloacetate. The oxaloacetate is reduced to malate, which is dehydrated to give fumarate. However, this reduction also requires NADH, which must be generated if all that produced in glycolysis has been used to reduce fumarate. What appears to happen is that the glycolytic NADH reduces the oxaloacetate, and more NADH is generated for the fumarate reductase step because half of the malate molecules are oxidatively decarboxylated to pyruvate. The net result is that an extra ATP molecule is generated for each pair of phosphoenolpyruvate molecules formed by glycolysis. The end products of this pathway appear to be pyruvate and succinate, but these are not excreted

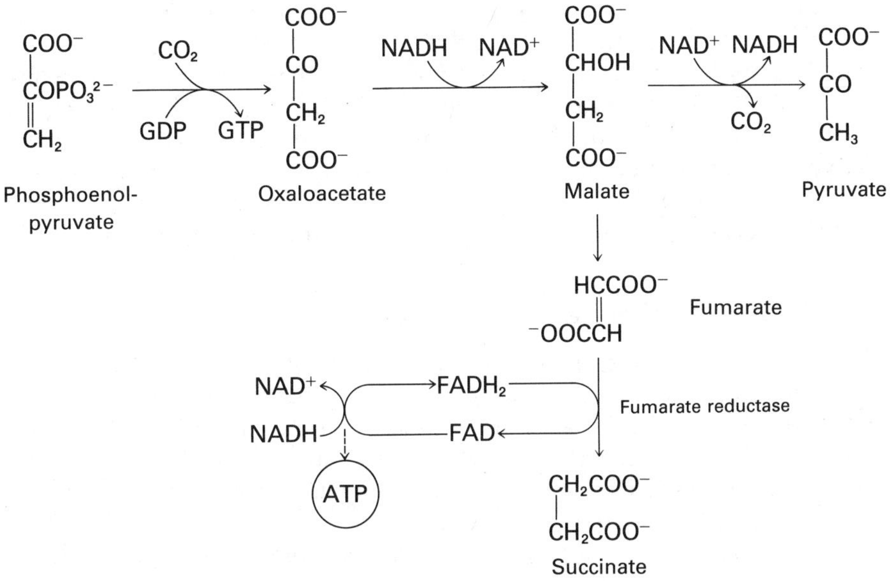

FIGURE 7.12 The pathway of anaerobic ATP synthesis from phosphoenolpyruvate in *Ascaris lumbricoides*.

as such. Perhaps in an attempt to form an excretory product that is less irritating to its host, these compounds are converted into methylbutyrate and methylvalerate by a series of balanced oxidations and reductions.

As described above, certain vertebrate muscles carry out anaerobic metabolism. But why should muscles of an aerobic organism become anaerobic? It appears that the rate of ATP synthesis in muscles required to contract very vigorously for short periods of time is limited by oxygen supply. Under these circumstances some advantage is apparently gained by switching over to oxygen-independent glycolysis, even though its rate must be very high to compensate for the low ATP yield of this pathway. Anaerobic metabolism predominates in the so-called white muscles, which owe their lack of color to the absence of cytochromes, together with the rest of the oxidative machinery. These muscles are adapted to perform relatively short bursts of intense contractile activity, in contrast with the red muscles, which are capable of steadier, more sustained, contractile activity. Many muscles contain both white and red fibers, but extreme cases in which only one type is present do occur. The psoas muscle of the rabbit, lying alongside the spine, is used in rapid locomotion and is almost entirely white. This contrasts with the semitendinosus muscle in the rabbit leg, which has a postural function and is deep red. The breast muscles of game birds, such as the pheasant, and of the domestic fowl are typically white, since these birds use flight primarily as an escape reaction. Other birds, like the pigeon, are capable of prolonged steady flight, and their breast muscles are characteristically red. Although the bulk of fish muscle is white, fishes frequently possess a small amount of lateral red muscle. Experiments with dogfish have shown that during normal swimming only the red muscle is used, but that during vigorous swimming (to catch prey or escape a predator) the large mass of white muscle becomes active. Apart from muscles, several other vertebrate tissues are largely anaerobic, including the lens, where blood vessels would impair vision, and fetal tissues, which can withstand transient anoxia during birth.

3.2   ATP Synthesis from Lipids

A second major source of ATP is the oxidation of fatty acids. These fatty acids are stored as triglycerides and made available by hydrolysis catalyzed by lipases. Many animals store and use both carbohydrate and lipids for the production of ATP, but in the absence of oxygen only the former can be used. Thus white muscles characteristically lack lipid stores, whereas red muscles frequently contain appreciable amounts.

The oxidation of fatty acids can again be considered as occurring in three stages. Only the first of these, the degradation of long-chain fatty acids to two-carbon acetyl-CoA, differs from carbohydrate oxidation. In the second stage the acetyl-CoA enters the tricarboxylic acid cycle, where it is completely oxidized and NADH is produced. Finally, ATP is generated from the reduced coenzymes formed in the first two stages. Since fatty acid oxidation occurs in the mitochondrial matrix, the long-chain fatty acid must first cross the inner mitochondrial membrane. This membrane is impermeable to fatty acids and their CoA esters, and so the fatty acyl group is transferred from its thioester with coenzyme A to an ester with carnitine $((CH_3)_3N^+ CH_2CH \cdot OH CH_2COOH)$. This ester can be transported across the inner mitochondrial membrane into the matrix where the CoA ester is reformed. Oxidation then proceeds in a stepwise manner, in which successive two-carbon units are split off from the carboxyl end of the fatty acid as acetyl-CoA. This process, known as β-oxidation, involves four reactions (Figure 7.13), during which the fatty acid derivatives remain attached to coenzyme A. Two of these reactions

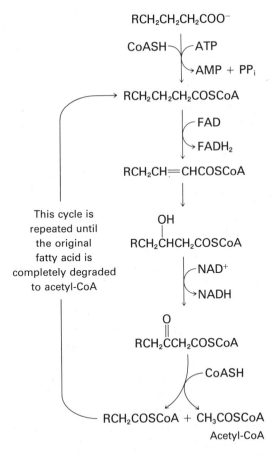

FIGURE 7.13   The β-oxidation pathway of fatty acid degradation.

TABLE 7.4   Summary of ATP Molecules Produced by the Oxidation of
One Molecule of Palmitic Acid

Process	Reduced Coenzyme	ATP Yield
Fatty acyl—CoA formation	—	−1
First cycle of β-oxidation	FADH$_2$	2
	NADH	3
Six further cycles	6 FADH$_2$	12
	6 NADH	18
Oxidation of one acetyl-CoA	FADH$_2$	2
	3 NADH	9
	—	1 (as GTP)
Oxidation of seven molecules of acetyl-CoA	7 FADH$_2$	14
	21 NADH	63
	—	7
	Total	130

are oxidations, the first involving a flavoprotein and the second NAD$^+$. In the final reaction of β-oxidation the β-keto fatty acyl-CoA undergoes cleavage, involving a second molecule of coenzyme A, to give acetyl-CoA and a fatty acyl-CoA with two fewer carbon atoms than the original fatty acid. This series of reactions is repeated until the whole molecule is degraded to acetyl-CoA. These reactions appear to be similar in all species studied. The yield of ATP from palmitic acid ($CH_3(CH_2)_{14} COOH$) is summarized in Table 7.4.

3.3   ATP Synthesis from Proteins

Although it is not normally considered a source of energy, protein may serve as such under a variety of conditions. The carbon skeletons of amino acids can, in principle, be used to generate ATP either directly by complete oxidation or indirectly via conversion to lipids or carbohydrates. The metabolic pathway followed varies according to the amino acid involved, but in general the amino acid is first deaminated to give the corresponding keto acid. This keto acid is channeled toward the degradative reactions already discussed in connection with carbohydrate and lipid oxidations. Most of the amino acid carbon skeletons are converted to intermediates of the tricarboxylic acid cycle, from which synthesis of carbohydrates is readily achieved. The production of glucose from amino acids (gluconeogenesis) is of particular importance since the ability of animals to store carbohydrates is limited. Glucose is required for metabolism not only in anaerobic tissues but also in nervous tissues, which, though aerobic, possess an obligatory glucose requirement. Not all amino acids are glucogenic—that is, capable of yielding glucose—since some are degraded to acetyl-CoA, from which glucose synthesis is not possible in animals. Administration of large quantities of this latter group of amino acids causes an increase in the concentration of circulating ketone bodies (acetoacetate and β-hydroxybutyrate); hence this group is referred to as the ketogenic amino acids. Only lysine and leucine are entirely ketogenic, but isoleucine, phenylalanine, tyrosine, and tryptophan are degraded in part to acetyl-CoA and in part to a gluconeogenic precursor.

Under what circumstances, then, will proteins constitute a significant energy source? One group of animals in which this occurs is the carnivores, whose diet will be rich in protein but relatively deficient in carbohydrate. Ruminants are a second group of animals likely to rely heavily on gluconeogenesis from amino acids. Any carbohydrate in the ruminant diet will be converted to short-chain fatty acids (for example, acetate, propionate, and butyrate) by the bacterial flora of the rumen, so the animal is chronically short of carbohydrate (see Chapter 8). Gluconeogenesis can be of importance too in omnivores, such as man, during starvation. Under these conditions muscle protein is broken down to provide glucose. This process is likely to be significant even during short periods of starvation, since the total glucose and glycogen reserves of a 70-kg man have been estimated at 245 g, whereas the glucose requirement is 180 g/day.

3.4 Storage and Transport of Fuel

The greatest rate of ATP utilization is found in cells performing contractile work. It can be calculated from oxygen uptake data that during flight the blowfly *Lucilia sericata* has an ATP utilization of over 2000 μmole/g muscle/min. Sacktor has measured the ATP concentration in fly flight muscle and obtained a value of 7 μmole/g. Thus if there were no resynthesis of ATP, flight could be sustained for only one-fifth of a second. In fact only slight changes in the amount of ATP in muscle can be detected during contractile activity, so ATP resynthesis must occur as soon as contraction begins. If ATP is not stored as a reserve, what is the source of this ATP? Some muscles contain appreciable amounts of phosphagens, which are N-phosphorylguanidine compounds capable of reversibly phosphorylating ADP (see Table 7.5). In vertebrate muscle the guanidine is creatine, but in most invertebrates it is arginine. In various species of the Annelida and related phyla, six other phosphorylated guanidines are found (Table 7.5). Although the reason for this diversity is unknown, it is assumed that all phosphagens have the same physiological function of rephosphorylating ADP during contraction. In a few muscles—for example, lobster (*Homarus*) abdominal muscle and scallop (*Pecten*) fast adductor muscle—the phosphagen constitutes a true "high energy" phosphate store since these muscles can use ATP at a rate greater than it can be generated

TABLE 7.5 Structure and Occurrence of Phosphagens

Phosphagen	Structure	Occurrence
Phosphocreatine	H_2O_3PNH—$C\cdot N\cdot CH_2COOH$, $\overset{\parallel}{N}H$ $\overset{\mid}{C}H_3$	Vertebrates (and some echinoderms, annelids, and sponges)
Phosphoarginine	H_2O_3PNH—$C\cdot NH(CH_2)_3CHCOOH$, $\overset{\parallel}{N}H$ $\overset{\mid}{N}H_2$	Most invertebrate phyla
Phosphoglycocyamine	H_2O_3PNH—$C\cdot NHCH_2COOH$, $\overset{\parallel}{N}H$	Some polychaete annelids
Phosphotaurocyamine	H_2O_3PNH—$C\cdot NHCH_2CH_2SO_3H$, $\overset{\parallel}{N}H$	Some polychaete annelids and related phyla

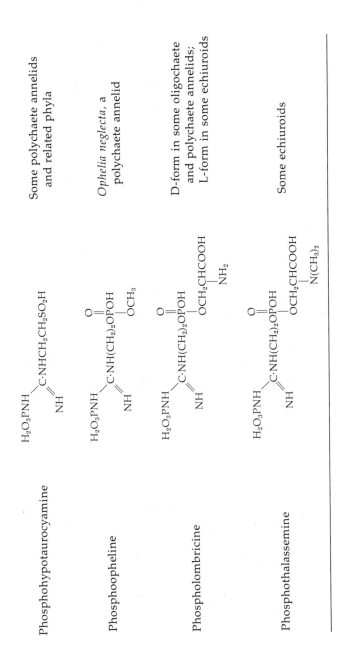

Phosphohypotaurocyamine	Some polychaete annelids and related phyla
Phosphoopheline	*Ophelia neglecta*, a polychaete annelid
Phospholombricine	D-form in some oligochaete and polychaete annelids; L-form in some echiuroids
Phosphothalassemine	Some echiuroids

by degradation of carbohydrates or lipids. These muscles are capable of only a limited number of vigorous contractions before the phosphagen is depleted and exhaustion occurs. In such muscles the phosphagen (phosphoarginine) content may exceed 40 μmole/g, and the enzyme catalyzing the rephosphorylation of ATP (arginine kinase) may exceed 13 percent of the soluble protein in the muscle. Much lower concentrations of phosphagens are present in those muscles that can sustain longer periods of activity. For example, the blowfly flight muscle referred to above has only 3 μmole of phosphoarginine per gram of muscle. In these muscles ATP must be rapidly synthesized from one of the fuels discussed in previous sections, and phosphagen acts only as a transient source of ATP as contraction is initiated.

Energy storage compounds (fuels) are in general either insoluble or have a high molecular weight. If this were not the case, large fluctuations in the concentration of these substances would cause deleterious changes in osmotic pressure (see Chapter 6). They must also be nontoxic and will be most useful if they have a high ratio of calorific value to weight. Fuels can be stored either in the cells where the ATP is required or in other tissues, in which case the fuel must be mobilized and transported when required. Polysaccharides and triglycerides are the most widely used storage compounds, and the advantages and disadvantages of these forms will be considered.

Carbohydrates.   In animals the predominant energy storage polysaccharide is glycogen, with a molecular weight of from 1×10^6 to about 2×10^8. It is mobilized for use within the cell by reaction with phosphate to produce glucose-1-phosphate, which is isomerized to the glucose-6-phosphate and thus enters the glycolytic pathway. Carbohydrate is generally transported in the blood in the form of glucose which is produced by the hydrolysis of the glucose-6-phosphate. However, in insects and some other invertebrates the transport carbohydrate is trehalose, a disaccharide composed of two glucose units linked $\alpha,\alpha(1 \rightarrow 1)$. The advantage of trehalose over glucose may be that its contribution to the hemolymph osmotic pressure is only half that of a calorifically equivalent amount of glucose. A disadvantage, however, is that its synthesis from glycogen (via UDP-glucose) requires a molecule of ATP not needed in glucose production.

Dried glycogen has a calorific value of 4.12 kcal/g, but it is extremely hydrophilic and under physiological conditions is highly hydrated. The presence of this associated water effectively reduces the calorific value to about 1.5 kcal/g. Despite this low calorific value, it is the only storage substance that can be used by anaerobic cells.

Lipids.   Triglyceride presents a much more economical proposition as a fuel. Not only does its calorific value (9.3 kcal/g) greatly exceed that of glycogen, but it is stored free from water. Triglycerides are particularly important when the energy/weight ratio must be kept high, as in migrating birds, for example. In one study several species of birds migrating from north to south across the Gulf of Mexico were captured after their flight, and their fat reserves were compared with those of nonmigrating birds of the same species. The fat index (the amount of ether-extractable fat material per gram dry weight of nonfat material) was between 2.5 and 3.5 in the premigration birds compared with values of between 0.2 and 0.4 in nonmigrating birds. Migrating insects too depend on stored triglyceride to sustain flight. An interesting example is the locust (*Schistocerca gregaria*), which uses carbohydrate for short flights but switches over to fat utilization after the first hour of flight. It has been calculated that this species can store sufficient triglyceride for about 12 hr of flight. Man also stores large quantities of triglyceride, and, although some individuals store much more than others, a typical 70-kg male will have 7 kg of lipid (and a 60-kg female will have 15 kg of lipid) which should enable him (and especially her) to survive many weeks of starvation. In mammals and birds triglyceride is stored predominantly in adipose tissue, the cells of which may contain 90 percent lipid. The distribution of adipose tissue varies according to the needs of different species. In man, for example, adipose tissue is distributed subcutaneously and around the viscera, where it probably subserves other functions such as thermal insulation and mechanical protection. Other animals, whose access to sources of food is irregular, have evolved special regions of fat deposition, such as the hump of the camel and the tail of the fat-tailed sheep. In aquatic mammals the subcutaneous adipose tissue is particularly extensive and forms blubber, which is important in thermoregulation (see Chapter 10). The lower vertebrates have little adipose tissue, but they do store triglycerides in the liver (hence cod-liver oil) or in their muscles (as in the so-called "oily" fishes, such as salmon, herring, and mackerel).

Despite its suitability as an energy storage substance, triglyceride has the disadvantage that it cannot be used to generate ATP in the absence of oxygen. If it were possible to synthesize carbohydrate from triglyceride, then this could occur when oxygen was available, and the carbohydrate would be used later under anaerobic conditions. However, with one exception, the roundworm *Ascaris*, animals are unable to carry out this conversion. No bypass reaction exists for the pyruvate dehydrogenase reaction, which is irreversible under physiological conditions. The oxaloacetate formed in the tricarboxylic acid cycle (Figure 7.8) cannot be converted to glucose since it is required for the continued operation of the cycle. Many plants and microorganisms are not subject to this restriction since they possess the enzymes of the glyoxylate shunt (Figure 7.14). In this pathway isocitrate is split to form succinate (which continues the cycle) and glyoxylate. The glyoxylate then condenses with a second acetyl-CoA

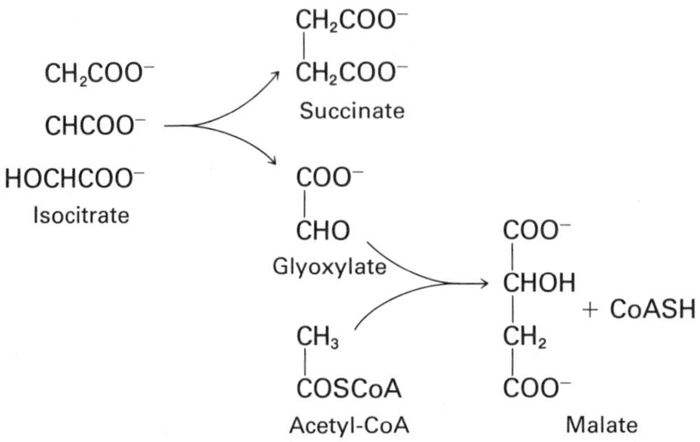

FIGURE 7.14 Pathway of the glyoxylate shunt.

molecule to form malate, which can be used synthetically. Despite the apparent advantages of being able to synthesize carbohydrate from fat, the existence of this pathway has been established in only one animal species, *Ascaris lumbricoides*. In this species the enzymes of the glyoxylate shunt are most active during the final larval stages, and it appears that the shunt functions to convert the triglyceride reserves of the aerobic larva to the glycogen stores of the anaerobic adult.

A second problem that must be overcome by organisms storing triglyceride is its efficient transport within the body. The insolubility of triglycerides precludes their mobility as such, but the problem is solved to some extent by the formation of lipoprotein emulsions. Lipoproteins of different composition, some containing cholesterol and phospholipids in addition to protein and triglyceride, are used for different transport functions. For example, in mammals chylomicrons are the lipoprotein complexes that transport triglyceride from the gut, and very low-density lipoprotein is involved in transport of triglyceride from the liver. Insects appear to transport a significant proportion of their fat as diglycerides, again in combination with proteins. In mammals not all lipid is transported as triglyceride since much is hydrolyzed, especially in adipose tissue, to liberate free fatty acids (FFA). These are not very soluble either, and they have a detergent action that can inactivate enzymes. Both problems are overcome by the transport of FFA in association with plasma albumin.

The availability of albumin in the plasma may well set an upper limit to FFA transport, and in times of increased lipid utilization (for example, starvation) ketone bodies make their appearance. The ketone bodies are acetoacetate ($CH_3CO \cdot CH_2 COOH$) and β-hydroxybutyrate

($CH_3CH \cdot OH\ CH_2\ COOH$). These four-carbon compounds are synthesized from long-chain fatty acids in the liver and have been described as "predigested lipids." They can be oxidized by many extrahepatic tissues. Their significance lies not only in their solubility and relative nontoxicity but also in the fact that they are the only form of lipid that can be utilized by the mammalian brain. The ability of the brain to utilize ketone bodies varies from species to species. The brains of ruminant animals have considerable ability to use ketone bodies, consistent with the nonavailability of glucose in these animals. The brain of the adult rat gains only a small proportion of its calorific requirement from ketone bodies, but in the newborn rat the brain has a much greater ability to utilize ketone bodies. In this situation the ketone bodies may provide material for phospholipid synthesis in the developing brain. An interesting situation occurs in man, whose brain does not normally use ketone bodies. During the first few days of starvation, however, enzymes for the utilization of ketone bodies are induced in the human brain, so after prolonged starvation as much as 70 percent of the calorific requirement of the brain is met by ketone bodies.

Proteins.   Any protein used for energy production in fed animals is derived from the diet. There is no protein with a specific energy storage function in animals. During starvation muscle protein is broken down to amino acids that are transported in the blood to the liver, where their fate is primarily gluconeogenic. Some amino acids, especially the branched-chain amino acids which are not readily converted to glucose in the liver, are partly degraded in muscle.

Although proteins are not stored specifically as energy sources, there is a single example of an amino acid, proline, being stored to provide energy for muscular contraction. Bursell found that the resting tsetse fly (*Glossina morsitans*) contained 150 μg proline and 20 μg alanine. The concentration of proline in the hemolymph was 180 mM. After a flight of 2 min the proline content had dropped to 25 μg, and the alanine content had risen to 115 μg. The probable metabolic pathway followed in the conversion of proline to alanine yields 14 molecules of ATP and is outlined in Figure 7.15. Proline utilization to power muscle contraction is not widespread among the insects, but it may also occur in some beetles. What can be the significance of this novel energy store? First, all arthropods have a relatively high total amino acid content in their hemolymph. Second, there is no change in osmotic pressure when proline is converted to equimolar amounts of alanine. If there is an advantage in having a soluble energy store, without attendant mobilization and transport problems, then proline would have an advantage since it is very soluble, would not disturb acid-base balance, has no other specific function, and gives a high

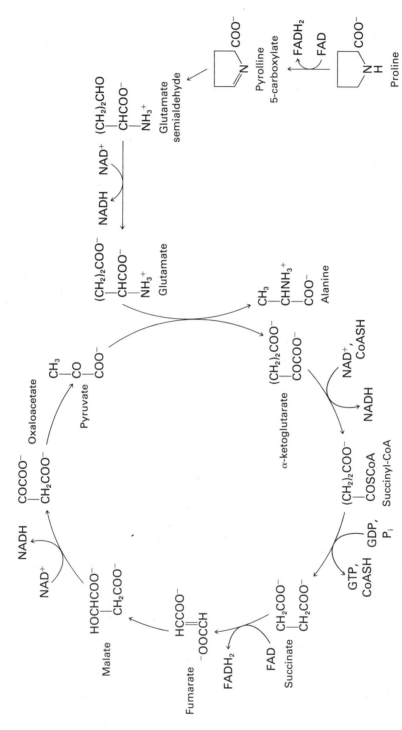

FIGURE 7.15 Pathway of proline oxidation in the flight muscle of the tsetse fly.

yield of ATP. During recovery the proline is presumably resynthesized from the alanine together with other substances from the diet. Why the tsetse fly has adopted this fuel strategy has yet to be explained.

Experimental Investigation of Fuel Utilization. Three types of method have been widely used to gain information on the fuels used by organisms and tissues. The first involves measuring the respiratory quotient (RQ), that is, the ratio of carbon dioxide produced to oxygen taken up. This can only be done for whole organisms if the respiratory gases can be collected or for tissues removed from an animal and incubated in vitro. The RQ is unity if carbohydrate alone is being used and 0.70 for lipid oxidation:

$$CH_3(CH_2)_{14}COOH + 23O_2 \longrightarrow 16CO_2 + 16H_2O$$

The respiratory quotient for protein oxidation depends on the amino acids involved but is intermediate between these values.

A more direct method, which is sometimes applicable, is the measurement of fuel reserves before and after activity. This approach normally requires killing the animal involved, but an extension of this method has been used to investigate fuel utilization in different vertebrate tissue that does not entail sacrifice of the animal. The concentration of fuel substances entering the organ in the arterial blood is compared with their concentration in the outflowing venous blood. Use of this method involves the assumption that there is no significant endogenous storage of fuel.

Finally, a less direct but more widely applicable method involves the determination of maximal activities of rate-limiting enzymes in different pathways. Thus phosphorylase activity would provide an index of glycogen-utilizing potential, hexokinase of glucose-utilizing potential, and phosphofructokinase an index of total glycolytic potential. This method can give information only on potential fuel utilization, but it is simple enough to be used to compare fuel utilization in a large number of tissues.

3.5 ATP Utilization

To consider all the ways in which ATP can be used by living organisms is beyond the scope of this chapter. Instead, two examples will be considered in detail. The utilization of ATP for muscle contraction is described in Chapter 4.

Carbohydrate Synthesis. Gluconeogenesis (from lactate, glycerol, and amino acids) in higher organisms occurs primarily in the liver. To give

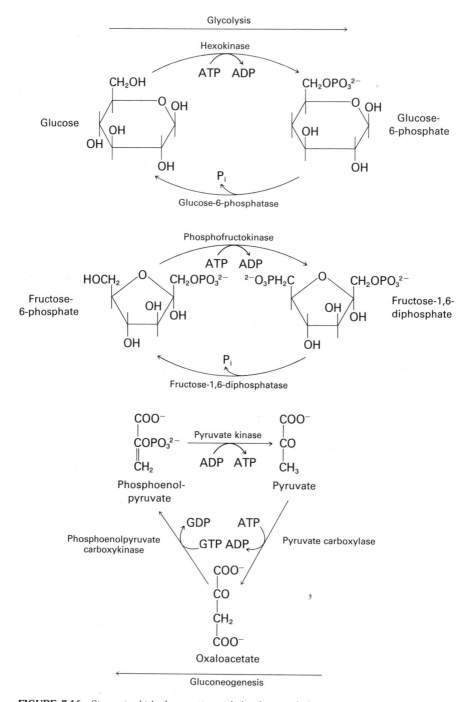

FIGURE 7.16 Steps at which the reactions of glycolysis and gluconeogenesis differ.

direction to carbohydrate metabolism in this organ, certain reactions are maintained away from equilibrium. Three such far-from-equilibrium reactions occur in glycolysis, namely, hexokinase, phosphofructokinase, and pyruvate kinase, so at these steps gluconeogenesis must involve different reactions. The remainder of the reactions in the pathway are close to equilibrium and can be used for both glycolysis and gluconeogenesis. The reactions specific to gluconeogenesis are outlined in Figure 7.16. Glucose-6-phosphatase catalyzes the hydrolysis of glucose-6-phosphate to glucose, whereas fructose diphosphatase catalyzes a similar hydrolysis of fructose-1,6-diphosphate to fructose-6-phosphate. It is at this stage that regulation can be effected, since phosphofructokinase and fructose diphosphatase can be independently controlled. Two reactions are involved in gluconeogenesis for the circumvention of the pyruvate kinase reaction. First, pyruvate is carboxylated to oxaloacetate in an ATP-requiring reaction involving pyruvate carboxylase. This oxaloacetate is then decarboxylated by phosphoenolpyruvate carboxykinase (PEPCK) in a GTP-dependent reaction to give phosphoenolpyruvate. Although the reactions of gluconeogenesis are similar in all species studied, there are interesting variations in the intracellular distribution of PEPCK. Pyruvate carboxylase is mitochondrial in all species studied, but in the rat and mouse PEPCK is mainly cytoplasmic, whereas in the rabbit and pigeon it too is mitochondrial. In the guinea pig and in man PEPCK occurs in both mitochondria and cytoplasm. These variations may impose restrictions on the means by which phosphorylation of pyruvate can be regulated.

Synthesis of glycogen involves the isomerization of glucose-6-phosphate to glucose-1-phosphate, which reacts with uridine triphosphate to give uridine diphosphate glucose. The glucosyl group can be transferred from this compound to extend a glycogen chain, as in the following:

$$\text{glucose-6-phosphate} \longrightarrow \text{glucose-1-phosphate}$$
$$\text{glucose-1-phosphate} + \text{UTP} \longrightarrow \text{UDP-glucose} + \text{PP}_i$$
$$(\text{glycogen})_n + \text{UDP-glucose} \longrightarrow (\text{glycogen})_{n+1} + \text{UDP}$$

The synthesis of each glucose unit in glycogen, from lactate, therefore requires a total of seven molecules of ATP.

Heat Generation. Not all of the energy change in metabolic reactions is conserved as chemical energy. Some is released as heat, and this might appear to be wasted. The heat produced will, however, raise the temperature of the animal above that of its environment and will therefore accelerate the metabolism of that animal. Some animals rely extensively on their metabolic heat for this purpose, whereas others possess mechanisms for absorbing radiant energy from their environment (see Chapter 10). The higher vertebrates have perfected means of regulating their body temperature at a constant level despite variations in the ambient temperature. In addition to the heat produced as a by-product of metabolism, some reactions appear to have the sole purpose of heat generation; two of these will be considered in detail.

Brown adipose tissue is found only in certain newborn and hibernating animals. Like white adipose tissue it contains large stores of triglyceride, but it has in addition a well-developed oxidative machinery for oxidizing the lipids in situ. The function of brown adipose tissue appears to be the generation of heat. The most likely way in which this is achieved is by the uncoupling of oxidative phosphorylation from electron transport. The energy change associated with the oxidation of NADH by oxygen is released as heat. The heat production by this tissue is under hormonal control.

A second mechanism has been proposed to explain the ability of bumblebees (*Bombus* spp) to raise their thoracic temperature before flight. It has been established that the bumblebee, like many other insects, must raise its thoracic temperature before flight can occur. Since this bee can fly at ambient temperatures down to 5°C and is frequently seen gathering nectar on cold days, it must have some means of maintaining body temperature while at rest. In fact several mechanisms may play a role. The insect is hairy and presumably loses heat very slowly. Second, it may generate appreciable amounts of heat by an almost isometric contraction of its flight muscles in a mechanism analogous to shivering. In addition to these mechanisms, Newsholme has suggested that the bumblebee may possess a metabolic heat-generating mechanism. This would depend on the simultaneous activity of phosphofructokinase and fructose diphosphatase (Figure 7.16) in the flight muscle. Both enzymes occur in this muscle, and since the muscle cannot carry out gluconeogenesis, the presence of fructose diphosphatase was at first surprising. If the two enzymes are simultaneously active, the net result is the hydrolysis of ATP and concomitant release of heat. By the use of radioactive isotopes it has been possible to measure the rate of this "substrate cycling" and show that at 5°C its rate is 10.4 μmole/min/g, but by 21°C the rate has dropped to 0.48 μmole/min/g, and cycling is not detectable at 27°C. Further evidence for the role of this cycle in heat generation comes from the observation that fructose diphosphatase is absent from the muscles of bees in the closely related genus, *Psithyrus*. These bees are brood parasites of *Bombus* and do not therefore need to forage to provide food for their offspring. *Psithyrus* is not seen on the wing on cold days.

In conclusion, it is apparent that metabolic data on species other than the laboratory rat and a few other mammals is relatively scanty. This is particularly true for the invertebrates. As our knowledge of biochemical pathways in different animal species improves, it should become increasingly possible to use comparisons between species to indicate the physiological significance of metabolic pathways.

REFERENCES

Beatty, C. H., and Bocek, R. M. Biochemistry of Red and White Muscle. *Physiology and Biochemistry of Muscle as a Food.* Vol. 2. E. J. Briskey, R. G. Cassens, and B. B. Marsh, eds. Madison: University of Wisconsin Press, 1970.

Benson, A. A., and Lee, R. F. The Role of Wax in Oceanic Food Chains. *Sci. Am.* 232:76–86, Mar. 1975.

Bursell, E. Aspects of the Metabolism of Amino Acids in the Tsetse Fly *Glossina* (Diptera). *J. Insect Physiol.* 9:439–452, 1963.

Cahill, G. F. Starvation in Man. *New Eng. J. Med.* 282:668–675, 1970.

Clark, M. G., Bloxham, D. P., Holland, P. C., and Lardy, H. A. Estimation of the Fructose Diphosphatase-Phosphofructokinase Substrate Cycle in the Flight Muscle of *Bombus affinis. Biochem. J.* 134:589–597, 1973.

Crabtree, B., and Newsholme, E. A. The Activities of Phosphorylase, Hexokinase, Lactate Dehydrogenase and Glycerol-3-Phosphate Dehydrogenase in Muscles from Vertebrates and Invertebrates. *Biochem. J.* 126:49–58, 1972.

Fairbairn, D. Biochemical Adaptation and Loss of Genetic Capacity in Helminth Parasites. *Biol. Rev.* 45:29–72, 1970.

Florkin, M. Biochemical Evolution in Animals. *Comprehensive Biochemistry.* Vol. 29B. M. Florkin and E. H. Stotz, eds. New York: Elsevier, 1975.

Gilmour, D. *The Metabolism of Insects.* Edinburgh: Oliver & Boyd, 1965.

Hanson, R. W., and Garber, A. J. PEP Carboxykinase: I. Its Role in Gluconeogenesis. *Am. J. Clin. Nutr.* 25:1010–1021, 1972.

Hochachka, P. W., and Somero, G. N. *Strategies in Biochemical Adaptation.* Philadelphia: Saunders, 1973.

Krebs, H. A. The Pasteur Effect and the Relations between Respiration and Fermentation. *Essays Biochem.* 8:1–34, 1972.

Newsholme, E. A., Crabtree, B., Higgins, S. J., Thornton, S. D., and Start, C. The Activities of Fructose Diphosphatase in Flight Muscle from the Bumble-bee and the Role of this Enzyme in Heat Generation. *Biochem. J.* 128:89–97, 1972.

Newsholme, E. A., and Start, C. *Regulation in Metabolism.* London: Wiley, 1973.

Odum, E. P. Adipose Tissue in Migratory Birds. *Handbook of Physiology, Section 5: Adipose Tissue.* A. E. Renold and G. F. Cahill, eds. Bethesda, Md.: American Physiological Society, 1965.

Sacktor, B., and Hurlbut, E. C. Regulation of Metabolism in Working Muscle *in Vivo*. II: Concentrations of Adenine Nucleotides, Arginine Phosphate and Inorganic Phosphate in Insect Flight Muscle during Flight. *J. Biol. Chem.* 241:632–638, 1966.

Saz, H. J. Anaerobic Phosphorylation in *Ascaris* Mitochondria and the Effects of Anthelmintics. *Comp. Biochem. Physiol.* 39B:627–637, 1971.

Thoai, N. V. and Roche, J. Sur la Biochimie Comparée des Phosphagènes et Leur Répartition chez les Animaux. *Biol. Rev.* 39:214–231, 1964.

Vague, J., and Fenasse, R. Comparative Anatomy of Adipose Tissue. *Handbook of Physiology, Section 5: Adipose Tissue.* A. E. Renold and G. F. Cahill, eds. Bethesda, Md.: American Physiological Society, 1965.

Weis-Fogh, T. Fat Combustion and Metabolic Rate of Flying Locusts (*Schistocerca gregaria* Forskål). *Phil. Trans. Roy. Soc. Ser. B* 237:1–36, 1952.

Williamson, D. H., and Hems, R. Metabolism and Function of Ketone Bodies. *Essays in Cell Metabolism.* W. Bartley, H. L. Kornberg, and J. R. Quayle, eds. London: Wiley, 1970.

DIGESTION AND NUTRITION

T. Richard Houpt

1 • DIGESTION

The nature of the molecules ingested by animals has been described in Chapter 7. The question of how these molecules gain access to the tissues of the animal, via the process of digestion, must now be considered. The basic processes are similar in the higher vertebrates and will be described before attention is given to the influence of diet on these processes in different species.

The body of an animal consists essentially of many cells surrounded by the extracellular fluid that supplies the needs of the cells and removes their waste products. This fluid is separated from the animal's environment by the specialized cells of the body surface: the skin or integument and the lining of the alimentary canal. In a sense, therefore, the lumen of the digestive tract can be considered as being outside the animal proper. The body surfaces act as barriers that prevent both the loss of compounds from the extracellular fluid and the entry of unwanted substances. Although few substances penetrate the integument, one of the major functions of the cells lining the digestive tract is to absorb nutritionally useful ions and molecules. The absorptive processes generally operate on small molecules that are formed by the degradation of larger molecules in the process of digestion. As well as rendering macromolecules more easily absorbed, digestion causes such molecules to become biologically inactive; thus, for instance, antibodies are not usually made against ingested proteins.

1.1 Ingestion

What appears in the lumen of the digestive tract is determined by what the animal ingests. There are three aspects of ingestion of interest to the comparative physiologist. First, the neural events, which determine when feeding will begin and when it will stop and thereby provide quantitative control of food intake; second, selection of food from the environment by the animal; and third, the neuromuscular processes by which the food is gathered and conveyed to the mouth.

The control of food intake involves matching the quantity of food ingested to the needs of the animal. Most attention has been focused on the adult animal, in which energy requirements are the primary concern. Among the parameters that are related to food intake, and so might participate in its control, are blood glucose concentration, amount of body fat, and distension of the stomach (see Figure 8.1). It is believed that all participate to some degree. A low blood glucose concentration is detected by certain areas of the brain (apparently in the region of the hypothalamus), causing a sensation of hunger, which leads to the ingestion of food. This glucostatic mechanism would presumably operate only over short periods of time—from meal to meal. Distension of the stomach by the ingested food is a most powerful inhibiting influence on intake and may often be the primary signal to end ingestion. It is believed that long-term control of food ingestion is related to the amount of fat in the body, and many animals, including man, seem able to maintain their body weight constant with an unchanging amount of fat present. The ways in which the amount

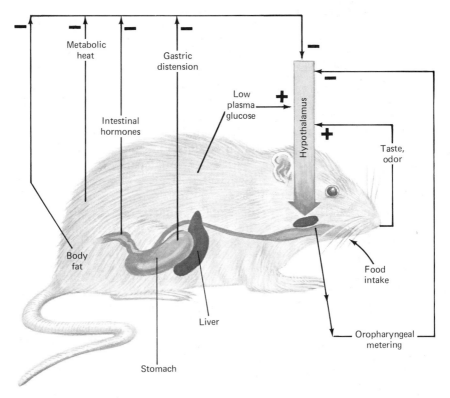

FIGURE 8.1 Control of food intake in the rat and some of the ways it may operate. Plus and minus signs indicate effect of increase of parameter on food intake.

of fat present in the body is sensed and influences ingestion are not known.

The selection of foodstuffs to be ingested involves many factors. The leopard is genetically programmed to hunt; it has the claws and teeth necessary to kill and ingest captured prey. The basic pattern imposed by heredity is certainly modified and elaborated by learned responses, but instinctive behavior and specific anatomy impose constraints on most animals with respect to their food selection. In making this selection the animal makes use of odor, taste, and visual clues. Carnivores are more likely to find all the required nutrients available in a single food item, but other animals may need to select a variety of foodstuffs. Herbivores will seek and eat specific nutrients in which they are deficient; for example, a goat will eat salt if it has a salt deficiency and will eat exactly the amount necessary to correct the deficit. This salt hunger seems to be as well developed and as well controlled in the ruminants as is the ingestion of water.

1.2   Physical Processes in Digestion

Once the animal has grasped the food and conveyed it to its mouth (prehension), the process of chewing, or *mastication*, follows. This is brief in carnivores but more protracted in herbivores. Swallowing, or *deglutition*, ensues, and as the masticated food is squeezed by the cheeks and tongue toward the pharynx, this physical distension initiates a reflex activity, causing a peristaltic wave of contraction that carries the food into the esophagus and down to the stomach. The movements of the digestive tract have three objectives: (1) to mix the food with digestive enzymes so as to facilitate chemical breakdown of complex food molecules, (2) to move the ingesta about so as to keep a flow over the absorptive membranes and thus facilitate absorption, and (3) to propel the ingesta onward to the next part of the digestive tract. The motility of the stomach is at first depressed, so ingested food is stored there until it can be digested. The ingesta will have already been mixed with the saliva, and salivary digestion may continue for a while in the stomach. Gastric juice, however, is secreted even before the food reaches the stomach, and under its influence liquefaction begins. As this occurs, stomach motility appears as shallow peristaltic waves passing toward the lower part of the stomach, the *antrum*. In the antrum peristaltic waves of considerable force carry the partly digested food, now called *chyme*, to the small intestine. It is here that breakdown of food molecules to their simple components is largely completed. Motility in the small intestine is predominantly of a mixing nature (Figure 8.2). The most frequently observed activities of the wall of the intestine are the segmentational contractions, which divide and stir the chyme about without propelling it down the intestine. As well as mixing food particles with

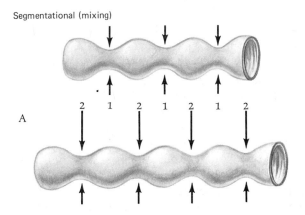

Segmentational (mixing)

A

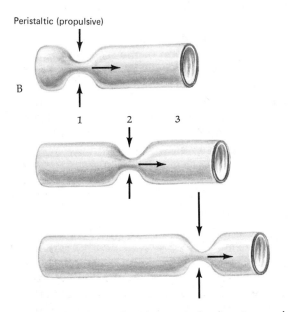

Peristaltic (propulsive)

B

FIGURE 8.2 Types of motility seen in the alimentary canal. (**A**) A contraction ring appears alternately at positions 1 and 2, churning contents back and forth without propulsion. (**B**) A contraction ring moves to the right, pushing the contents onward.

digestive enzymes and bringing digested material in contact with the absorptive surfaces, this motion helps the emulsification of fatty substances for digestion. This emulsification is aided by the bile salts, which are secreted into the intestine.

As digestion approaches completion, propulsive, peristaltic contraction waves appear in the small intestine and gently move the contents toward the large intestine. The ileocecal sphincter, or valve, plays an important part in preventing the premature movement of chyme into the large intestine. In the large intestine, water and sodium chloride are absorbed. Motility here resembles that of the small intestine, but in slow motion.

A new process appears in the large intestine: the growth of bacteria. The upper digestive tract harbors few microorganisms, but in the large intestine large populations of a variety of bacteria exist. For their nutrition these microorganisms depend upon food residues, digestive secretions, and substances leaking into the large intestine from the blood. This results in continuous bacterial growth, with much of the colony continually being removed in the feces, which are largely composed of microorganisms.

Slow propulsive waves occurring in the large intestine move the contents toward and into the rectum. Distension of the rectum produces sensory impulses, which travel to reflex centers in the sacral part of the spinal cord and initiate defecation.

One form of motility of the alimentary canal, so far unmentioned, is vomiting. Although usually thought of as an abnormal form of gastrointestinal motility, in some animals it might be considered normal. Carnivores, in particular, eat hastily and readily reject, by vomiting, any piece of food that proves to be irritating. Many other animals, such as rats and horses, rarely vomit under any circumstances.

1.3 Enzymatic Digestion

During digestion peptide, glycoside, and ester bonds are split by stepwise hydrolysis catalyzed by enzymes secreted into the alimentary canal. The control of the secretory glands is such that the secretions containing digestive enzymes appear when food arrives or is about to arrive in the part of the tract served by the corresponding gland. The general scheme of control of secretion is illustrated in Figure 8.3.

Digestion of proteins begins in the stomach with the action of pepsin, which cleaves peptide bonds in which the amino function is contributed by either an aromatic or acidic amino acid. It is curious that only vertebrates consistently have this step in protein digestion; in many invertebrates enzymes similar to pancreatic trypsins accomplish the same cleavage. Pepsin is secreted as pepsinogen, an inactive form of the enzyme that is activated by the hydrochloric acid of the stomach. Both pepsinogen and this acid are secreted in response to the hormone gastrin and to vagal stimulation. In addition to activating pepsinogen and providing

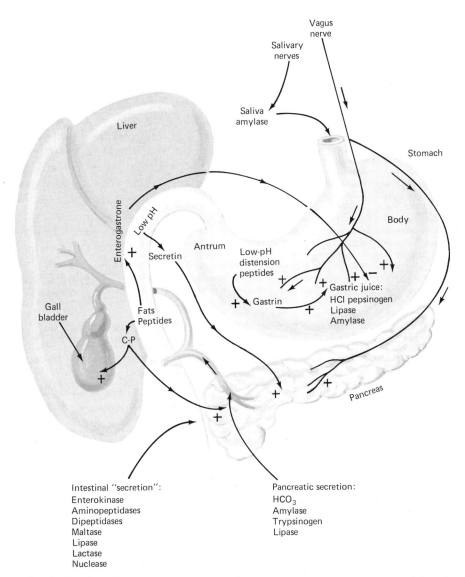

FIGURE 8.3 Control of gastric, pancreatic, and intestinal secretions and contraction of the gall bladder. C-P is cholecystokinin-pancreozymin, which causes an enzyme-rich pancreatic secretion and contraction of the gall bladder.

the low pH necessary for pepsin action, the acid also destroys bacteria and breaks down cell walls.

As the chyme enters the small intestine, its low pH triggers the release of secretin from the intestinal mucosa. This hormone travels in the bloodstream to the pancreas, where it causes a copious secretion of a bicarbonate-rich solution that neutralizes the acid. This secretion contains other proteolytic enzymes, including trypsinogen, which is activated to trypsin by enterokinase, also secreted into the intestinal lumen. Trypsin has a marked specificity for peptide bonds whose carbonyl groups are contributed by arginine or lysine. Also present in the pancreatic secretion is chymotrypsinogen, which is activated to chymotrypsin by the action of trypsin. Chymotrypsin has a broader specificity than trypsin but preferentially cleaves those peptide bonds whose carbonyl function is contributed by an aromatic amino acid. Pepsin, trypsin, and chymotrypsin are endopeptidases; that is, they can cleave peptide bonds in the middle of polypeptide chains. The exopeptidase, carboxypeptidase, is also secreted by the pancreas and will cleave single amino acids from the end of polypeptides bearing a free carboxyl group. Its action is complemented by aminopeptidases, secreted into the intestine, which cleave amino acids from the amino end of polypeptides. The combined action of these enzymes results in a mixture of absorbable amino acids and dipeptides. The latter are split by dipeptidases in the microvilli of the epithelial cells as the dipeptides are absorbed.

Carbohydrate digestion similarly follows a sequence of progressive hydrolysis. In a few species, including man, this begins in the saliva where an α-amylase, ptyalin, causes a limited amount of starch hydrolysis. More effective and widespread is a similar α-amylase, present in the pancreatic secretion, which produces mainly disaccharides. Disaccharides are hydrolyzed by enzymes such as maltase, lactase, and sucrase. This hydrolysis occurs in large part as the sugars are absorbed through the epithelial membrane with which the disaccharidases are associated.

The enzymatic digestion of lipids is complicated by the water-insoluble nature of fatty substances. In the absence of some mechanism to render the lipid susceptible to attack by lipases, little hydrolysis would occur. The problem is effectively solved by the dispersion of lipid particles into the aqueous phase. This emulsification is aided by the churning motion of the intestinal contents and the detergent action of the bile salts. The surface area of the lipid is greatly increased by this emulsification, and the rate of lipid hydrolysis by pancreatic lipase is correspondingly increased. As lipid hydrolysis proceeds, the emulsification is carried further by the action of the generated fatty acids, monoglycerides, and diglycerides, all of which are themselves surface active agents (detergents). Eventually the cloudy mixture of lipids, lipases, and bile salts clears and takes on the appearance of a true solution due to the formation of micelles (molecular aggregates 3 to 10 nm in diameter). From these micelles, fatty acids and monoglycerides are absorbed into the intestinal epithelial cells (Figure 8.4). The bile salts are also absorbed, but from a lower part of the small intestine.

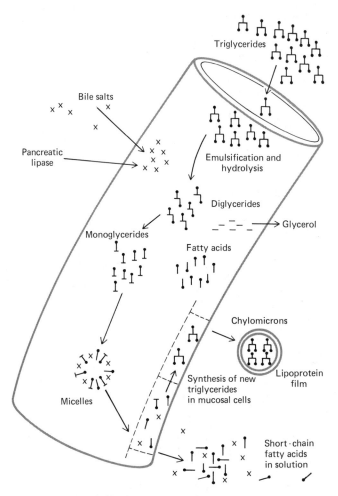

Triglycerides

Bile salts

Pancreatic lipase

Emulsification and hydrolysis

Diglycerides

Glycerol

Monoglycerides

Fatty acids

Chylomicrons

Micelles

Synthesis of new triglycerides in mucosal cells

Lipoprotein film

Short-chain fatty acids in solution

FIGURE 8.4 Enzymatic digestion and absorption of lipids. (Adapted from K. J. Isselbacher, *Fed. Proc.* 26:1420, 1967.)

1.4 Absorption

Absorption of the end products of enzymatic digestion is greatly facilitated by the enormous increase in surface area of the intestinal mucosa due to the various infoldings of its surface (Figure 8.5). There are specific transport mechanisms for many substances that can move the molecule from the intestinal lumen, across the mucosal cell, and into the blood draining the intestine. The different transporting systems share a number of common characteristics. They are believed to involve a transient combination with a carrier molecule in the cell membrane (Figure 8.6); they are dependent on a source of ATP, since substances are being transported up

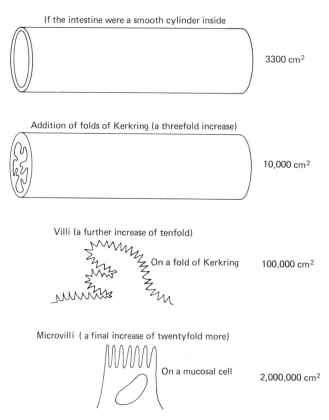

FIGURE 8.5   How the surface area of small intestine is increased by folds and projections into the lumen (the human species). (Adapted from T. H. Wilson, *Intestinal Absorption.* Philadelphia: Saunders, 1962.)

a concentration gradient, and they are believed to require the simultaneous transport of sodium ions in the same direction. Most amino acids are transported by such mechanisms, and there are at least three transporting systems, each shared by a set of amino acids. Glucose and galactose share another active transport system.

Many small, water-soluble molecules appear to be absorbed simply by diffusion across the cell membrane. The water-soluble vitamins, fatty acids with fewer than 10 carbon atoms, and water itself are examples of such substances. An exception among the water-soluble vitamins is cyanocobalamin (B_{12}), which has a significantly more complex structure than that of the other B vitamins. It is absorbed by a special transport system in the small intestine after first being bound to a glycoprotein, called *intrinsic factor*, which is released by the gastric mucosa.

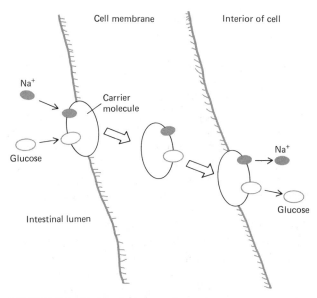

FIGURE 8.6 Model of glucose absorption system across the intestinal cell membrane. (Adapted from R. K. Crane, *Fed. Proc.* 24:1000, 1965.)

The absorption of the water-insoluble fatty acids and of monoglycerides presents a problem since they must be dispersed in a form suitable for transport in the aqueous plasma. As shown in Figure 8.5, the products of fat digestion are resynthesized into triglycerides in the intestinal cells. This triglyceride combines with protein in the intestinal cells to form particles called *chylomicrons,* which are about 1 μ in diameter. The lipoprotein in the chylomicrons acts to prevent coalescence of the fat into larger particles, which could block blood vessels. During fat absorption the lymph draining the intestine becomes milky in appearance due to the high concentration of chylomicrons. Many other fat-soluble substances are absorbed and transported in the blood with the triglycerides. Notable among these are cholesterol and the fat-soluble vitamins, A, D, E, and K.

2 • NUTRITIONAL ADAPTATIONS

In the first part of this chapter the features of digestion common to all vertebrates were discussed. The success of a species must depend on adaptations that permit it to exploit new sources of food, to utilize existing sources more efficiently, and to withstand periods of nutritional stress. It is now appropriate to consider the nature of these adaptations. The emphasis on vertebrates, apparent in the first part of this chapter, will be maintained since the higher animals have been extensively studied in this respect.

One could expect that an animal species might cope with problems of nutritional scarcity in two ways. First, the basic pattern of metabolism might be modified so that certain of the required nutrients (for example, vitamins and essential amino acids) would be synthesized in the body. This possibility seems to have been little used as an adaptive mechanism to nutritional stress, and the basic needs of body cells seem to differ little across the whole range of vertebrate forms (see Chapter 7). Second, adaptations might involve modification and elaboration of organ systems. Those modifications seen in present vertebrates appear to be directed toward the solution of two nutritional problems: the digestion of cellulose in plant material and the reutilization of waste products. A widely employed solution to both problems has been the development of symbiotic relationships between the vertebrates and microorganisms, usually occurring in some special compartment of the alimentary canal. Bacteria are the most important microorganisms involved in these relationships, but protozoa and yeasts also contribute to the microbial population.

The pattern of flow of nutrients between the vertebrate groups is indicated in Figure 8.7. Animals eat what is available and what they are

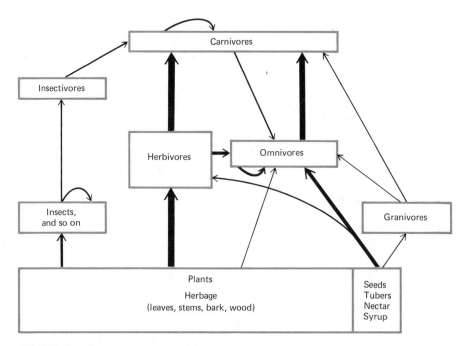

FIGURE 8.7 Pattern of biological energy and nutrient flow: vertebrates.

adapted to ingest and digest. Some are herbivores and others carnivores; some are specialist feeders and others more catholic. How have animals approached the problems associated with getting the essential nutrients for the metabolism of their cells?

2.1 Carnivores and Insectivores

The simplest solution is to eat other animals, for the tissues of any animal contain all the nutrients needed by another animal in a compact form. However, the proportions of nutrients in the body of one animal are different from what another animal ideally requires in its diet. Most important, the food of a carnivore contains a far higher proportion of protein than is needed for the maintenance of adult tissues. The fat content may also be high if the prey is well fed, but the carbohydrate content is likely to be very low. Two metabolic processes assume particular importance when the proportion of protein in the diet is high. These processes are the synthesis of glucose from amino acids (gluconeogenesis), described in Chapter 7, and the disposal of the nitrogen removed in this process.

The first step in the utilization of amino acids as an energy source consists of deamination. The most important route of deamination involves, first, the transamination of the amino acid with α-ketoglutarate to form the keto acid corresponding to the amino acid, together with glutamate. Second, this glutamate is deaminated under the catalytic influence of glutamate dehydrogenase to reform α-ketoglutarate and release ammonia; that is,

$$HOOC \cdot CH_2CH_2CO \cdot COOH + R \cdot CH(NH_2)COOH \longrightarrow$$
$$HOOC \cdot CH_2CH_2CH(NH_2)COOH + R \cdot CO \cdot COOH$$

$$HOOC \cdot CH_2CH_2CH(NH_2)COOH + NAD^+ \longrightarrow$$
$$HOOC \cdot CH_2CH_2CO \cdot COOH + NADH + NH_4^+$$

This ammonia must be rapidly removed from the body fluids because of its toxic actions, most dramatically on the central nervous system. The excretion of nitrogen is dealt with in detail in Chapter 6. Mammals convert the ammonia to urea, a relatively nontoxic, inert, water-soluble compound that is excreted in the urine. The presence of urea in body tissues is not normally detrimental, but excessive accumulation may occur in animals with malfunctioning kidneys. Under these circumstances urea diffuses into the large intestine, where it is hydrolyzed to ammonia by the bacteria present. This ammonia passes back into the blood and is carried to the liver, thus increasing the demand for urea synthesis in this organ. The energy cost of the synthesis of urea must be taken into account when assessing the energy yielded from ingested proteins.

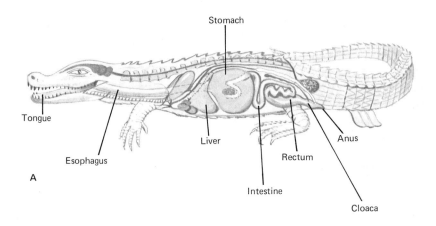

A

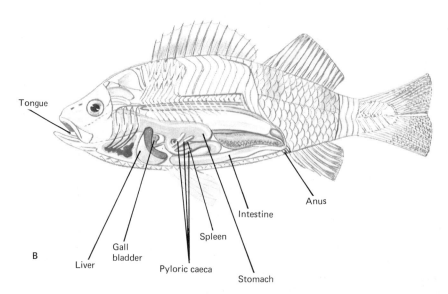

B

FIGURE 8.8   Digestive systems of typical vertebrate carnivores. (**A**) American alligator. (**B**) Yellow perch.

The metabolic consequences of dependence on a high-protein diet have had little influence on the evolution of the digestive tract in carnivores. Basically there is a mouth, pharnyx, esophagus, stomach of limited capacity and great acidity, small intestine into which flows pancreatic juice and bile, and a relatively short large intestine with rudimentary cecum (Figure 8.8). The specialized teeth, so effective for cutting and crushing animal tissues, are the outstanding structural adaptation. The minimum requirement for protein in carnivores is determined more by behavioral characteristics than by biochemical need. Although there appears to be no metabolic reason why carnivores like dogs and cats should not thrive on a low-protein diet, dogs, for example, will not ordinarily eat food containing less than 20 percent protein. Cats have an even stronger desire for animal protein.

Insectivores are presumed to cope with their nutritional needs in ways similar to those employed by carnivores, but some additional adaptations are seen. The polysaccharide, chitin, present in insect exoskeleton presents a potentially significant energy source, and a chitinase system, for its breakdown, has been found in the stomach of bats as well as in certain insectivorous birds, reptiles, and fish.

The starting point in the evolution of each class of vertebrates appears to have been an unspecialized carnivore of small size. Some of these carnivore types have persisted relatively unchanged to the present time, whereas others have evolved into omnivores and herbivores. Many birds are carnivorous (such as eagles, owls, cormorants, and hawks) and many others insectivorous (such as flycatchers, nighthawks, woodpeckers, and robins). As in carnivorous mammals, the digestive tract of these birds is unspecialized, except for the bird's beak. Again, most present-day amphibians and reptiles are carnivores or insectivores, and the same pattern of alimentary canal structure is seen.

2.2 Herbivores

Quantitatively, the greatest need of the adult animal is for energy-yielding foodstuffs. The carnivore, in the absence of prey, is surrounded by tremendous quantities of energy-containing compounds that it cannot use. The energy is locked up in cellulose molecules, which make up a large proportion of plant material. Energy, energy everywhere but not a calorie to burn! As described in Chapter 7, cellulose is a polymer of glucose units linked $\beta(1 \rightarrow 4)$, but vertebrates lack the enzyme (cellulase) necessary to hydrolyze this bond. Herbivores have solved the problem by evolving a cooperative venture on a grand scale with a wide variety of microorganisms. The vertebrates gather plant material and break it down into small fragments before supplying it to the microorganisms that exist in special enlargements of the digestive tract. These regions are warm and an-aerobic, and the microorganisms rapidly hydrolyze the cellulose there and ferment the glucose thus produced. In this way the microorganisms obtain the energy for their survival and excrete the products of the fermentations, mostly a variety of short-chain fatty acids (see Figure 8.9). These fatty acids (for example, propionic, acetic, and butyric acids) are absorbed by the host and oxidized to provide energy for the host's tissues (see Chapter 7). Ruminants have probably gone furthest in developing this symbiosis and harbor their microorganisms in extensions of the stomach. Other

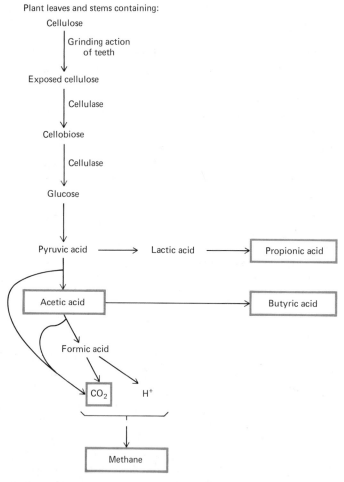

FIGURE 8.9 General pathways of cellulose digestion by symbiotic micro-organisms of the alimentary canal of herbivores. The usual major end prod-ucts of this fermentative breakdown are indicated in boxes.

mammals, for example the rodents, culture their microbial flora in enlargements of the cecum.

Ruminants.   The dictionary definition of a ruminant is an animal that chews a cud. Observation of a ruminant (for example, a cow, sheep, camel, or giraffe) will soon reveal the phenomenon of chewing the cud. From time to time something comes up the esophagus as the result of a slight coughlike motion; this "cud" is then chewed for a few minutes before being reswallowed. This cycle of regurgitation, rechewing, and reswallowing is known as rumination. The anatomy of the digestive tract of a typical ruminant is shown in Figure 8.10. The most obvious feature is the enlargement and compartmentalization of the stomach. Chewed herbage travels down the esophagus to enter the reticulum and rumen. These chambers are only partially separated by a partition, and because the contents are continually mixed back and forth, the reticulorumen acts functionally as a single compartment. The reticulorumen is never empty, and the swallowed food slowly mixes with the ingesta already in this chamber. The contractions of the stomach wall in ruminants are more complex than in a simple-stomached animal. Each cycle of contraction begins with a brief, double contraction of the reticulum, which projects the ingesta back into the rumen. The most recently ingested material tends to float on top of the mass in the rumen and is returned to the vicinity of the esophageal opening as the rumen sacs contract twice. At this point there may be an-

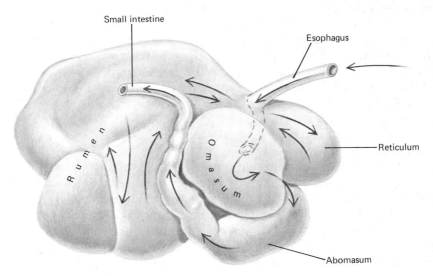

FIGURE 8.10   The stomach of the ruminant. (Esophageal groove position indicated by dashed lines leading to omasal entrance.)

other brief contraction of the reticulum and a brief inspiratory effort with the glottis closed. This activity sucks some of the floating ingesta into the esophagus, whence it is carried by muscular contractions back to the mouth. Rechewing, with the addition of more saliva, proceeds, and the bolus is reswallowed. It has been estimated that a cow may make 45,000 chewing motions every day. As a result of this rumination, food particles are ground more and more finely, thereby hastening the bacterial and protozoan action in the rumen. Particles not completely digested sink in the rumen. During contractions of the rumen, these heavier particles tend to move into the reticulum in the region of the omasal opening. This opening relaxes at the time of the second reticular contraction, and a few milliliters of fluid, with solid particles and microorganisms, flows into the omasum.

Much of the water, electrolytes, and fatty acids are absorbed from the omasum, where contractions facilitate this process and then push the ingesta into the abomasum. The abomasum is analogous to the simple stomach of the dog or man; it secretes acid and digestive enzymes, including pepsin. The absorptive action of the omasum may be necessary to prevent too great a dilution of the abomasal secretions. The acid probably serves to destroy the cellular structure of microorganisms entering this chamber since few escape destruction. The microbial proteins, carbohydrates, and fats are thus exposed to the digestive enzymes of the abomasum and small intestine. Digestion in the small intestine appears to differ little from that in other animals. The cecum, however, is also enlarged and contains a microbial colony, so bacterial action resumes and processes begun in the rumen continue in the cecum and colon.

In the young ruminant the esophageal groove leads from the junction of the esophagus and the reticulum directly into the omasum, near its exit. This is indicated in Figure 8.10. Suckling, or the presence of fluid in the pharnyx, initiates reflexes that cause the groove to partially close and thus carry the milk from esophagus to omasum without a stop in the rumen. As the young ruminant matures, this reflex weakens and finally ceases to operate. When the milk reaches the abomasum, it is coagulated by rennin, a digestive enzyme that has some proteolytic action at the low pH of the abomasum.

The proportion of cellulose present in the diet that is digested will depend on the type of plant ingested. Pure cellulose may be completely digested, and all of its chemical energy used by the microbe-mammal symbiotic combination. The cellulose from young, fast-growing plants may be 90 percent digested, but that from mature, woody plants may be only 50 percent digested. In the latter case the cellulose is partly protected from microbial action by other complex plant substances, such as lignin, which are resistant to bacterial breakdown. In this case the grinding action of the rumination cycle is particularly important in physically exposing cellulose to bacterial action.

Next in importance to the continual need for energy-containing compounds is the requirement for certain amino acids. Adult ruminants, like other animals, hydrolyze and resynthesize a significant proportion of their proteins each day. Many bacteria can, if supplied with a source of nitrogen, synthesize all of the biologically useful amino acids. Ruminants take advantage of this synthetic ability by arranging for some of the circulating urea in the blood to diffuse into the rumen. This urea can be utilized by the bacteria present for the synthesis of amino acids, which become incorporated into bacterial proteins. When the bacteria are digested in the abomasum and small intestine, the newly synthesized amino acids are made available to the host. Thus ruminants can survive even when the nitrogen content of their food falls to low levels, and it has been estimated that this amino acid regeneration cycle (Figure 8.11) may double the survival time of sheep on very low-protein diets.

Microorganisms also serve their ruminant hosts in another way: by the synthesis of vitamins. Ruminants show no dietary requirements for vitamins of the B group and vitamin C. Only vitamin B_{12}, cyanocobalamin, may present a problem since its molecule contains a cobalt atom, but ruminal microorganisms can synthesize cyanocobalamin if cobalt is present in the diet.

In arid regions the most conspicuous nutrient shortage is that of

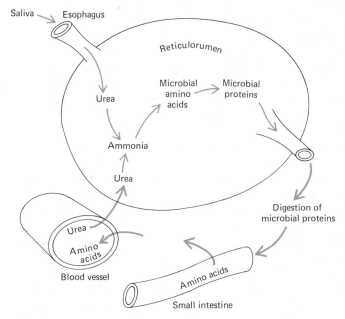

FIGURE 8.11   Reutilization of body urea for amino acid synthesis in the ruminant.

water, and it is notable that ruminants are the predominant large mammals in such regions. The ability of the camel to survive without drinking is legendary. The camel cannot store water, either in its hump or in its rumen, but it does have considerable ability to minimize water loss from its body. The volume of urine excreted is low, not because of an exceptional ability to concentrate urine but because of the lower rate of production of excretory solutes. This is in part due to the utilization of urea described above and partly because of the lower rate of metabolism per unit body weight characteristic of large animals. The thick, tough hide slows water loss through the skin, and although camels do sweat, they do so less profusely and at a lower rate than does, for example, the donkey. Because less water is expended in cooling the animal, its body temperature rises during the day and later falls at night. The wool on the dorsal surface of the animal is retained in summer and is an effective insulator, slowing the gain of heat from intense solar radiation. Finally, although these strategems will slow water loss, the camel's last defense is an ability to withstand a degree of dehydration that would kill many mammals of its size. Thus the camel can lose one-third of its body weight as water before showing signs of discomfort.

Rodents and Lagomorphs.   Some rodents are strict herbivores, whereas others (such as the rat) are omnivores; however, all possess enlarged ceca, which support large microbial populations. The lagomorphs, including the rabbit, hare, and pika, are all herbivores. They too have a large cecum, and the rabbit will be used as an example to illustrate this adaptation.

The general arrangement and proportions of the digestive tract of the common rabbit is shown in Figure 8.12. The stomach of the rabbit is somewhat enlarged, and some microbial digestion occurs there, but the most striking feature is the large cecum. Like the ruminant, the lagomorph, with the aid of its microorganisms, is able to digest considerable quantities of cellulose, reutilize significant amounts of its urea, and synthesize various vitamins. The positioning of the region of microbial digestion after the small intestine, which is the major site of absorption, presents the problem of how to move the nutrients produced by the bacteria into the body proper. The products of cellulose digestion and fermentation, together with synthesized vitamins, are absorbed directly from the cecum. Amino acids, however, are not absorbed directly. In the lagomorphs and in many rodents the simple solution to this problem is coprophagy. The animal ingests a special form of fecal pellet, taken directly from the anus at certain times of the day. In rabbits coprophagy occurs at dawn and dusk and so is rarely observed. It is believed that the fecal pellet is moved from the cecum to the anus by a special motion of the large intestine. It appears that the rabbit has solved the problem of how to gain the benefit of microbial synthetic activity by transporting the cecal contents to the beginning of the alimentary canal.

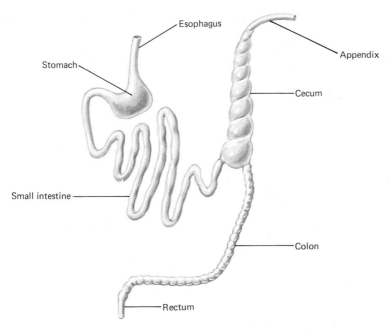

FIGURE 8.12 The alimentary canal of the rabbit. (From H. M. Kaplan, *The Rabbit in Experimental Physiology*, 2d ed. New York: Scholar's Library, 1962.)

In arid regions the ruminants are the most common large animals and the desert rodents the most common small ones. Smallness confers the great advantage that a heat stress may be escaped. A camel must remain exposed to the sun, but a kangaroo rat can enter the subterranean stratum of cooler and moister air during the heat of the day. Generally, soil temperatures a few centimeters below the surface are lower than body temperatures, so the desert rat in its burrow can lose heat by conduction, convection, and radiation. No water need be expended to cool a rodent resting below ground. Furthermore, as warm moist air is exhaled, it is cooled in the nasal passages, and thus much of the water vapor in the exhaled air is removed by condensation within the body. Losses from the skin and from feces are also very low. Finally, urinary water loss is minimized by the most effective concentration mechanism found in any vertebrate kidney. Solutes are excreted in urine with very little water loss. This superb combination of water conservation systems is so effective that the small water loss can usually be balanced by the water produced as a by-product of metabolism, and no drinking water is required even when these animals are fed on dry food.

The Horse and Its Relatives.   The equids (horses, asses, and zebras) also have relatively large ceca, but it is something of a mystery as to how the products of microbial synthesis are transported to the body proper. Co-prophagy is not seen as a normal and quantitatively important aspect of behavior. The horse has a lower ability than a typical ruminant to digest cellulose, conserve nitrogen, and synthesize vitamins.

Birds.   In recent geological time the mammals appear to have had the most success with a herbivorous diet. Although many species of birds eat seeds and some vegetation, strict herbivores are few in number. The flightless, or almost flightless, birds of arid regions such as the ostrich, rhea, and tinamou are herbivores. The latter does possess a sacculated cecum, but whether or not it serves a function similar to that seen in the herbivorous mammal is not known.

Lower Vertebrates.   There are few herbivorous amphibia and reptiles. It is interesting to note, however, that the tadpole of frogs is vegetarian and possesses a longer digestive tract than does the adult. A few lizards and some turtles and tortoises are herbivores. The Galapagos tortoise can digest cellulose, but little is known of the details of its digestive tract function. Among extant lizards the curious fact exists that almost all herbivorous species are large (heavier than 300 g), whereas all smaller species are carnivores or insectivores. It has been suggested that the larger lizards have evolved in the direction of herbivory because they were less mobile and unable to compete with small lizards in the catching of elusive prey. Fortunately, their lower metabolic rate per unit weight can be sustained by lower-energy-yielding, but more reliably available, vegetation.

Reptiles cannot be left without some reference to their glorious past. For 100 million years the dominant land vertebrates were reptiles: the dinosaurs. The evolutionary radiation that occurred in the mammals was preceded by a similar pattern in the reptiles during the Mesozoic era. From small carnivorous or insectivorous species a variety of larger omnivores, carnivores, and herbivores evolved. Presumably the lush vegetation and warm moist climates of that time supported this fauna of giants. Evolution reverses with difficulty, and the inability of those highly specialized reptilian herbivores to change in keeping with the changes in vegetation, as climates became cooler and drier, probably accounts for their disappearance at the end of the Mesozoic. The carnivores would, of necessity, follow their prey into extinction.

2.3   Omnivores, Granivores, and Fence Straddlers

This group has some of the characteristics of both herbivores and carnivores, and it avoids the extreme specializations that seem to preclude carnivores from eating vegetation and herbivores from eating meat.

Pigs and people are rather alike in their digestive apparatus. Teeth of both species can be used for a moderate amount of grinding but are even better for tearing up flesh. In both animals the cecum is fairly small, but the colon is long, wide, and sacculated (Figure 8.13A and B). The

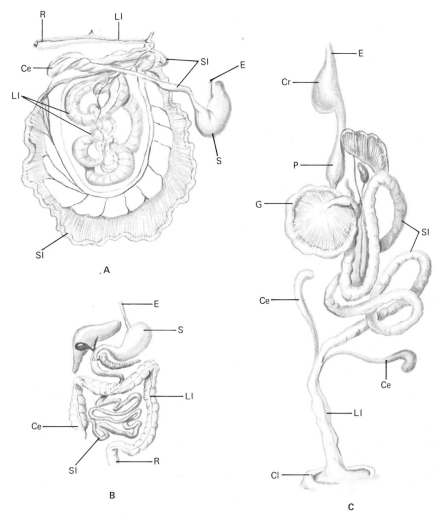

FIGURE 8.13 Digestive tracts of the (**A**) pig, (**B**) human being, and (**C**) chicken, somewhat schematically represented. Code: E—esophagus; S—stomach; SI—small intestine; Ce—cecum; LI—large intestine; R—rectum; Cr—crop; G—gizzard; P—proventriculus; Cl—cloaca. (Sources: (**A**) R. Nickel, A. Schummer, E. Seiferle, and W. O. Sack, *The Viscera of the Domestic Mammals.* Berlin: Verlag Paul Paray, 1973; (**B**) B. A. Schottelius and D. D. Schottelius, *Textbook of Physiology,* 17th ed. St. Louis: Mosby, 1961; (**C**) P. D. Sturkie, *Avian Physiology,* 2d ed. First edition copyright © 1954 by Cornell University. Second edition copyright © 1965 by Cornell University. Used by permission of Cornell University Press.)

greatest advantage to the omnivores is the increased range of foodstuffs available; meat, insects, grains, tubers, pulses, and fruits can all be ingested and digested. Intake of herbage remains somewhat limited since the ability to digest cellulose is modest.

The chicken is a good example of an avian omnivore. Here, however, the primitive vertebrate digestive tract has been modified, as seen in Figure 8.13C. The chicken in nature has only a limited intake of fibrous vegetation and feeds mainly on insects and seeds. The major adaptations seen appear to facilitate the digestion of hard seeds. The chicken eats rapidly, filling its crop, and slowly moves the ingesta on to the proventriculus and gizzard. Foodstuffs are softened in the crop and subjected to some degree of bacterial action. The proventriculus secretes acid and pepsin into the food mass as it moves through the gizzard. The thick muscular walls of the gizzard contract rhythmically to grind the ingesta. The effectiveness of this grinding is increased by the presence of small pieces of stone in the gizzard, deliberately ingested for this purpose. Gastric digestion continues into the first part of the small intestine, where the usual pancreatic enzymes and bile are added. The two ceca appear to have a limited ability to digest cellulose by bacterial action, and vitamin synthesis also occurs. Under agricultural conditions the chicken and the pig share the distinction of converting feed into meat more efficiently than any other domestic animals.

For man the greatest adaptations are behavioral rather than structural. The first important step was the acquisition of the ability to cultivate plants and thus increase the supply of plant starch and protein—a feat surely comparable with the digestion of cellulose by herbivores. It is possible, however, that another cultural technique was fully as important as agriculture, and that was the use of heat to prepare food. Although most meats and fruits can be digested uncooked, albeit more slowly in the case of meats, raw plant starches are highly indigestible. This is because starch granules are surrounded by a cellulose wall that is little affected by mastication. By causing the starch to swell, cooking breaks this cellulose wall, and the starch is exposed to the action of digestive enzymes. It appears that man, once again through the superior nature of his central nervous system, has overcome the limitations of an organ system—in this case the digestive tract.

REFERENCES

Bell, D. J., and Freeman, B. M., eds. *Physiology and Biochemistry of the Domestic Fowl.* New York: Academic Press, 1971.

Chatterjee, I. B. Evolution and the Biosynthesis of Ascorbic Acid. *Science* 182:1271–1272, 1973.

Crawford, M. A., ed. *Comparative Nutrition of Wild Animals*. New York: Academic Press, 1968.

Davenport, H. *Physiology of the Digestive Tract*. 3rd ed. Chicago: Year Book Medical Publishers, 1972.

Grassé, P. P., ed. *Vertebrates*. Traité de Zoologie, vols. XII–XVI. Paris: Masson et Cie, 1950–1973.

Halver, J. E., ed. *Fish Nutrition*. New York: Academic Press, 1972.

Houpt, T. R. Urea Utilization by Rabbits Fed a Low-Protein Ration. *Am. J. Physiol.* 205:1144–1150, 1963.

————. Utilization of Blood Urea in Ruminants. *Am. J. Physiol.* 197:115–120, 1959.

Houpt, T. R., and Houpt, K. A. Nitrogen Conservation by Ponies Fed a Low-Protein Ration. *Am. J. Vet. Res.* 32:579–588, 1971.

Jeuniaux, C. On Some Biochemical Aspects of Regressive Evolution in Animals. *Biochemical Evolution and the Origin of Life*. E. Schoffeniels, ed. Amsterdam: North-Holland Publishing, 1971.

Mitchell, H. H. *Comparative Nutrition of Man and Domestic Animals*. Vols. I and II. New York: Academic Press, 1962, 1964.

Moir, R. J., Somers, M., and Waring, H. Studies on Marsupial Nutrition. I. Ruminant-like Digestion in a Herbivorous Marsupial (*Setonix brachyurus* Quoy and Gaimard). *Austral. J. Biol. Sci.* 9:291–304, 1956.

Novin, D., Wyrwicka, W., and Bray, G. A., eds. *Hunger: Basic Mechanisms and Clinical Implications*. New York: Raven Press, 1976.

Pough, F. H. Lizard Energetics and Diet. *Ecology* 54:837–844, 1973.

Robinson, D. W., and Slade, L. M. The Current Status of Knowledge on the Nutrition of Equines. *J. Animal Sci.* 39:1045–1066, 1974.

Schmidt-Nielsen, B., Schmidt-Nielsen, K., Houpt, T. R., and Jarnum, S. A. Water Balance of the Camel. *Am. J. Physiol.* 185:185–194, 1956.

Schmidt-Nielsen, K. *Desert Animals: Physiological Problems of Heat and Water*. London: Oxford University Press, 1964.

Swenson, M. J., ed. *Dukes' Physiology of Domestic Animals*. 9th ed. Ithaca, N.Y.: Comstock Publishing Associates, 1977.

Vonk, H. J. Comparative Biochemistry of Digestive Mechanisms. *Comparative Biochemistry*. Vol. 6. M. Florkin and H. S. Mason, eds. New York: Academic Press, 1963.

ENDOCRINOLOGY

P. J. Bentley

Hormones are chemical messengers that work in harmony with nerves to coordinate the physiological functions of animals, both vertebrates and invertebrates, that are described in this book. The physiological processes that utilize hormonal control mechanisms include nutrition and energy metabolism in all their many aspects, such as digestion and interconversions between fats, carbohydrates, and proteins. Hormones also regulate such diverse processes as growth and temperature regulation. They are the principal controllers of osmoregulation and are vital for the regulation of calcium and phosphate levels in the body. The term "hormone" has the popular connotation of reproductive activity attached to it. This characteristic is true, and these chemicals also have a great influence on embryonic differentiation and nurture of the young.

The vertebrates are a diverse group of animals, with many different and specific needs. It is thus not surprising that hormones have very special roles to play in certain species. Two examples of such special functions are control of color change of the skin and metamorphosis of tadpoles into frogs.

To exert their numerous and varied actions, hormones ultimately act to alter the physicochemical properties of cells. They do this in a variety of ways, including changes in cellular permeability, electrical activity, and energy metabolism. This chapter describes such hormone-mediated processes among the vertebrates. Invertebrate physiological processes are also regulated by hormones, but the hormones of invertebrates are sufficiently different and specialized to necessitate separate description, which space does not allow. It is hoped that this chapter will help the reader to appreciate how hormonal actions have allowed different vertebrates to adapt to the wide variety of environmental conditions that they have encountered.

1 • INTRODUCTION

Coordination of the activities of cells in different parts of an animal is an integral part of its life. It may be performed in different ways, usually involving the transmission of messages by electrical and chemical changes. The nervous system, as described in Chapter 3, sends its cablelike fibers to all regions of the body. Most of the information collected is transmitted and interpreted in the brain and spinal cord. The message for the initiation of an appropriate response may be sent along the nerve fibers to another nerve or to an effector tissue, such as muscle or glands. Transmission along nerves is fast and of a transitory nature, so the response is rapid and not usually sustained. All cells are bathed in extracellular fluids with which the blood is contiguous. The latter thus offers another pathway for the transfer of information in the form of special chemical compounds called *hormones*. Compared to transmission along nerve fibers, this process is a slow one, reflecting the time taken for the blood to circulate and for the hormones to cross the capillaries and diffuse to their sites of action. The effects may, however, be of a more sustained nature and often result in prolonged stimulation of a tissue. Both nervous and hormonal pathways are essential, and each is specifically suited to the integration of certain physiological processes. However, hormones and nerves work in close collaboration and are interdependent.

Cells are complex structures arranged in such a way as to be able to undergo certain reactions with extraneous chemical compounds. Certain regions on their external and internal surfaces utilize this reactivity to form compounds with molecules that are part of the body's information network. Ultimately, nerve transmission depends on the release of chemical compounds such as acetylcholine and norepinephrine. These compounds, however, only travel for the very short distances from the nerve endings to the tissues effecting the response. They are rapidly destroyed at the site of action. Hormones, which travel in the blood, survive much longer, though for a finite time. As the blood and the fluid bathing the cells can ultimately come into contact with all the tissues in the body, hormones may gain access to many more cells than are necessary for the ultimate response. Thus there is a sorting and translation problem with regard to the correct cells getting the appropriate message. This is solved in two ways. First, in contrast to the nerve transmittors, a large number of chemically distinct hormones are produced. Second, the cells themselves can react only with certain of these hormones and then respond only in a specific manner.

The tissues that make and store hormones are called *endocrine glands*. They release their products directly into the extracellular fluids from which they pass into the blood vessels that supply them, unlike exocrine glands, which secrete their products through discrete tubular ducts. There are at least eight such glands in a mammal, and they produce more than 40 hormones (Table 9.1). Anatomically and physiologically homologous endocrine tissues exist in all vertebrates, ranging from the jawless cyclostome fishes, like lampreys and hagfishes, to the mammals (Figure 9.1). The glands and their secretions are, however, not identical in all vertebrates. Many differences exist in their morphology and particularly in the chemical structures of the hormones that they secrete. Some vertebrates may lack certain endocrine glands that are present in mammals, or they may even possess novel ones.

All the endocrine glands are not essential for life, but they help. Thus in the absence of a functioning neurohypophysis mammals form large volumes of urine due to an inability to reabsorb water across the kidney tubules. Life is not incompatible or even shortened by this defi-

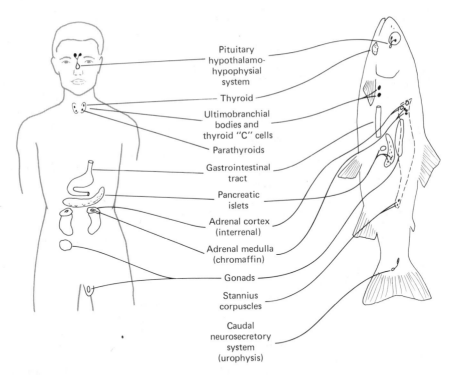

Pituitary
hypothalamo-
hypophysial
system

Thyroid

Ultimobranchial
bodies and
thyroid "C" cells

Parathyroids

Gastrointestinal
tract

Pancreatic
islets

Adrenal cortex
(interrenal)

Adrenal medulla
(chromaffin)

Gonads

Stannius
corpuscles

Caudal
neurosecretory
system
(urophysis)

FIGURE 9.1 Comparison of the endocrine system of man and a teleost fish. Notable differences are the lack of parathyroids in the fish. The adrenal tissues (adrenocortical and chromaffin) are separated in the fish, and two putative endocrine glands, the corpuscles of Stannius and urophysis, are present in addition. The precise roles of the latter two tissues in fish are not clear. (From H. A. Bern, *Science* 158:455–462, 1967. Copyright © 1967 by the American Association for the Advancement of Science.)

TABLE 9.1   The Secretions of the Endocrine Glands

Gland	Hormones	Target Tissues
Pituitary		
Adenohypophysis		
Pars distalis	Follicle stimulating hormone, FSH	Ovary and testis
	Luteinizing hormone, LH (also called interstitial-cell stimulating hormone, ICSH)	Ovary and testis
	Thyrotrophic hormone, TSH	Thyroid
	Corticotrophic hormone, ACTH (adrenocorticotrophic hormone)	Adrenocortical tissue
	Growth hormone, GH (somatotrophic hormone)	Liver; forms somatomedin, which alters tissue metabolism (liver, muscle, adipose tissue)
	Prolactin (luteotrophic hormone, LTH)	Mammary glands, fish gills, tadpole metamorphosis, corpus luteum, skin, and so on
	β-lipotrophin	Adipose tissue
Pars intermedia	Melanocyte stimulating hormone, MSH	Melanocytes, melanophores (pigmentation and color change)
Neurohypophysis		
Pars nervosa	Vasopressin (ADH), vasotocin	Kidney, amphibian skin and urinary bladder
	Oxytocin	Mammary gland, uterus
Hypothalamus	Pituitrophins, LH/FSH-RH, TRH, and so on	Adenohypophysis
Thyroid gland	Thyroxin (T_4) Triiodothyronine (T_3)	Tissue metabolism and differentiation, calorigenic (mammals), morphogenetic (amphibians)
Parathyroid glands	Parathormone, PTH	Bone, kidney, and gut (?)

TABLE 9.1 (*Continued*)

Gland	Hormones	Target Tissues
Ultimobranchial bodies ("C" cells in mammalian thyroid)	Calcitonin, CT (also called thyrocalcitonin)	Bone and kidney (?)
Adrenal glands		
Cortex (interrenals in sharks and rays)	Cortisol, corticosterone, cortisone, 1α-hydroxycorticosterone	Liver and muscle; proteins to amino acids, gluconeogenesis. Intestine and gills in teleosts; increased Na$^+$, K$^+$-activated ATPase
	Aldosterone	Kidney, sweat and salivary glands, gut, amphibian skin and bladder; Na and K metabolism
Medulla (chromaffin tissue)	Norepinephrine (noradrenaline) Epinephrine (adrenaline)	Liver, muscle and adipose tissue; glycogenolysis, mobilization fatty acids, calorigenic, smooth muscle constriction and relaxation
Islets of Langerhans		
α-cells	Glucagon	Liver (glycogenolysis), adipose tissue (fatty acid release)
β-cells	Insulin	Liver, muscle, and adipose tissue; amino acids to protein, glucose to fat and glycogen
Gonads		
Ovary		
Graafian follicle	Estrogens (estradiol)	Female sex organs and characters, mammary glands, brain
Corpus luteum	Progestins (progesterone)	Uterus, mammary glands and brain
Testis		
Interstitial tissue (Leydig cells)	Androgens (testosterone)	Male sex organs and characters, sperm maturation, brain
Sertoli cells(?)	Androgens	Sperm maturation

TABLE 9.1   (*Continued*)

Gland	Hormones	Target Tissues
Placenta (pregnant eutherian mammals)	Estrogens (estriol), progesterone	Uterus, mammary glands, fetus
	Chorionic gonadotrophin, HCG	Corpus luteum
	Placental lactogen, HPL (somatomammotrophin, HCS)	Mammary glands
Gut		
Stomach (pyloric mucosa)	Gastrin	Stomach; stimulates secretion of gastric juices
Duodenum (mucosa)	Enterogastrone (?)	Stomach; inhibits secretion of gastric juices
	Secretin	Exocrine pancreas; stimulation of pancreatic juices
	Cholecystokinin-pancreozymin	Exocrine pancreas and gall bladder; enzyme secretion and release of bile
	Enteroglucagon	As for glucagon
Kidney		
Tubular cells	1,25-dihydroxy-cholecalciferol	Bone and intestine; resorption of calcium
Juxtaglomerular cells	Renin	Plasma α_2 globulin; angiotensin, stimulates adrenal cortex and smooth muscle
Putative Endocrines		
Pineal gland	Melatonin	Hypothalamus (inhibits release MSH and gonadotrophins), melanophores (larval amphibia and cyclostomes)
Corpuscles of Stannius (some bony fishes)	Hypocalcin	Calcium metabolism, osmoregulation (?)
Urophysis (some fishes)		Smooth muscle contractions, gill (?) osmoregulation

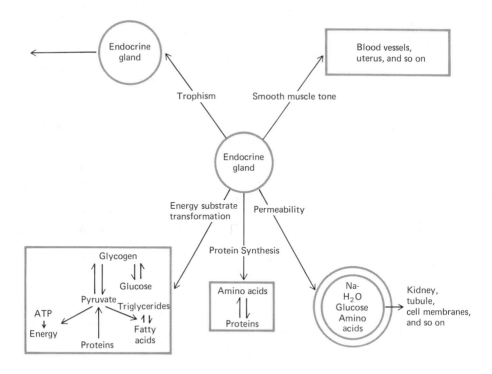

FIGURE 9.2   Some basic effects of the secretions of the endocrine glands. (From P. J. Bentley, *Comparative Vertebrate Endocrinology.* New York: Cambridge University Press, 1976.)

ciency if enough drinking water is available. However, the production of up to 20 liters of urine each day may be inconvenient and constitute, at least, a social problem. In other instances endocrine malfunction may result in physiologically and anatomically important changes. For instance, a lack of sufficient thyroid hormone may lead to the inadequate early development of mammals (cretinism), and life itself may be made precarious, as when the adrenal cortex is not functioning (Addison's disease). Problems may arise that are not only associated with a lack of adequate endocrine function but also with an overproduction of hormones. Some of the difficulties that may occur when the endocrine system is not working properly are summarized in Table 9.2.

Hormones from the endocrine glands can influence all the tissues in the body. Some of their effects, such as the contraction of the uterus or an increase in blood pressure, are more obvious than others. We can classify their actions into several major groups, which are shown in Figure 9.2. (1) They may have a *trophic effect* and stimulate other endocrine glands, such as when the gonadotrophins act on the gonads. (2) They may contract

TABLE 9.2   Some Effects of Endocrine Dysfunction in Man

Gland	Secretory Activity	Abnormality	Principal Effects
Pituitary			
Adenohypophysis	↑	If growth hormone: Giantism, acromegaly	Excessive growth
	↓	Dwarfism	Retarded growth
Neurophypophysis	↓	Diabetes insipidus ADH	Excessive loss of water in urine
	↑	Schwartz–Bartter syndrome ADH	Low plasma Na
Thyroid gland	↑	Graves' disease	High metabolic rate and nerve cell activity
	↓	Myxedema	Low metabolic rate
	↓	Cretinism	Inadequate development and growth
Parathyroids	↑	Hyperparathyroidism	Hypercalcemia, polyuria, reduced bone calcium
	↓	Hypoparathyroidism	Muscle tetany
Islets of Langerhans β-cells	↓	Diabetes mellitus	Hyperglycemia, muscle wasting
Adrenal cortex	↓	Addison's disease	Renal Na loss and K retention (low plasma Na, high plasma K), low blood pressure, muscle weakness
	↑	Cushing's syndrome	High blood pressure, obesity, retarded growth
Adrenal medulla	↑	Phaeochromocytoma	Hyperglycemia, high blood pressure
Ovaries	↓		Sterility, failure to develop or maintain secondary sex characters
Testis	↓		As above

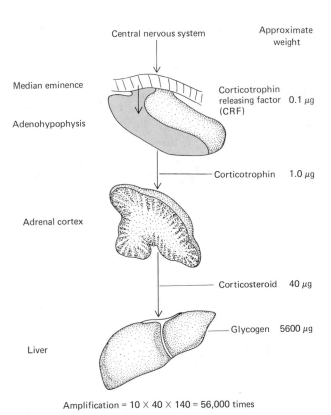

Amplification = 10 × 40 × 140 = 56,000 times

FIGURE 9.3 An example of "biological amplification" illustrated by the action of the hypothalamo-adenohypophysial-adrenocortical axis. The total amplification (or magnification) given in terms of product weight from the release of corticotrophin-releasing factor (CRF) to the deposition of glycogen in the liver is 56,000. (From P. J. Bentley, *Comparative Vertebrate Endocrinology*. New York: Cambridge University Press, 1976.)

smooth muscle, as does oxytocin on the uterus. (3) The *permeability* of membranes can be altered by hormones, such as ADH acting on the renal tubule. (4) The *metabolism of cells* may be altered, as when thyroxin exerts its calorigenic effect.

Hormones occur only at very low concentrations in the animal's body fluids. Antidiuretic hormone, ADH, from the neurohypophysis, is active when present at a concentration less than 10^{-10} M, or 0.1 μg in 1 liter of plasma. Hormones can nevertheless have far-reaching effects, and they provide an excellent example of the process of biological amplification (or magnification). Thus in the instance (Figure 9.3) of the effects of corticotrophin-releasing factor (CRF) from the hypothalamus, 0.1 μg of this local releasing hormone leads to the generation of 5600 μg of glycogen in the liver, representing (in terms of weight) an amplification of 56,000 times. If one knew the quantities of the neural transmittors associated with CRF release, the amplification would be even more astounding.

2 • MECHANISMS OF HORMONE ACTION

Until about 10 years ago little was known about just how hormones initiated their effects, and knowledge is still far from complete. It was once considered likely that hormones had a direct action in their effector tissue. Thus a change in cell membrane permeability could conceivably be pro-

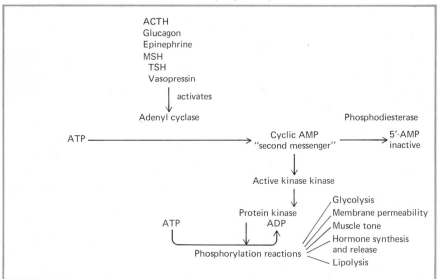

FIGURE 9.4 The role of adenosine-3',5'-monophosphate (cyclic AMP) in hormone action: its structure, synthesis, metabolism, and some hormones whose actions are mediated by it.

TABLE 9.3 Relations between the Actions of Some Hormones and Cyclic AMP

Hormone	Actions Shared by Cyclic AMP
ACTH	Steroid hormone synthesis in adrenal cortex
Epinephrine	Glycolysis in liver and muscle, lipolysis in fat cells, contractility of heart muscle
Glucagon	Glycolysis
Luteinizing hormone	Steroid hormone synthesis by corpus luteum
Melanocyte stimulating hormone	Dispersion of melanin granules in melanophores
Parathormone	Mobilization of Ca from bone, increased Ca resorption across renal tubule
Vasopressin (ADH)	Increased water transfer across renal tubule

duced by a hormone molecule interacting directly with some structural component of the cell. This combination could alter a molecule's shape and make a membrane more or less permeable or even result in some mechanical movement. Unequivocal examples of this type of process are not currently available, the events being far more complex and involving at least several intermediate steps. Hormones initially may interact with some site on the cell's surface and either enter the cell or from this position initiate a chain of chemical events. The position where the hormone is so bound is called its *receptor site*. Receptors can react only with certain molecules, and these (not necessarily all) will initiate a response. This has often been likened to the alignment of a key (the hormone) in a lock (the receptor). If the lock can then be turned, the cell acquires its ability to discriminate between different hormones. However, different cells often respond to the same hormone in different ways, depending on the particular series of chemical events with which the receptor is aligned. For instance, epinephrine can promote glycogenolysis in skeletal muscle, dilatation of the bronchi, and contraction of blood vessels. Taking the key-lock analogy one step further, the door when opened may lead into different rooms.

Two major types of mechanism mediating the actions of hormones have been described. One of these (Figure 9.4 and Table 9.3) involves the formation of the nucleotide *adenosine-3',5'-cyclic monophosphate* or *cyclic AMP* (see Chapter 2, Section 3). The other requires the *synthesis of new proteins*, which often have an enzymatic role that can speed up certain processes in the cell.

2.1 Cyclic AMP

Glycogen is broken down to glucose in the liver under the influence of the hormone's glucagon and epinephrine. This process involves the activation of the enzyme *phosphorylase* that is converted from the inactive to the active form. While studying the steps in this process, Earl Sutherland discovered that a factor was formed, in the presence of either of these hormones, that activated this enzyme. The substance was *cyclic AMP*. It is formed as a result of the activation of *adenyl cyclase* in the cell wall, and this speeds the formation of this nucleotide from ATP. The cyclic AMP so formed activates the phosphorylase following a series of reactions that initially involves stimulation of a *protein kinase* in the liver. The destruction of cyclic AMP is promoted by *phosphodiesterase*. These processes are summarized in Figure 9.4. Cyclic AMP thus acts as a "second messenger" in the hormone's action. This role of the nucleotide is not limited to glucagon and epinephrine but is widespread in the body and associated with the actions of many hormones. Some of its effects are summarized in Table 9.3. These include the initiation of diverse responses, such as synthesis of adrenocorticosteroids by corticotrophin, expansion of melanophores by MSH, increase in membrane permeability to water by ADH, and mobilization of bone calcium by parathormone. Cyclic AMP is not confined to animals; it even plays a role in bacterial physiology. Its importance may be compared to that of the role of ATP in the transformation of energy. Sutherland was awarded the 1972 Nobel Prize for his role in its discovery.

2.2   Induction of Protein Synthesis

Several hormones act in a manner that does not appear directly to involve cyclic AMP or the cell membrane. Instead, a change in the nuclear gene expression results in the induction of new proteins. Such effects mediate the actions of hormones from the gonads and adrenal cortex, which are steroids, as well as those that induce complex metabolic changes and include thyroxin, growth hormone, prolactin, vitamin D_3, and possibly the hypothalamic-releasing and -inhibiting hormones. The steps in this process have been studied using the female sex hormones, estradiol-17β and progesterone, thyroxin, aldosterone, and cortisol. The pattern of their effects in the cells seems to be basically similar. The hormone enters the cell (Figure 9.5) and in combination with a *cytoplasmic receptor* crosses to the nucleus, where it combines with the chromatin and is closely associated (possibly even bound) to DNA. Genetic transcription occurs as a result of an increased action of RNA polymerase, so messenger RNA is formed (see Chapter 7, Section 2.4). The newly synthesized messenger RNA initiates and programs the formation of new proteins in the cytoplasm that in turn initiate the hormone's effect. This response may be a change in membrane processes, as is probably the case with aldosterone or, more generally, a metabolic process associated with growth, differentiation, or the transformation of substrates providing energy.

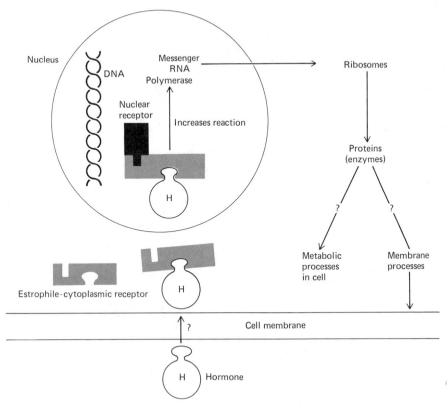

FIGURE 9.5 The mechanism of action of hormones that alter nuclear genetic translation and transcription in the cell. This could apply to the actions of estrogens, progestins, thyroid hormones, and corticosteroids. (From P. J. Bentley, *Comparative Vertebrate Endocrinology.* New York: Cambridge University Press, 1976.)

3 • EVOLUTION OF THE ENDOCRINE GLANDS

The gross structural arrangements of the endocrine tissues may differ considerably among the vertebrates. Homologous tissues are usually found throughout the vertebrate series, but there are exceptions. Thus parathyroid glands have not been identified in fishes, whereas urodele amphibians (newts and salamanders) and cyclostome fishes lack the alpha cells of the islets of Langerhans. Putative endocrine glands, like the urophysis and corpuscles of Stannius (see Section 8), are confined to certain fishes.

There is an evolutionary tendency for the endocrine tissues to assume a more discrete and compact form as one ascends the phyletic scale. Thus in mammals the adrenal gland is made up of a separate compact

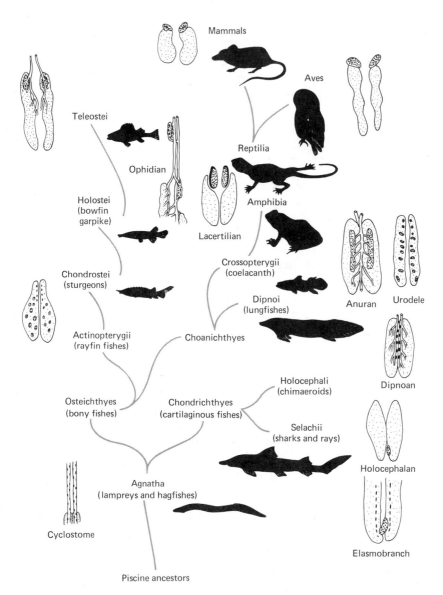

FIGURE 9.6   The evolution of the adrenal glands. The adrenal tissues are shown in association with the ventral surface of the kidneys (light stippling). Black indicates chromaffin tissue (catecholamines), dark stippling adrenocortical (interrenal) tissue. The adrenals are shown as if in cross section.

mass of chromaffin tissue surrounded by an equally distinct piece of adrenocortical tissue. However, in the fishes these tissues usually exist as relatively isolated islets of tissue lying along the major blood vessels and the kidneys. As one ascends the phyletic tree (Figure 9.6) the adrenal tissues assume a more compact form and become increasingly associated with each other. The reasons for the close association of the chromaffin tissue, which produces epinephrine, and the adrenocortical tissue, which secretes corticosteroid hormones, have for a long time been a mystery. Embryologically they have diverse origins, being derived respectively from neuroectodermal and coelomic epithelium. They also produce chemically very different hormones: catecholamines and steroids. It appears that in mammals high local concentrations of corticosteroid hormones are necessary for the maintenance of adequate amounts of the enzyme phenylethanolamine-N-methyltransferase (PNMT), which converts norepinephrine to epinephrine. The corticosteroids pass relatively undiluted through local portal blood vessels from the cortex directly to the medulla of the adrenal gland, where they exert these actions.

Similar symbiotic evolution of the endocrine tissues is seen in the instance of the pituitary gland and hypothalamus. The pituitary is basically composed of several types of tissues held in varying degrees of association with the hypothalamus at the base of the brain. The neurohypophysis is neural tissue and maintains direct nervous connections with the brain. The adenohypophysis, on the other hand, is ectodermal. The two tissues lie in close juxtaposition to each other and in apposition to the hy-

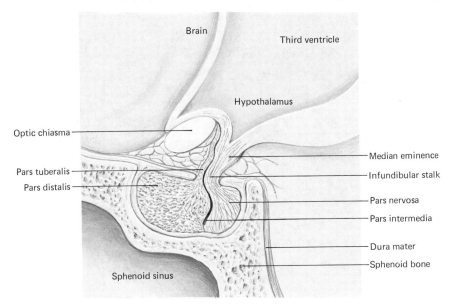

FIGURE 9.7 The mammalian pituitary gland. The adenohypophysis includes the pars tuberalis, pars distalis, and pars intermedia (intermediate lobe). The neurohypophysis includes the median eminence, pars nervosa, and neural lobe. The pars tuberalis and pars distalis are sometimes called the anterior lobe and the pars intermedia and pars nervosa the posterior lobe. (Modified from R. Guillemin and R. Burgess. Copyright © November 1972 by *Scientific American*. All rights reserved.)

pothalamus (Figure 9.7). The latter merges into the median eminence (nervous tissue) that lies parallel to the adenohypophysis. This latter tissue contains several distinct regions or cell types that are responsible for forming its different hormonal secretions. In the cyclostome fishes these tissues are loosely connected, but their morphological association becomes closer in higher vertebrates (Figure 9.8). Different parts of the tissues differentiate in a manner that can be related to their function. Thus the neurohypophysis enlarges posteriorly in tetrapods, so a relatively larger and more distinct neural lobe becomes apparent in the amphibians. The hormonal secretions of this tissue (for instance ADH and vasotocin)

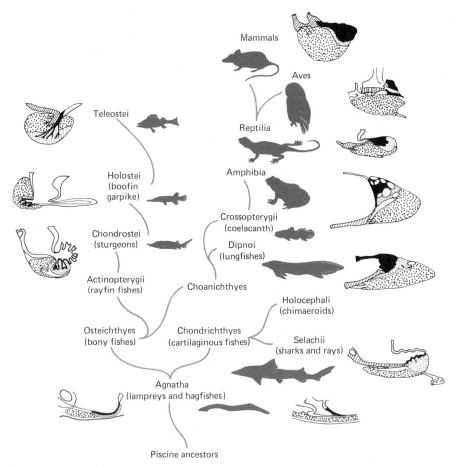

FIGURE 9.8 Evolution of the pituitary gland. Black—neural lobe; crosshatching—median eminence; large dots—adenohypophysis; small dots—pars intermedia. (Based on K. Wingstrand, *The Pituitary Gland*, vol. 1. Berkeley: University of California Press, 1966.)

play a distinct role in osmoregulation of such terrestrial vertebrates, and the amount of stored hormone is much greater than that found in fishes. A most important interrelationship in the pituitary is that between the adenohypophysis and the median eminence, and this affords a direct connection for the reception, by the adenohypophysis, of information collected and transcribed by the nervous system. In the teleost fishes the adenohypophysis and median eminence are separated by a plexus of capillaries, and thus their interconnections are rather tenuous. In the tetrapods and other fishes a distinct portal system arises into which secretions from the median eminence can be discharged and then pass directly to the adenohypophysis. These secretions are hormones that originate in nerve cells by a process called *neurosecretion* (see Chapter 3, Section 8). Such nerves discharge their secretions directly into blood vessels rather than into the region of another nerve, muscle, or gland. Such neurosecretions from the median eminence influence the formation and release of seven hormones made by the adenohypophysis.

There are other examples of morphological differences in the endocrine glands, the physiological significance of which, if any, is not understood. For example, the ultimobranchial bodies (which secrete calcitonin) are discrete glands in birds, reptiles, amphibians, and fishes, but in mammals this tissue is embedded in the thyroid gland and makes up the "C" cells.

4 • PHYLETIC DISTRIBUTION, POLYMORPHISM, AND EVOLUTION OF THE HORMONES

Many hormones from different vertebrate species exhibit a considerable polymorphism of chemical structure. On the other hand, others are remarkably conservative and have few variants throughout the vertebrates.

Thyroxin and epinephrine have been identified in members of all the vertebrate groups from cyclostome fishes to mammals. Natural analogs are scarce, and they consist, respectively, of triiodothyronine and norepinephrine. The relative proportions of each may vary within a species, but the identical molecular structures persist. The steroid hormones, such as those secreted by the gonads and adrenal cortex, are a little less conservative. Several corticosteroids exist that have different biological activities and exhibit a fairly precise pattern in their phyletic distribution (Figure 9.9). Their presence in the cyclostomes is in some doubt, but cortisol, cortisone, and possibly corticosterone are present in the Osteichthyes (bony fishes). The Selachii (sharks and rays), on the other hand, secrete predominantly 1α-hydroxycorticosterone. Aldosterone, a hormone that has an im-

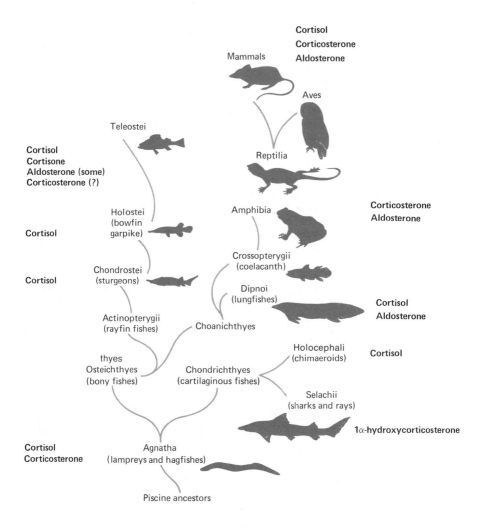

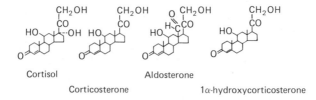

FIGURE 9.9 The principal adrenocorticosteroid hormones in vertebrates and their phyletic distribution. (From P. J. Bentley, *Comparative Vertebrate Endocrinology*. New York: Cambridge University Press, 1976.)

portant role in the regulation of sodium and potassium balance, is present in all the major tetrapod groups, together with corticosterone or cortisol or even both. It has recently been shown that the South American lungfish, *Lepidosiren paradoxa,* in contrast to most other bony fishes, also secretes aldosterone. This is particularly interesting because this group of vertebrates, the Dipnoi, is considered to be close to the original line of tetrapod evolution. The Dipnoi are a living "missing link" in evolution, and, as we shall see, several other of their hormones (including prolactin, growth hormone, and mesotocin) also show closer affinities to tetrapods than to other fishes.

It is among the protein and polypeptide hormones that we see the greatest polymorphism. There is insufficient space to give a complete account of this variation here. In some instances there appear to exist almost random differences in the amino acid composition of the hormones, whereas in others one can seemingly trace the path of the hormone's evolution. Differences in structure of such homologous hormones from different species became apparent even before their precise amino acid sequences could be determined. This was revealed by differences between the relative sizes of responses elicited by hormones from various species when their actions are compared in biological preparations. Thus the neurohypophysial peptide isotocin (see Figure 9.10) from teleost fishes behaves in many respects like mammalian oxytocin, and both contract the rat uterus. However, when equiactive amounts as measured on the uterus are compared for their abilities to increase sodium transport across frog skin, isotocin is much less effective. In isotocin this reflects the substitution of two different amino acids: serine at position 4 and isoleucine at position 8. Additional evidence of polymorphism among hormones is sometimes seen when a hormone preparation made from one species is administered to a different species. A progressive decline in the response can occur due to the formation of antibodies as a reaction to the foreign hormone. Such results of hormone polymorphism are of practical, as well as theoretical, significance, because endocrine glands from animals usually provide the hormonal preparations that are used to correct endocrine deficiencies in man. *Insulin* is the earliest example of such a source, and this hormone, which is used for the treatment of diabetes mellitus (inadequate insulin), is derived from the pancreas of pigs and cattle. Pig and cattle insulins differ from human insulin by one and three amino acid substitutions, respectively. Although their activities in lowering blood sugar levels in man are basically similar, the continual administration of either one of these animal hormones can lead to a decline in the response, due to development of an immunity. If one then substitutes injections of the other animal's hormone, sensitivity will be regained since antibodies have not been formed to neutralize this particular insulin. The construc-

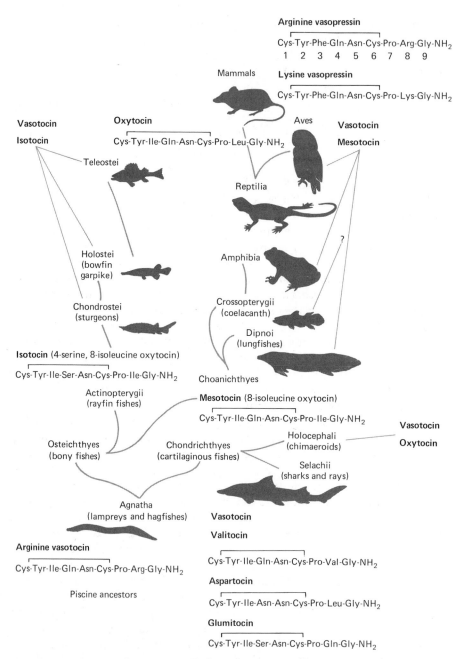

FIGURE 9.10 The principal neurohypophysial hormones in vertebrates and their amino acid sequence and phyletic distribution. (From P. J. Bentley, *Comparative Vertebrate Endocrinology*. New York: Cambridge University Press, 1976.)

tion of evolutionary pathways for such molecules can sometimes be difficult; thus although pig insulin differs from that in chicken by the substitution of six amino acids, it differs from that of guinea pigs (whom most of us would consider a closer relative) by 17 such replacements.

Similarities and differences in the pituitary *growth hormone* and *prolactin* are particularly interesting. These molecules show a considerable crossover in their biological activities. Growth hormone has a prolactin-like activity and vice versa. In addition, interspecific variations in their respective structures result in differences in their biological potency when they are administered to other species.

An extreme example of species specificity to a hormone is growth hormone in man. Some children suffer from a deficiency in growth hormone, which can result in dwarfism. Other mammalism growth hormones are completely ineffective for the treatment of this condition. Human pituitaries thus must be collected from cadavers and the growth hormone extracted for administration to these children. In other species, cross responsiveness to growth hormone does occur, although this is dependent on the closeness of the relationship. Fish growth hormone is usually not effective in tetrapods, but a notable exception is that from the lungfish, *Protopterus aethiopicus.*

In the instance of prolactin a possible pattern of the hormone's evolution has been proposed. This hypothesis is based on the differences in the biological activities of prolactins from different groups of vertebrates. Prolactin has several effects when injected into different species. These include promotion of milk secretion in mammals (milk letdown), the formation of "pigeon's milk" from the crop sac of pigeons and doves, the stimulation of certain newts to return to water and breed (eft water-drive effect), and a reduction in the permeability of the gills of some fishes to sodium (teleost sodium-retaining activity). When the activities of prolactin from different major vertebrate groups are compared with respect to these responses, a pattern, presumably reflecting differences in the hormone's structure, emerges (Figure 9.11). Cyclostomes seem to lack prolactin, but all other species exhibit the eft water-drive activity; lungfishes (but not other fishes) and the tetrapods all have pigeon crop sac activity; and milk secretion only occurs in response to the tetrapod hormones. That we are dealing with homologous molecules in the fishes and tetrapods is confirmed by their immuno-cross-reactivity. Antisheep prolactin serum reacts with fish prolactin, while antifish prolactin serum reacts with sheep prolactin.

It has recently been confirmed that the overlap in the biological effects of prolactin and growth hormones depends on their chemical relationships. The precise structural differences are difficult to define, but it was shown that antibodies to human growth hormone do not interact

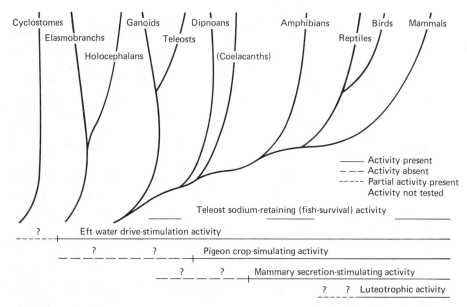

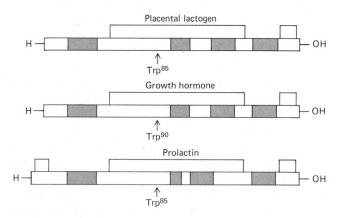

FIGURE 9.11 Phyletic distribution in vertebrates of some of the actions exhibited by prolactin and prolactinlike (paralactin) hormones. (From P. J. Bentley, *Endocrines and Osmoregulation.* New York: Springer-Verlag, 1971; and H. A. Bern.)

FIGURE 9.12 Diagrammatic representation of the structure of placental lactogen, growth hormone, and prolactin molecules. The cross-hatched areas represent internally homologous sequences of amino acids in each. Other similarities can be seen in the presence of disulfide bridges (lines above) and the tryptophan (Trp) residues at position 85 in placental lactogen and growth hormone and 90 in prolactin. (From H. D. Niall et al., *Proc. Nat. Acad. Sci. U.S.A.* 68:866–869, 1971.)

with human prolactin. Another hormone, produced by the human placenta during pregnancy (human placental lactogen, HPL, also called human chorionic somatomammotrophin) also has both growth hormone and prolactin activities. Its structure has also been described, and the three are compared in Figure 9.12. Considerable homologies exist between the molecules, and it has been suggested that they may have evolved by gene reduplication from a common ancestral peptide.

Comparable evolution may also have occurred among the pituitary glycoprotein hormones FSH, LH, and TSH. The structures of the two latter hormones in mammals have been compared. Each consists of two subunits called α- and β-chains. It was found that the α-chains of LH and TSH are virtually identical, and differences in their activities reflect variations in the β-chains. The β-chains can be chemically dissociated from their α counterparts but alone exhibit no activity. However, if the β-chain from one hormone, say, TSH, is combined with the α-chain of the other (for example, β-TSH with α-LH), the TSH activity is restored. Mammalian TSH, as well as LH and FSH, have a thyrotrophic action in fishes. This crossover presumably reflects the similarities between the three molecules and the piscine TSH. All of these hormones may have evolved from a common ancestral glycoprotein.

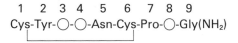

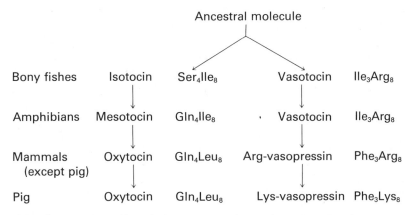

FIGURE 9.13 A hypothetical scheme of the evolution of neurohypophysial hormones. One gene duplication and a series of subsequent single substitutions at positions 3, 4, or 8 produce two molecular "lines." (From R. Acher *Proc. Roy. Soc. B* 170:7–16, 1968.)

The neurohypophysial hormones are peptides consisting of a ring of five amino acids joined to a side chain containing three more. The possibility of genetic change is far more limited than in the larger protein hormones, and a clearer picture of their possible evolution can be seen. So far eight such peptides have been described among the vertebrates. Usually two exist in a single species. In mammals these are arginine-vasopressin (AVP or ADH) and oxytocin; in birds, reptiles, amphibians, and lungfish, vasotocin (the ring of oxytocin and side chain of ADH) and mesotocin; in teleost fish, vasotocin and isotocin, and so on (Figure 9.10). Anomalies may occur within a group. Thus in mammals lysine-vasopressin (LVP), instead of arginine-vasopressin (AVP), is found in domestic pigs, whereas other members of the Suinidae, like the warthog and peccary, may contain either LVP or AVP or both. A proposed scheme for the evolution of the neurohypophysial hormones from an ancestral molecule (which could be vasotocin) and involving single-step genetic changes is shown in Figure 9.13.

5 • HORMONE SYNTHESIS, RELEASE, TRANSPORT, AND METABOLISM

The effector's response to a hormone in the animal's body is the fulfillment of its physiological role, but this is only part of a complex series of events that begins with its synthesis (Figure 9.14) and includes the storage of the hormone, its release into the blood, its transport to the site of action, and finally its destruction or removal from the body.

The endocrine glands *synthesize* and then store the products that they ultimately release into the blood. The diversities in the synthesizing processes reflect the differences in the chemical nature of the hormones. Some of the processes involved in the formation of the steroid hormones, the catecholamines, and the thyroid hormones are summarized in Figures 9.15 and 9.16. In glands, like the adenohypophysis, that secrete many different hormones the syntheses are not directly linked to each other and are carried out by distinctly different types of cells that compose the tissue. The peptide hormones of the neurohypophysis are formed in nerve cells in the hypothalamus, pass down the nerve fibers, and are stored in the neural lobe. The process of hormone synthesis may be controlled by a feedback control system (see Figure 9.17 and Chapter 1, Section 4) that involves a trophic hormone. Thus corticotrophin initiates steroidogenesis in the adrenal cortex, and the trophic hormone in its turn is regulated by the concentrations of corticosteroids in the blood. Products of the hormone's action, or the lack of them, might also influence synthesis; for instance, high

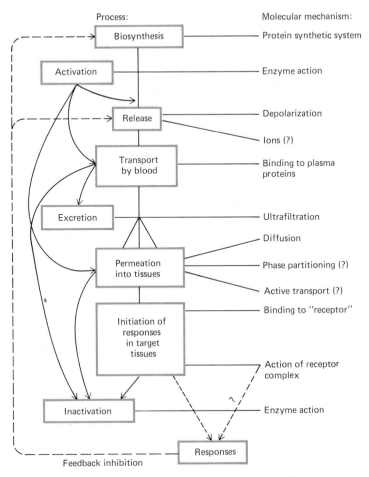

FIGURE 9.14 The life history of a hormone in the body. The dashed lines represent possible pathways for a "feedback" inhibition. (From J. Rudinger, *Proc. Roy. Soc. B* 170:17–26, 1968.)

glucose levels in the blood passing through the hypothalamus decreases the release of growth hormone. Such effects may be mediated by other hormones—in this instance, an increased release of growth hormone-release inhibiting hormone (somatostatin) from the hypothalamus.

The mode of *storage* of the newly synthesized hormones differ. In some glands, such as the pituitary and thyroid, substantial reserves are usually retained as compared with the adrenal cortex, for instance. When substantial storage of hormones occurs, special structures are utilized for this purpose. The thyroid hormones are stored outside the cell attached to

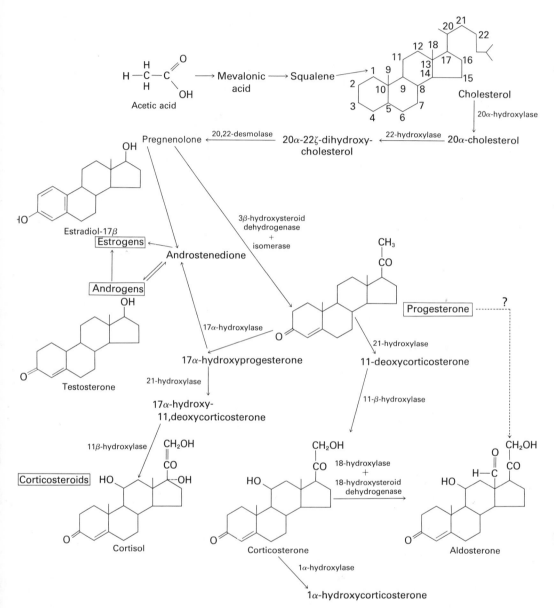

FIGURE 9.15 Some of the steps and interrelationships in the biosynthesis of steroid hormones. (From P. J. Bentley, *Comparative Vertebrate Endocrinology.* New York: Cambridge University Press, 1976.)

large protein molecules called *thyroglobulin.* Catecholamines, neurohypophysial peptides, and adenohypophysial hormones are contained in small

TABLE 9.4   **Physiological Factors Influencing the Release
of Hormones in the Body**

Hormone	Releasing Stimuli
Aldosterone	Low Na plasma concentration, angiotensin
Calcitonin	High plasma calcium levels
Cortisol and corticosterone	Corticotrophin
Epinephrine	Neural stimuli
Estrogens	FSH
Follicle stimulating hormone (FSH)	External stimuli, rhythms, low estrogen levels
Glucagon	Low plasma glucose levels
Growth hormone	Sleep, exercise, apprehension, hypoglycemia
Insulin	High plasma levels of glucose and amino acids, glucagon, growth hormone; in ruminants, high propionic and butyric acid levels
Luteinizing hormone (LH)	External stimuli, sexual excitement (male) rhythms, estrogen (female), low progesterone or testosterone levels
Melatonin	Light inhibits, diurnal rhythm
Melanocyte stimulating hormone (MSH)	Light on retina
Oxytocin	Suckling, stimulation of reproductive tract
Parathormone	Low plasma calcium levels
Progesterone	LH, chorionic gonadotrophin
Prolactin	Diurnal rhythms, suckling, parturition, low plasma osmotic concentrations (in some fish), estrogens
Renin	Low plasma Na, reduced renal blood flow, β-adrenergic stimuli
Testosterone	LH (ICSH)
Thyroid Hormones	TSH
Thyrotrophic hormone (TSH)	Low thyroxin and temperature
Vasopressin	Increased osmotic concentration in plasma

granules inside the cell. Within these storage granules the hormones are attached to specific protein. For example, oxytocin and vasopressin are bound to neurophysin, and this protein even differs in its structure between species secreting arginine- and lysine-vasopressin.

The *release of hormones* may take place in response to stimuli that reflect the physiological imbalance experienced by the animal and which the hormone tends to correct (see Table 9.4). They may in addition be released by nonspecific and unpleasant stimuli, some of which are termed

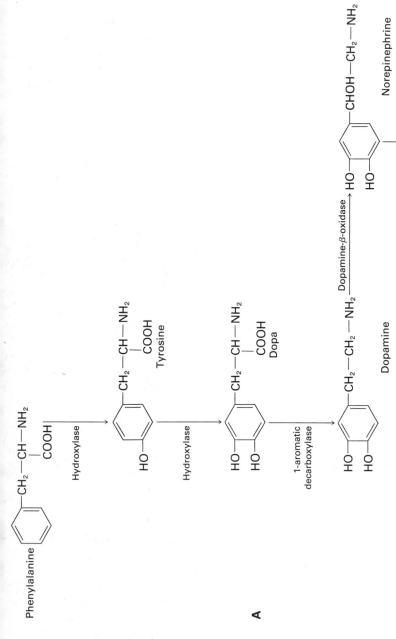

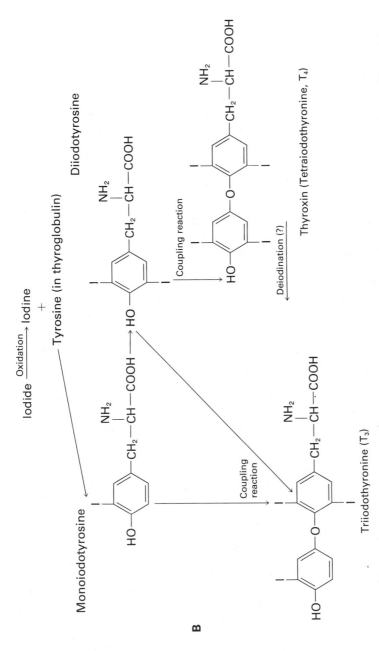

FIGURE 9.16 The synthesis of two groups of hormones from amino acids. (**A**) Catecholamines: nonrepinephrine and epinephrine. (**B**) Thyroid hormones: thyroxin and triiodothyronine. (From P. J. Bentley, *Comparative Vertebrate Endocrinology*. New York: Cambridge University Press, 1976.)

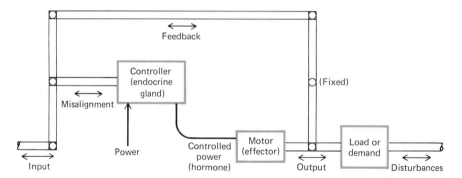

FIGURE 9.17 The analogies between a negative feedback servosystem and endocrine coordination in the body. The double arrows indicate that the various links can move in either direction. (From P. J. Bentley, *Endocrines and Osmoregulation.* New York: Springer-Verlag, 1971.)

"stress." The control of the circulating levels of a hormone through the initiation or inhibition of their release from the endocrine glands, involves a series of feedback controls that engineers call a servosystem (see Chapter 1). The analogies between these are shown in Figure 9.17. The hormone's release may be initiated by a physiological imbalance involving a deficiency or excess, which when the hormone acts on its effector system will then tend to be corrected. This adjustment will in its turn inhibit the further release of the hormone. Thus as a result of water loss by evaporation the osmotic concentration of the blood rises, and this will initiate the release of vasopressin (ADH) from the neurohypophysis. The ADH in its turn reduces water loss from the kidney (the urine becomes more concentrated), so the osmotic concentration of the blood declines, thereby reducing the release of ADH. Conversely, excessive water intake inhibits release of ADH; thus water excretion by the kidney is enhanced by this process.

The intimate mechanisms involved in the release of hormones from cells is a field of intensive research. The hormones, which are stored in granules, appear to be discharged, together with the proteins to which they are bound. Calcium is essential for the coupling of this event to the initiating stimulus. In the instance of the release of thyroxin and cortisol by TSH and corticotrophin, respectively, cyclic AMP is involved, whereas in such other glands as the adrenal medulla a nerve impulse may carry the signal (synaptoid control). Rhythmical, periodic release of hormones is a common and important phenomenon. This may occur during the course of the day (diurnal and circadian) rhythms, as seen with corticosteroids, prolactin, and growth hormone. In man cortisol levels are lowest at night, whereas prolactin and growth hormone are released during sleep (Figure 9.18). This pattern may differ somewhat among species. Reproduction is

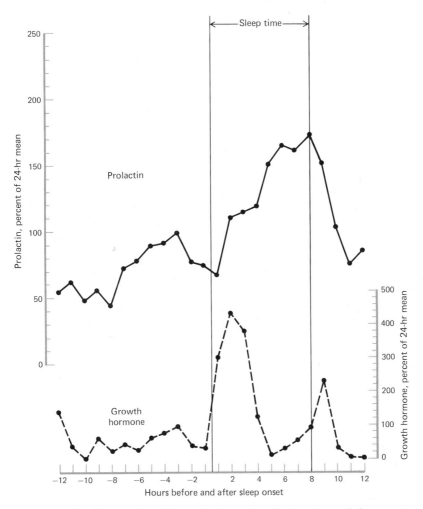

FIGURE 9.18 The circadian rhythmical release of prolactin and growth hormone in man. (From J. F. Sassin et al., *Science* 177:1205–1207, 1972. Copyright © 1972 by the American Society for the Advancement of Science.)

particularly dependent on such periodically timed releases of hormones. Successful reproduction, consistent with the generation and survival of the young, will depend to a large extent on suitable environmental conditions, such as adequate food, water, and temperature. The timing of the breeding season is thus critical and will depend on the reception of external stimuli that include light, temperature, food, sound, and even smell. Such stimuli by impinging on an internal "biological clock" can influence the release of the pituitary gonadotrophic hormones. The "clock" appears to be situated somewhere in the brain and initiates neural and/or hormonal messages that control hormone release from the adenohypophysis.

In the blood the hormones may circulate in a free, unattached form, or they may associate and be bound to various plasma proteins. Thyroxin is bound to albumins and, in mammals (but not lower vertebrates), to a specific alpha-globulin called *thyroxin-binding-globulin* (*TBG*). Steroid hormones may also be bound to plasma proteins, and the degree of such binding appears to be greater as one ascends the phyletic scale. In mammals cortisol is bound to a specific globulin, called *transcortin*, that has not been identified in nonmammals. Binding of steroid hormones to albumins also occurs. The linkage of hormones to plasma proteins may influence their actions by reducing the rate of their excretion and inactivation and influencing the ease with which they cross the capillaries and gain access to their receptors.

Finally, the hormone's action needs to be terminated. The efficiency of the control system will depend on the speed with which such an effect can be terminated. The adjustment will partly depend on the cessation of the hormone's release and sometimes on the appearance of another hormone with an opposing action. The *inactivation* of a hormone is also important and results from its metabolic conversion to another chemical substance. This process usually occurs in the liver or kidney or both, where the hormone may be converted into a form that facilitates its excretion in the urine and bile (for example, lipid- to water-soluble form) or is broken down into smaller fragments. Some hormones may also be excreted in a relatively unchanged form. The persistence of a hormone in the body varies considerably: in terms of its half-life, from a few minutes for vasopressin to several hours for thyroxin. This is consistent with the nature of its action and the effects it has in the body.

6 • CONTROL OF THE ENDOCRINE SYSTEM: THE PITUITARY GLAND

The central control of the endocrine system and its interrelationships with the nervous system are principally mediated by the pituitary gland. As we have seen, this is situated at the base of the brain close to the hypothalamus (see Figure 9.19 for an overall summary of this section). Neural connections from the brain control the release of hormones from the neurohypophysis and intermediate lobe. The median eminence (a part of the hypothalamus) is also under neural control. This region of the brain is connected to the adenohypophysis by small blood vessels, the *hypothalamo-hypophysial portal system*. Neurosecretory products from the nerve

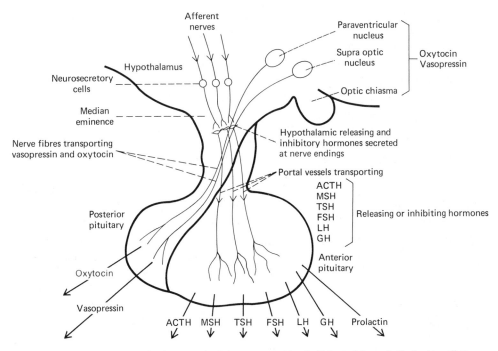

FIGURE 9.19 Hypothalamic control of the pituitary gland. (Adapted from J. S. Jenkins, *Brit. Med. J.* 2:99–102, 1972.)

endings in the median eminence are discharged into these vessels and are carried to the adenohypophysis, where they play a dominant role in its control. Thus neural information from both the internal and external environments and the brain itself can be transferred to the pituitary gland.

The importance of this association between the brain and the pituitary can be demonstrated by transplanting the pituitary to another part of the body (an ectopic transplant)—for example, to a kidney or to an eye. The normal synthesis and release of its various hormones is then severely impaired. Retransplantation back to its original site results in a restoration of control, following a brief period of time during which blood vessels connecting it to the median eminence are regenerated. Severing the nerve tracks connecting the hypothalamus and the pituitary results in diabetes insipidus due to the degeneration of the neurohypophysis and, in frogs, a persistent expansion of the melanocytes resulting from a continuous, uncontrolled release of MSH. If the tissue containing the connecting blood vessels, the pars tuberalis, is cut, only a temporary incapacity in control of the adenohypophysis results as regeneration of the vessels occurs. However, if a piece of waxed paper is inserted between the median eminence and the adenohypophysis, this reconnection cannot take place and the effects are equivalent to transplantation of the adenohypophysis.

Ectopic transplantation of the adenohypophysis varies in its effects. The activity associated with most adenohypophyseal hormones declines, though to a varying extent. Control of secretion of the thyrotrophic hormone seems to be less compromised than, for instance, that of the gonadotrophins, the release of which is inhibited. On the other hand, prolactin secretion in mammals is increased, and, as we shall see, this reflects the lack of inhibitory control from the hypothalamus. In birds, however, prolactin secretion declines, and this may reflect a different (stimulating) mode of hypothalamic control for this hormone. Only sporadic experiments of this kind have been carried out in lower vertebrates. In amphibians the hypothalamic control of the pituitary seems to be similar to that in mammals, but this has not been described in the cyclostome and selachian fishes. In teleost fishes hypothalamo-hypophysial portal vessels do not occur regularly, which makes it difficult to conceive a physiological role for such systems in these fishes. However, the story should be considered as incomplete.

Information or messages are transmitted from the median eminence to the adenohypophysis by a series of hormones that are carried in the connecting portal blood vessels. The compounds that have been chemically characterized are peptides, but catecholamines may also be involved. They are called pituitary-releasing or -inhibiting hormones (sometimes "factors") or *pituitrophins*. Thus TSH is controlled by TSH-releasing hormone (TRH), prolactin by prolactin-release inhibiting hormone (P-R-IH), and so on. The release of these pituitrophins is controlled by the brain and a feedback control system that usually reacts to circulating concentrations of the hormones they ultimately control. The site for the feedback is thought to be in the hypothalamus, including the median eminence and possibly the pituitary. The circulating levels of thyroid hormone, the corticosteroids, the gonadal steroids, and possibly prolactin are influenced by a negative feedback control. In other instances the metabolic products, such as fatty acids and glucose, may be involved. When the negative feedback inhibition is due to hormones (or metabolites) derived from the peripheral endocrine glands, such as the gonads and the adrenal cortex, it is termed a "long loop" feedback. It is also suspected that some adenohypophysial hormones, such as prolactin and growth hormone, may exert more direct effects on the release of pituitrophins, and this is called a "short loop" feedback. Pituitrophins appear to control the synthesis and release of all the adenohypophysial hormones. Their total number is not yet clear because in some instances both an inhibitory and releasing hormone may control a single adenohypophysial hormone. The chemical structure of several of these hormones has been described (Figure 9.20). The peptides

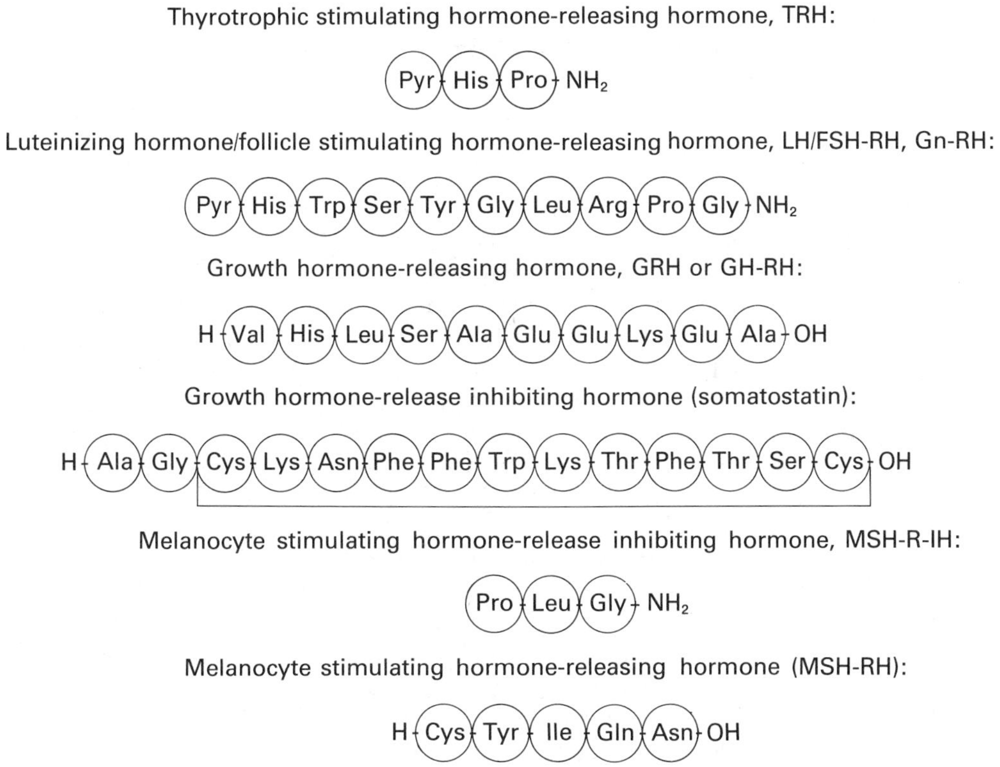

Thyrotrophic stimulating hormone-releasing hormone, TRH:

$$\text{Pyr}-\text{His}-\text{Pro}-NH_2$$

Luteinizing hormone/follicle stimulating hormone-releasing hormone, LH/FSH-RH, Gn-RH:

$$\text{Pyr}-\text{His}-\text{Trp}-\text{Ser}-\text{Tyr}-\text{Gly}-\text{Leu}-\text{Arg}-\text{Pro}-\text{Gly}-NH_2$$

Growth hormone-releasing hormone, GRH or GH-RH:

$$H-\text{Val}-\text{His}-\text{Leu}-\text{Ser}-\text{Ala}-\text{Glu}-\text{Glu}-\text{Lys}-\text{Glu}-\text{Ala}-OH$$

Growth hormone-release inhibiting hormone (somatostatin):

$$H-\text{Ala}-\text{Gly}-\text{Cys}-\text{Lys}-\text{Asn}-\text{Phe}-\text{Phe}-\text{Trp}-\text{Lys}-\text{Thr}-\text{Phe}-\text{Thr}-\text{Ser}-\text{Cys}-OH$$

Melanocyte stimulating hormone-release inhibiting hormone, MSH-R-IH:

$$\text{Pro}-\text{Leu}-\text{Gly}-NH_2$$

Melanocyte stimulating hormone-releasing hormone (MSH-RH):

$$H-\text{Cys}-\text{Tyr}-\text{Ile}-\text{Gln}-\text{Asn}-OH$$

FIGURE 9.20 The chemical structure (amino acid sequence) of some of the *pituitrophins* from the hypothalamus. Note that the structures of GH-RH and MSH-RH are tentative ones, and some doubt even has been expressed about MSH-R-IH. (From P. J. Bentley, *Comparative Vertebrate Endocrinology.* New York: Cambridge University Press, 1976.)

are small molecules containing as few as 3 amino acids in TRH or 10 in LH/FSH-RH. These hormones are the products of neurosecretion from the nerve fibers that terminate in the median eminence and originate somewhere in the hypothalamus. They may represent fragments of molecules that were originally larger; thus MSH-R-IH is identical to the side chain of oxytocin, and an enzyme that can cleave this fragment from oxytocin has been demonstrated in the median eminence. TRH, on the other hand, is assembled from its three constituent amino acids under the influence of TRH synthetase found in the hypothalamus and median eminence. While homologous releasing and inhibiting hormones undoubtedly also exist in nonmammals, their chemical structure is unknown. It will be particularly interesting to know what variations can occur in the structure of such small, active peptide hormones.

7 • ENDOCRINES AND THE CONTROL OF INTERMEDIARY METABOLISM, RESPIRATION, AND GROWTH

Cells are in a continual state of metabolic perturbation. This is due to their various special functions that require energy in order to do work and various substrates, principally amino acids, to maintain structure and growth. The activity of such processes is usually a continuous one over extended periods of time. The coordination of these metabolic pathways is predominantly by means of hormones (Figure 9.21). The role of hormones in digestion is described in Chapter 8 (Section 1.3 and Figure 8.3).

The metabolic transformations and transpositions that occur in an animal are described in Chapter 7. Briefly, such changes involve three major groups of chemical compounds: fats, carbohydrates, and proteins. These furnish the principal substrates used to supply energy and support the major structural requirements for growth. Transformations continually occur between these compounds, and the processes involved concern their constituent subunits (see Chapter 7). In fats (or triglycerides) these are fatty acids and glycerol; proteins are made up of amino acids and sometimes some sugars; carbohydrates can be broken down into simple sugars such as glucose and fructose. The general nature of the interchanges that occur in the liver, muscle, and adipose tissue are shown in Figure 9.22. More notably, glucose can be formed from amino acids in the liver, a process called *gluconeogenesis;* glucose also can be converted to glycogen or to fatty acids and glycerol and vice versa. The pathways for such transformations are far more complex than indicated in Figure 9.22, which only serves as a summary.

Various hormones can be shown to influence the processes of intermediary metabolism as reflected by changes in the levels of glucose, fatty acids, and glycogen in blood, liver, and muscle tissues (Table 9.5). The manner in which they do this is only partly understood but involves two types of mechanisms. First, the various substrates must be able to enter and leave the cell, and this provides the gate that may influence subsequent reactions in the cell. Insulin is notable for its actions at this site, and it promotes the uptake of substrates such as glucose, amino acids, and fatty acids by certain cells. Second, hormones may influence the rates of enzymatically controlled transformations that take place within the cell. They alter the rate of enzyme formation or possibly act as activators, cofactors, or inhibitors in such reactions. Some notable effects of hormones or intermediary metabolism are the following. (1) Amino acids are mobilized from muscle and transformed in the liver to glucose (gluconeogenesis). This is mainly increased by corticosteroids but also by growth hormone. (2) Insulin has widespread effects and increases conversion of glucose to

TABLE 9.5 Effects of Hormones on Intermediary Metabolism as Reflected by Changes in the Concentrations of Energy Substrates

Injection of Mammalian Hormone Preparations	Blood		Liver	Muscle
	Glucose	*Fatty Acids*	*Glycogen*	*Glycogen*
Insulin	↓	↓	↑	↑
Glucagon	↑ *	↑	↓	0
Epinephrine	↑	↑	↓	↓
Corticosteroids	↑	↑ †	↑	Variable
Growth hormone	↑	↑	—	—

↑ = increase; ↓ = decrease; 0 = no change.
* *Not* in cyclostome and chondrichthyean fishes.
† *Not* in chondrichthyean fishes.

glycogen and fat and of amino acids to protein. (3) Glycogen is converted to glucose under the influence of glucagon (in the liver) and epinephrine (in liver and muscle). (4) Triglycerides are broken down to fatty acids and glycerol in the presence of epinephrine, growth hormone and corticosteroids, and high concentrations of glucagon.

These processes are basic ones, so it is not surprising to find that they generally occur throughout the vertebrates, from cyclostomes to mammals. There appear to be one or two exceptions, such as the inability of the cyclostome fish and possibly the Chondrichthyes (sharks and rays) to respond to glucagon. However, some of the anomalies may reflect the fact that mammalian hormones are usually injected experimentally, and these may differ from the homologous ones in a particular species.

The hormones that control metabolism and growth appear to be present in all the vertebrate groups but, as discussed earlier, they may exhibit considerable variation in their chemical structure. The relative importance of their effects may differ between species and may vary with the seasons of the year, depending on the physiological condition of the animal. Poikilotherms are not as responsive to such hormones as homoiotherms, and they may be particularly sensitive to seasonal temperature changes. The animal's nutritional state (which reflects the availability of food), the onset or awakening from estivation and hibernation, reproduction, lactation, and growth in the young may all influence the role of hormones in intermediary metabolism. Some of these factors will be discussed later.

Carnivorous and omnivorous mammals and birds are more responsive to insulin than are herbivorous species. In addition, pancreatectomy generally has a greater hyperglycemic effect in carnivores than in herbi-

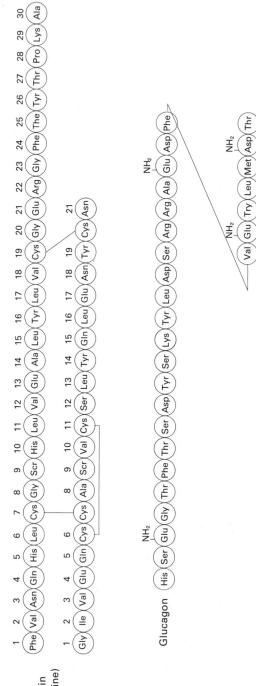

Insulin (bovine)

Glucagon

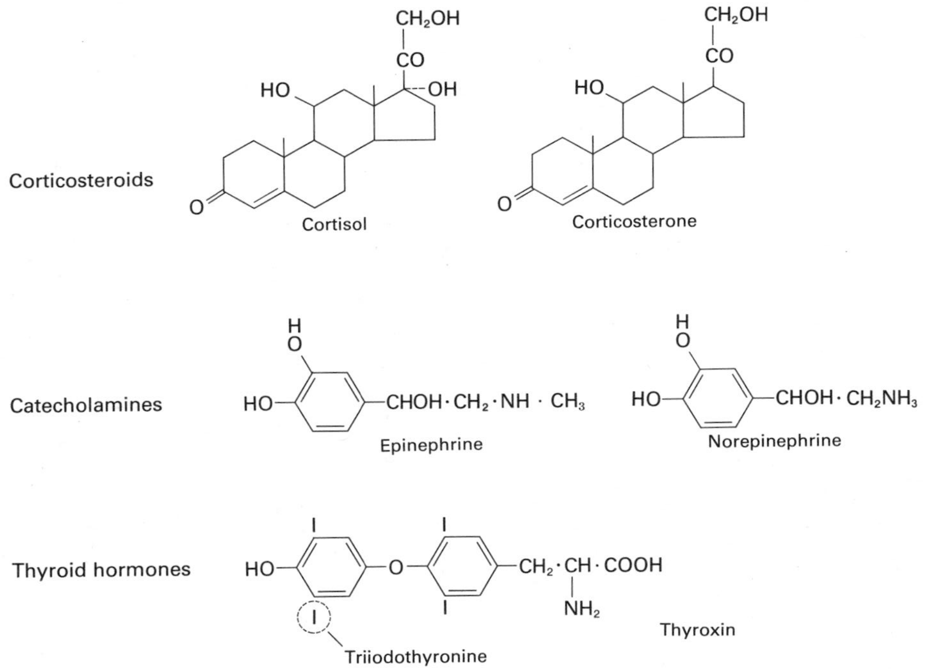

Corticosteroids

Cortisol

Corticosterone

Catecholamines

Epinephrine

Norepinephrine

Thyroid hormones

Triiodothyronine

Thyroxin

Growth hormone: protein containing 190 amino acids

FIGURE 9.21 The principal hormones that influence intermediary metabolism in vertebrates.

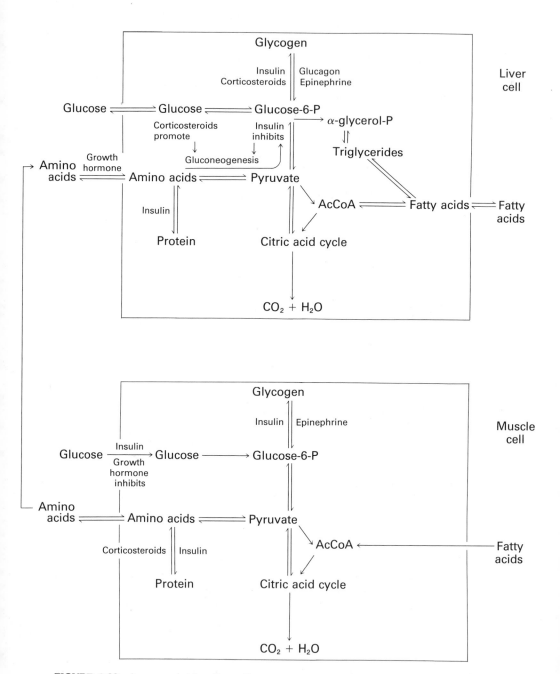

FIGURE 9.22 A summary of the effects of hormones on intermediary metabolism in liver, muscle, and adipose tissue. (From P. J. Bentley, *Comparative Vertebrate Endocrinology.* New York: Cambridge University Press, 1976.)

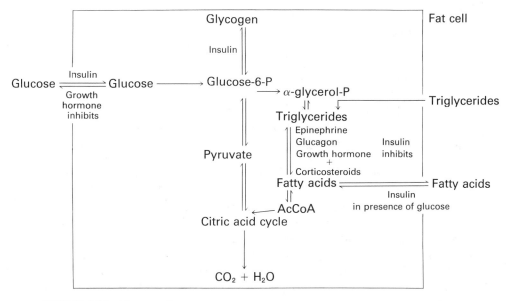

FIGURE 9.22 *(Continued)*

vores. It has been suggested that these differences reflect the relative importance of energy-storing tissues and their abilities to respond to insulin in animals that only eat periodically as compared to those (the herbivores) that nibble continually (see Chapter 8).

One of the end processes of intermediary metabolism is the oxidation of the products to carbon dioxide and water with the release of energy. Homoiotherms continually produce heat in order to maintain their body temperatures, so such oxidative processes occur very much more rapidly in these animals than in poikilotherms. The activity of the thyroid gland and its hormones, thyroxin and triiodothyronine, are vital if this production of energy is to occur optimally. Such is the added need for oxygen under the influence of the thyroid gland that the potency of its hormones can be estimated by comparing the survival times of mice placed in a closed jar. Those mice injected with thyroid gland extract die much more rapidly than normal mice, reflecting quicker asphyxiation due to their increased rate of metabolism. Although many hypotheses have been forthcoming, the precise mechanism by which thyroid hormones increase metabolism of energy substrates is unknown. They can act on the cell nucleus to alter genetic transcription and promote protein synthesis. This action could have widespread effects and may be important in the maintenance of the overall metabolic integrity of the tissues. Thyroid hormones do not exert prominent effects on oxygen consumption in poikilotherms. There is evidence that in some such species, under certain conditions such as a high body temperature, oxygen consumption may be stimulated. The effects are, however, equivocal and usually difficult to reproduce. The calorigenic effect of thyroid hormones may represent a specialization of their role or possibly an increased importance for such effects, which become more manifest at high metabolic rates. As we shall see, thyroid hormones also have other roles to play in controlling the proper functioning of certain organs, such as the heart, and they also influence growth and differentiation, particularly of nervous tissue. These effects are consistent with the concept that the thyroid hormones have a general role in maintaining optimal metabolic conditions in tissues.

As described in Chapter 10, the control of heat production is an important part of the process of thermoregulation in homoiotherms. The role of the thyroid in long-term chronic adjustment is not clear, but its level of activity may be greater at lower ambient temperatures. In acute situations, when animals are subjected to sudden large drops in environmental temperature, the catecholamines from the sympathetic nervous system and the adrenal may play an important role in increasing the production of heat. Injections of epinephrine and norepinephrine increase the rate of oxygen consumption in mammals and probably birds, as well as in some poikilotherms, such as frogs. These hormones can also reduce the loss of body heat as they produce piloerection (hair raising) and diminish peripheral blood flow (an α-adrenergic effect). In addition, the catecholamine hormones mobilize glucose and fatty acids. Oxygen consumption is stimulated, which reflects the increase in available energy substrates, but there is also a direct stimulation of metabolism. This calorigenic response is particularly well seen in brown fat. Such effects of catecholamines are β-adrenergic responses which are mediated by cyclic AMP. Hypothyroid animals are not as responsive to such β-adrenergic stimulation by catecholamines, and it seems that the thyroid increases the tissue's sensitivity to these hormones. If baby mice are subjected to a sudden drop in the ambient temperature to 0°C, they cannot survive if they have been preinjected with drugs that block β-adrenergic activity. This is not related to processes of heat transfer that are α-adrenergic effects (*not* mediated by cyclic AMP, possibly an inhibition of adenyl cyclase is involved) but is due to a reduction in the calorigenic metabolic responses. The catecholamines also may play a role in thermoregulation in mammals; in many species they stimulate the sweat glands to secrete and so facilitate evaporation.

Hormones play an important role in the control of growth. As we have seen, the endocrine glands can regulate the formation and breakdown of protein. An excess of cortisol indeed reduces the rate of growth in young animals, and this reflects a catabolic effect on protein. Growth hormone, on the other hand, in conjunction with insulin increases the incorporation of amino acids into proteins. Similarly, some steroid hormones, notably androgenic ones (like testosterone, among others), also promote the formation of protein. The overall effects of these hormones also depend on adequately functioning thyroid tissue. Growth is thus a multihormone discipline.

The effects of growth hormone are very dramatic, and they are readily apparent in man. Undersecretion may lead to dwarfism and oversecretion to gigantism. Prolactin also has a growth-promoting activity, which is more apparent in some nonmammals. Growth hormone, apart from facilitating protein anabolism, also promotes the growth of cartilage and bone and the accumulation of various structural components including sulfate. Incorporation of sulfate into cartilage can be seen in vitro, but this is not a direct effect of the hormone. Growth hormone initiates the formation, by the liver, of an intermediary called *sulfation factor* or *somatomedin*. This substance is present in the blood, and it appears that most, if not all, of the actions of growth hormone are mediated by it.

8 • THE REGULATION OF CALCIUM METABOLISM

Calcium plays a major part in the processes of life. Apart from its more obvious structural role in the skeleton of most vertebrates, it is vital for the adequate function of all tissues. The presence of calcium is necessary in many enzyme-mediated reactions within the cell, and it helps to maintain the permeability of the surrounding plasma membrane. It is also essential for the integration and coupling of the processes associated with nerve conduction, muscle contraction (see Chapter 4, Section 9), and the responses to hormones. Several endocrine glands regulate calcium levels in the body: the parathyroids, the ultimobranchial bodies, the ovaries, and the corpuscles of Stannius in some bony fishes.

The parathyroid glands usually lie in juxtaposition to the thyroid. There are exceptions, however, such as in sheep where they are associated with the thymus. These glands are derived from the branchial pouches (Numbers III and IV) and are typically four in number, though this can also vary. They produce a polypeptide hormone (Figure 9.23) called *parathormone* that elevates blood calcium levels. Parathyroids are not present in the fishes and make their phyletic debut in amphibians.

Tissues that lie near the last pair (VI) of gill pouches give rise to the *ultimobranchial bodies,* and these are present in all vertebrates with the single exception of the cyclostome fishes. The anatomical site of this tissue varies somewhat, and in nonmammalian vertebrates it is situated in the pericardial region. In mammals the ultimobranchial tissue is associated with the thyroid gland, where it makes up the "C" (or parafollicular) cells. These tissues, in species ranging from sharks to man, contain another polypeptide hormone, which opposes the effect of parathormone and decreases blood calcium level. It was formerly called thyrocalcitonin but is now called *calcitonin*.

A third substance, *vitamin D*, plays a vital role in calcium homeostasis. Two forms, vitamin D_2 and D_3—ergocalciferol (a synthetic material) and cholecalciferol, respectively—are mainly obtained in the diet. However, vitamin D_3 is also formed, as a result of ultraviolet radiation, in the skin, while both substances are converted to the more active forms 25-hydrooxyergocalciferol and 25-hydroxycholecalciferol in the liver. The latter precursor is converted under the trophic action of parathormone in the kidney to an even more active substance, 1,25-dihydroxycholecalciferol $(1,25(OH)_2D_3)$, which is considered to be a hormone. 1,25-dihydroxycholecalciferol stimulates the absorption of dietary calcium across the intestinal wall and also plays a vital role in controlling calcium resorption from bone, where it permits the action of parathormone.

The interactions of these three hormones in the calcium metabolism of mammals are described in Figure 9.24. In most vertebrates that have a bony skeleton, an equilibrium exists between the calcium present in the bones and that in the tissue fluids. Bone acts as a substantial storage site that accumulates calcium and from which it can be reabsorbed into the extracellular fluid. Resorption of calcium is increased by parathormone, thus increasing blood calcium levels. The initiation of this process depends on a permissive effect exerted by $1,25(OH)_2D_3$ that induces a "Ca^{2+}-carrier" protein in the bone cells. Parathormone can act only in the presence of this protein. Calcitonin inhibits the resorptive action of parathormone

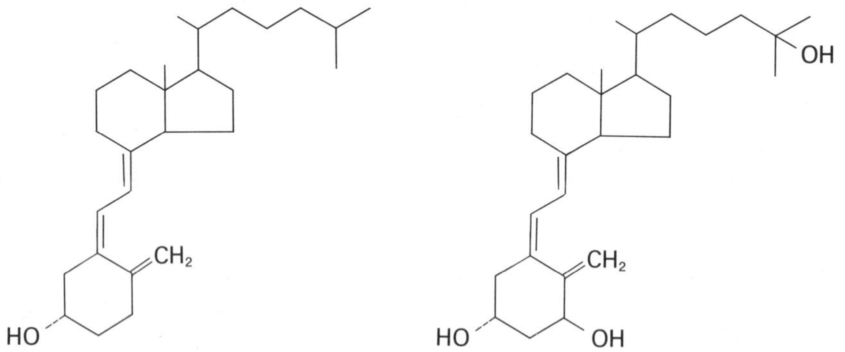

Vitamin D_3
(cholecalciferol)

1,25-dihydroxycholecalciferol
$(1,25\text{-}(OH)_2\text{-vitamin } D_3)$

FIGURE 9.23 The principal hormones that influence calcium metabolism in vertebrates.

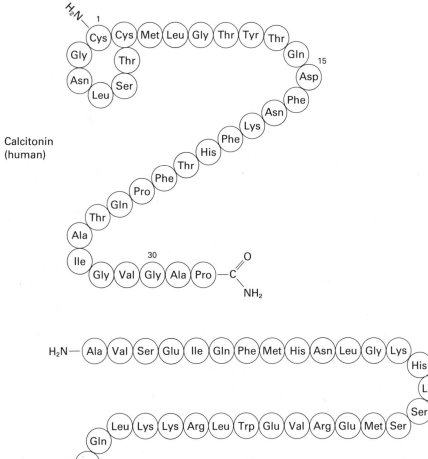

Calcitonin
(human)

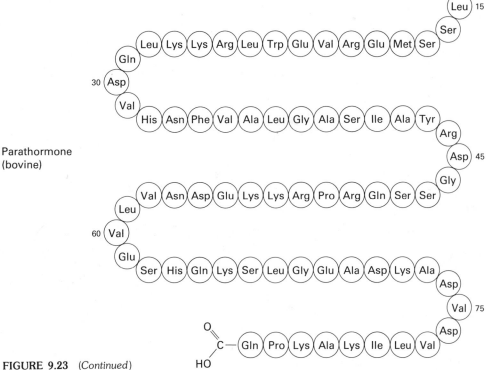

Parathormone
(bovine)

FIGURE 9.23 (*Continued*)

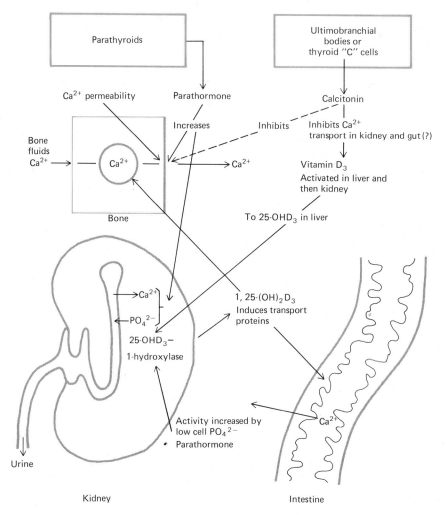

FIGURE 9.24 A summary of the effects of hormones on calcium metabolism. (From P. J. Bentley, *Comparative Vertebrate Endocrinology.* New York: Cambridge University Press, 1976.)

on bone. Parathormone also acts on the kidney, promoting excretion of phosphate and the conversion of vitamin D_3 to $1,25(OH_2)D_3$ and decreasing the loss of calcium in the urine. Thus in mammals three hormones inextricably interact with one another to maintain calcium levels in the body.

The regulation of Ca^{2+} metabolism in nonmammals shows a number of differences from that in mammals, and this is not altogether surprising. As we have seen, the associated endocrine glands themselves are not present in all vertebrates, and neither is an important effector tissue, a bony skeleton. The cyclostome and chondrichthyean fishes lack a calcified skeleton, while in many teleosts the bone is acellular and not as responsive to the initiation of ionic interchanges with the extracellular fluid. There is no evidence indicating that parathormone, calcitonin, or vitamin D_3 are active either in cyclostomes and chondrichthyeans or in some bony fishes such as the lungfish *Lepidosiren paradoxa* and the killifish *Fundulus heteroclitis*. The parathyroids undoubtedly play a role in calcium regulation in amphibians, reptiles, and birds, but the role of calcitonin is not clear. Although there are histological changes in the ultimobranchial bodies associated with imbalance of calcium, the injection of calcitonin is usually not very effective in lowering calcium in the blood of these vertebrates. In addition, removal of the ultimobranchial bodies does not appear to compromise significantly the animal's calcium balance. The precise role of calcitonin in nonmammals remains to be defined.

The chemical structures of parathyroid hormone and calcitonin differ somewhat in various species. Some of these variations in the calcitonin molecule are shown in Figure 9.25. Extensive differences exist between human and salmon calcitonin, and it is somewhat surprising to find that the fish hormone is very much more effective (about 10 times)

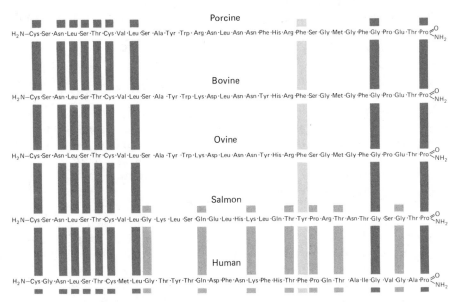

FIGURE 9.25 Comparative structures (amino acid sequences) of calcitonin molecules from five vertebrate species. The dark bars indicate amino acid positions that are homologous in all species. It can be seen that extensive differences exist especially in the central parts of the molecules. The short tinted bars indicate homologies between salmon and humans; the tall, lightly tinted bar indicates comparable hydrophobic residues. (From J. T. Potts et al., *Gen. Comp. Endocrin.*, Suppl. 3, 405–410, 1972. Copyright © 1972 by Academic Press.)

in man than is the human one. The metabolic machinery for inactivating salmon calcitonin in man is not as efficient as for the homologous hormone, so it persists much longer in the body.

Teleost and holostean fishes possess islets of tissue called *corpuscles of Stannius*. These vary in their number from two in some teleosts to 40 or 50 in the bowfin *Amia calva*. Their role in the body is uncertain, but histological changes have been observed under various conditions, including some associated with skeletal defects. When these tissues are removed from eels, the plasma calcium levels rise. The injection of extracts from the corpuscles of Stannius lower plasma Ca^{2+} levels in several teleost fishes. These tissues appear to secrete a hormone that depresses the calcium concentration in the body fluids of teleosts. This "new" hormone has been called "hypocalcin," and it may be especially important in fishes living in seawater where the Ca^{2+} levels are high.

Finally, mention should be made of the effects of female ovarian estrogenic hormones on blood calcium. It has been observed in many vertebrates that the onset of reproduction and egg laying is associated with considerable increases in blood calcium concentrations. This can be mimicked in oviparous species but not in ovoviviparous or viviparous ones (the Chondrichthyes and mammals) by the injection of estrogens. The estrogens initiate the formation in the liver of a protein (vitellin) that binds calcium in the blood. This is preparatory to its incorporation into the egg yolk.

9 • ENDOCRINOLOGY OF THE SKIN

Several hormones play an important role in maintaining the integrity of the skin and its appendages, such as scales, feathers, and hair. In addition, pigmentary changes associated with breeding cycles and with responses to different shades and colors in the environmental background are also mediated by the endocrine glands.

The thyroid hormones play a basic role in maintaining the optimal physiological condition of the skin. If thyroid secretion is deficient, the skin in man becomes very dry and brittle, and excessive accumulations of mucopolysaccharides arise (myxedema). Many animals periodically shed parts of their integument, including scales, feathers, or the outer epidermal covering. This is called molting, and it occurs in amphibians, reptiles, birds, and mammals, where it may be associated with several events, including growth, seasonal changes in the environment, and the breeding cycle. In the absence of the pituitary, normal molting is usually prevented. This response reflects deficiencies of several different hormones. The thyroid (via TSH) usually stimulates molting, and its activity is important for the successful accomplishment of this process. Other hormones that may also play a role include ACTH and corticosteroids (in anuran amphibians) and prolactin. The latter hormone has been observed to reduce the period between successive molts (or shedding) in hypophysectomized reptiles, and it accelerates molting in urodele amphibians. The postnuptial molt, following mating, in birds may reflect a decline in gonadal activity.

The skin, fur, and feathers of vertebrates display considerable pigmentary diversity. Species, racial, and sexual differences are common. Changes may occur that are associated with the breeding season and the onset of summer or winter. Color change, as a response to the background coloration of the environment, can occur in representatives of all groups of cold-blooded vertebrates. These differences and changes reflect alterations in the amounts and distribution of pigments in the epidermis and dermis. The pigments include pteridines, carotenes, and melanin, and they are often formed in special cells, the melanocytes that are present in the epidermis and dermis. The melanocytes are derived from nerve cells, whose shape they resemble. The melanin that they may contain is synthesized from tyrosine within the cell. Different concentrations of this pigment may accumulate, resulting in variation of the hue of the skin, hair, and feathers. In some circumstances high levels of MSH from the pituitary can lead to excessive synthesis of melanin in the skin of mammals. This process involves an increased activity of tyrosinase in the melanocyte, but it does not appear to be a normal physiological process. Estrogen and LH can increase the melanin content in the feathers of some birds.

Many reptiles, amphibians, and fishes change their color quite rapidly to conform, or blend, more closely with the shades and hues of their background. This process involves the dispersion (darker skin color) or aggregation (lightening) of melanin granules in pigment cells called *melanophores*. Such changes can result from nervous stimulation, the release of MSH from the pituitary, and possibly also from the secretion of melatonin by the pineal gland. In some animals both humoral and neural coordination of color change appears to coexist. The contrast in the colors of two dogfish in their dark and pale color phases is shown in Figure 9.26. Equally dramatic changes in color can be seen in bony fishes, amphibians, and reptiles.

Events involved in lightening and darkening of the skin in cold-blooded vertebrates are summarized in Figure 9.27. The phyletic distribution of neural and humoral control of the melanophore is shown in Figure 9.28. The chondrichthyeans (elasmobranchs) and the amphibians have an exclusively humoral control, as do some reptiles and teleost fish. The latter two groups sometimes have a neural control mechanism, and in teleosts both may coexist in the same animal. Nerves initiate an aggregation of melanin in the melanophore through the action of the transmitters nor-

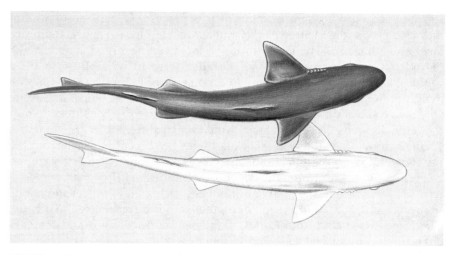

FIGURE 9.26   Two dogfish (*Squalus*) in their dark and pale phases. (From H. Waring, *Proc. Roy. Soc. B* 125:264–282, 1938.)

epinephrine and epinephrine. MSH, on the other hand, disperses melanin, an effect mediated by cyclic AMP. It is also possible that some nerves have a "dispersing" action on melanin granules, and this could represent a β-adrenergic effect.

Melanophores may also respond directly to changes in light and temperature, but the most important effects are initiated by light falling on different parts of the retina. In aquatic species the incidence of light on the basal part of the retina stimulates the release of MSH from the pars intermedia. This secretion is under an inhibitory control from the hypothalamus that may involve inhibitory nerve pathways and also the presence of an MSH-release inhibiting hormone from the median eminence. The interrelationships of these two inhibitory processes are not clear. The stimuli arising from the retina overcome this inhibition so that MSH is released into the blood.

In some cyclostome fishes removal of the pineal gland abolishes the animal's usual diurnal variation in color: dark in the day and pale at night. The pineal is a tissue which is considered a putative endocrine gland. It is present in all vertebrates but has undergone extensive phylogenetic changes. The pineal is associated with the brain, and in fishes and amphibians it is photosensitive—hence its description as a "third eye." This ability to respond to light may account for the observations in cyclostomes. However, the picture has become more complex since the original observations were made. The pineal in vertebrates has been shown to synthesize a substance called *melatonin* (Figure 9.29). This has been identi-

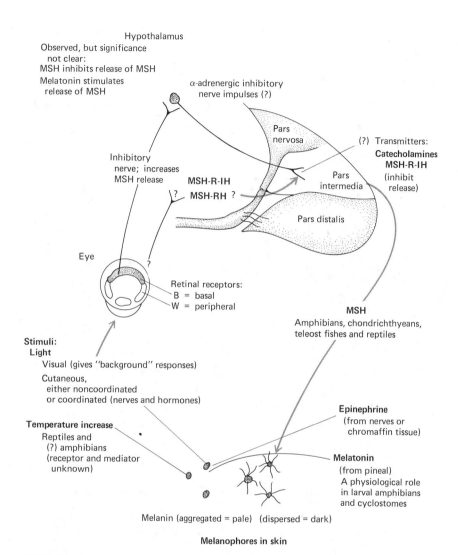

FIGURE 9.27   A summary of the role of hormones in vertebrate color change. (From P. J. Bentley, *Comparative Vertebrate Endocrinology*. New York: Cambridge University Press, 1976.)

fied in the blood, and it has a potent action in producing skin lightening in larval amphibians and some cyclostomes. It thus could oppose the effect of MSH. As we shall see later, melatonin may also modify the processes of reproduction by an action in the hypothalamus.

Like all the polypeptide hormones, MSH displays considerable polymorphism (Figure 9.29). Two forms, α-MSH and β-MSH, usually coexist

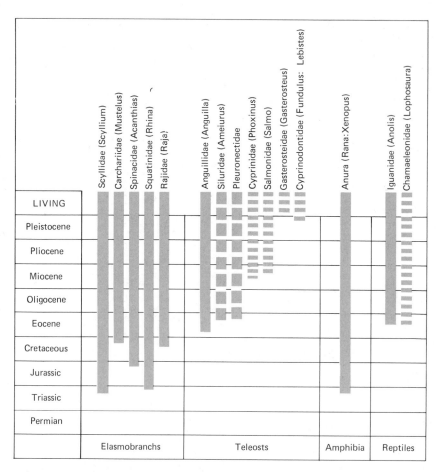

FIGURE 9.28 The phyletic distribution of neural and hormonal coordinating mechanisms for color change in relation to the geological age of the vertebrate groups. Code: solid bar—predominantly hormonal; broken bar—mixed hormonal and neural; small squares—predominantly neural. (From H. Waring, *Biol. Rev.* 11:120–150, 1942. By permission of Cambridge University Press.)

in a single species, though in some instances three or four different but related peptides have been identified. Apart from the variants that have been structurally delineated in mammals, chemical and immunological studies indicate an even more widespread polymorphism among fishes, amphibians, reptiles, and birds.

The biological significance of color change and the MSHs has been widely debated. The potential survival value of an ability to blend in with the surroundings is obvious. In addition, it is possible that changes in color may influence the absorption of radiant heat and could influence the control of body temperature in some poikilotherms. Even though not all

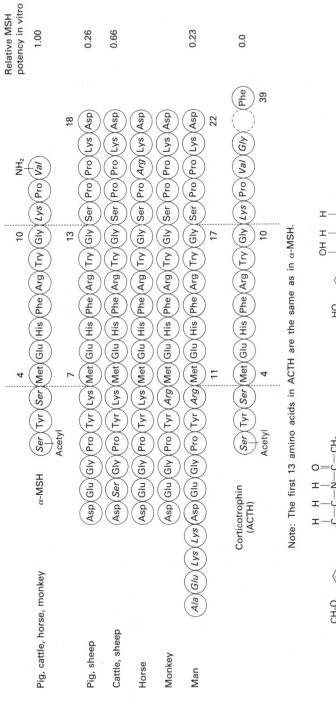

FIGURE 9.29 Hormones that influence color change in vertebrates.

vertebrates exhibit such an ability to change color, they all possess MSH. What it does, if anything, in those species lacking a melanophore response is unknown.

10 • ENDOCRINES AND OSMOREGULATION

The control of water and solutes in animals has been described in Chapter 6. Life can clearly be successful (if not happy) in diverse osmotic environments that may potentially provide too much or too little water and salts. The ability to maintain an equilibrium in such different conditions depends on the responses and the function of several organs and tissues that promote the conservation or excretion of water and solutes. In many instances the integrity of these organs and their appropriate responses are mediated by hormones. The effectors include the kidneys, gut, skin (in amphibians), urinary bladder (in some amphibians, reptiles, and teleost fishes), gills, and salt glands. The ability of such an effector organ to respond to a particular hormone varies considerably among the vertebrates. Vasotocin thus increases water reabsorption from the urinary bladders of frogs and toads. This effect is not seen in the bladders of other vertebrates or, for that matter, even in most newts and salamanders.

The principal endocrine secretions involved in osmoregulation (see Figure 9.30) are: (1) the *neurohypophysial hormones,* principally vasopressin (or ADH) in mammals and vasotocin in many other vertebrates, the main effects of which are on water movement, though sodium transfer may also be facilitated; (2) the *corticosteroids,* cortisol, corticosterone, and aldosterone (from the adrenocortical tissue), which have widespread effects on sodium and potassium metabolism; and (3) *prolactin,* or a prolactinlike hormone, which has a number of interesting effects, especially that limiting permeability of the gills of some euryhaline teleost fishes to sodium.

Other hormones may play less direct roles or may exert actions at a pharmacological level. The *urophysis,* a caudal neural tissue in some fishes, is a putative endocrine gland that may influence osmoregulatory activity in some fishes. The *corpuscles of Stannius* have received similar consideration but probably have a primary effect on Ca^{2+} metabolism, as discussed earlier. It is often difficult to decide whether the effect of an injected hormone or tissue preparation really reflects a physiological role, since many tissues exhibit an innate responsiveness to diverse chemical materials, especially in high concentrations. A response to an administered tissue or glandular extract need not indicate a physiological role. Such an action could, however, reflect a potential role or a "long lost" effect that is, or was, manifest in some ancestral or related forms. This

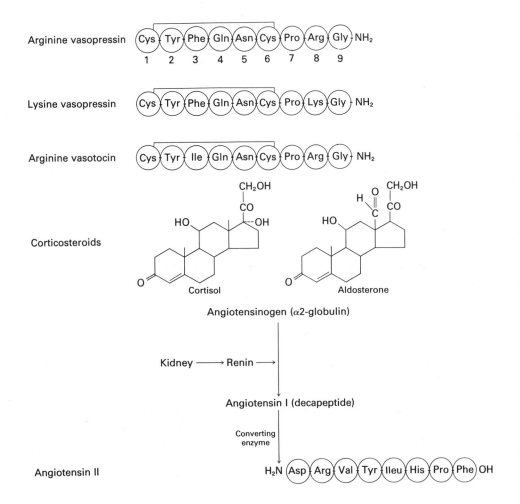

Prolactin: a protein containing 198 amino acids

FIGURE 9.30 The principal hormones that influence osmoregulation in vertebrates.

problem of physiological versus pharmacological effects arises in all vertebrates, and especially in the fishes, where it is further emphasized by considerable specific variations.

Neurohypophysial hormones can act on the kidney of tetrapod vertebrates and reduce the flow of urine. In mammals the absence of a functioning neurohypophysis results in excessive losses of water in the urine.
Vasopressin and vasotocin reduce renal water loss in two ways: (1) *by
increasing the absorption of water* from the kidney tubules and (2) in some

birds, reptiles, and amphibians also by *reducing the rate of filtration at the glomerulus*. In many instances it is doubtful that the latter effect is a physiological one. In a few tetrapods, especially aquatic species of amphibians, the kidney responds very weakly to neurohypophysial peptides. In some fishes these hormones have the opposite effect to that seen in mammals; that is, urine flow is increased, as seen in the African lungfish, *Protopterus aethiopicus,* and the goldfish, *Carassius auratus.* However, in most fishes these peptides do not affect the kidney. The renal antidiuretic response thus seems to have been first acquired by the tetrapods and may be related to a need for water conservation in their usual terrestrial environment. Vasotocin not only influences water conservation by a renal action in the tetrapods but also may exert comparable effects on other tissues (Table 9.6). Thus water uptake across the skin of anuran (but not urodele) amphibians is increased. Water reabsorption from the capacious anuran urinary bladder is accelerated by this hormone as well. The neurohypophysial peptides also exert other actions in the body, including an increase or decrease in blood pressure, increased blood sugar levels, and an accelerated rate of sodium uptake across the skin and bladder of some amphibians. These effects, however, appear to be pharmacological ones. Vasotocin also contracts the uterus and oviduct, and, as we shall see, this may be useful in reproduction.

The *corticosteroids,* apart from their effects on intermediary metabolism, also promote sodium conservation and potassium excretion in the urine of mammals. The most physiologically active is aldosterone, which influences transport of these ions (sodium reabsorption and potassium secretion) in the renal tubules. Whether or not this hormone has the same renal effects in other vertebrates is not clear, but it appears likely in birds and reptiles though not in amphibians. Most fishes (with the exception of lungfishes and some teleosts) lack aldosterone, and their principal corticosteroid, cortisol, does not appear to influence renal electrolyte excretion. The corticosteroids may influence sodium and potassium metabolism by actions on other nonrenal target tissues (Table 9.6). In mammals these hormones reduce sodium and promote potassium loss in sweat secretions and saliva, and they promote sodium reabsorption from the colon. The latter effect also occurs in the gut of birds and in amphibians. Corticosterone is necessary to maintain the optimal secretory activity of the nasal salt glands in birds, and a similar dependence also appears to be present in reptiles. Sodium accumulation or reabsorption across the skin and urinary bladder of many amphibians, and the bladder of some chelonian reptiles, is increased in the presence of aldosterone. Thus the same types of hormones may exert similar general effects on salt metabolism, but this may involve actions on different effector tissues. In some fishes, such as eels, two types of response can be seen in a single effector; that is, injected cortisol increases salt uptake across the gills in fresh water but promotes its excretion again across the gills in seawater. Fishes drink seawater, and the absorption of this fluid from the gut is promoted by cortisol. These effects of corticosteroids in fishes are due to their ability to increase the amounts of sodium-potassium-activated ATPase in the tissues (see Chapter 2, Section 3).

The secretion of aldosterone in mammals and amphibians can be stimulated by an octapeptide called *angiotensin*. This is formed from an α-2 globulin (angiotensinogen) in the plasma as a result of the action of a renal enzyme (or hormone?) called *renin*. This enzyme is released in response to changes in renal blood flow that reflect small decreases in the volume of the body fluids resulting from sodium deficiency or hemorrhage. Stimulation of the renal nerves also promotes secretion of renin. The released corticosteroid, by conserving sodium, will tend to correct the deficiency and so complete the inhibitory feedback. Renin has been identified in the kidneys of representatives of all the vertebrate groups with the exception of chondrichthyeans and cyclostomes. Thus the renin-angiotensin system first appears on the phyletic scale in bony fishes. Whether or not it has a similar function in stimulating the release of corticosteroid hormones in all the vertebrates that possess it is uncertain. Evidence has recently been presented that angiotensin may even exert a direct stimulating action on sodium transport across the colon and kidney tubules in mammals and the skin of frogs and stimulate drinking by an action on the thirst center in the brain.

Many fishes can readily survive in either fresh water or the sea—contrasting circumstances that they may experience during seasonal or breeding migrations. Many such euryhaline fishes can readily survive after their pituitary is removed, but only if they are kept in seawater. In fresh water they soon die because of a rapid loss of sodium. Injections of prolactin, but no other pituitary hormones, can prolong their survival indefinitely. This hormone acts mainly by limiting losses of sodium, by diffusion, across the gills, but effects at several other sites have also been described. Sticklebacks, *Gasterosteus aculeatus,* normally spend their winters in the sea, but in the spring they make an annual nuptial migration up freshwater rivers. If winter sticklebacks from the sea are placed in fresh water, they soon die; if they are first injected with prolactin, they survive. Increased levels of prolactin (or a prolactinlike hormone) appear to protect these fishes and help them adapt during their migration into fresh water. Prolactin is also concerned with maintaining milk secretion in mammals and in some birds the secretion of a milky paste from the crop sac. This is a far cry from the osmotic effects of this hormone in fishes, though both actions could be the result of changes in the permeability epithelial membranes.

TABLE 9.6 The Effects of Various Hormones on Osmoregulatory Processes in Vertebrates

Vertebrate Group	Neurohypophysis		Adrenocortical Tissue	
	Hormone	*Effector*	*Corticosteroid*	*Effector*
Tetrapods Mammals	Arginine vasopressin or lysine vasopressin	Kidney: antidiuresis	Aldosterone	Kidney, sweat and salivary glands, colon: decreased Na, increased K in secretions; increased resorption Na
Birds	Vasotocin, AVT	Kidney: antidiuresis	Aldosterone	Kidney: Na conservation, K excretion Nasal salt gland: maintains ability to secrete Na
Reptiles	AVT	Kidney: antidiuresis	Aldosterone	Kidney: decreased Na loss in urine (?) Cephalic salt glands: Na and K secretion (?)

Amphibians	AVT	Kidney: antidiuresis Skin and bladder: increased water absorption	Aldosterone	Skin, bladder and colon: increased Na absorption
Fishes Dipnoi (lungfishes)	AVT	Kidney: diuresis	Aldosterone	?
Teleosts*	AVT	Kidney: diuresis (sometimes)	Cortisol	Gills: Na uptake in fresh water and output in seawater Gut: increased Na absorption in seawater
Chondrichthyes	Various		1α-hydroxycorticosterone	*Note:* ACTH plus prolactin may maintain high branchial osmotic permeability
Cyclostomes	AVT		Cortisol (?) Corticosterone (?)	Gills: Na uptake (?)

* In some teleost fishes (in fresh water) prolactin restricts the permeability of the gills to sodium and so prevents its loss.

11 • HORMONES AND REPRODUCTION

Reproduction commences with gametogenesis when the female produces ova, and the male sperm. This is a complex procedure, but hardly less so than the subsequent nuptial events and arrangements that ensure the union of the sexes and the mixing of their genetic material. Following fertilization, and as a result of cell proliferation, the transfiguration of the gametes to a new and viable member of the species will be accomplished.

Pituitary gonadotrophins:

Follicle-stimulating hormone, FSH

Luteinizing hormone, LH (glycoprotein)

Prolactin (sometimes called luteotrophic hormone, LTH)

Gonadal steroids:

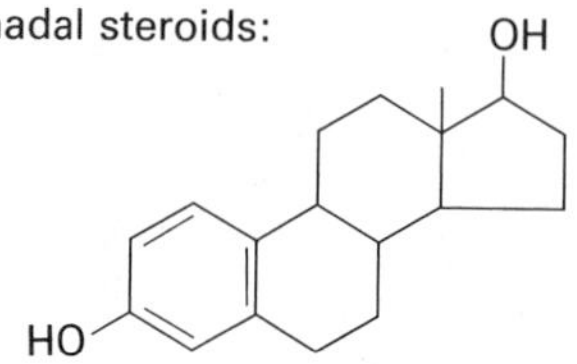

Estradiol 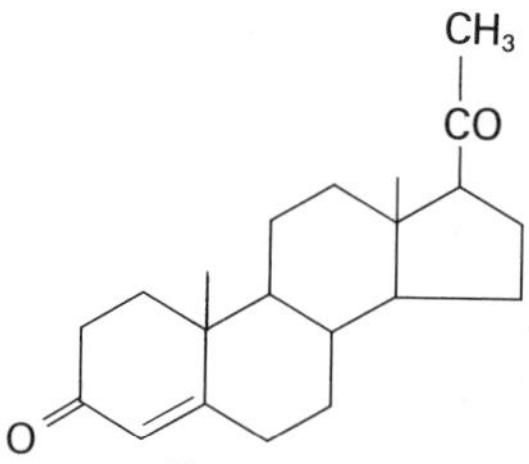   Progesterone 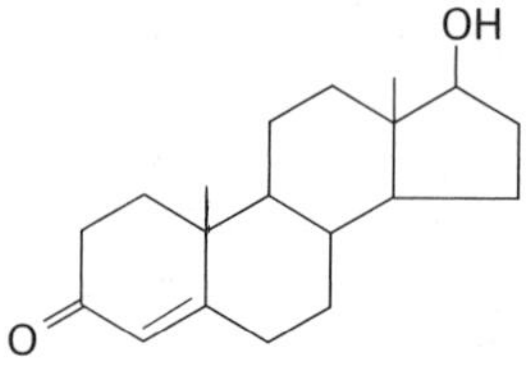   Testosterone

Oxytocin: Cys-Tyr-Ile-Gln-Asn-Cys-Pro-Leu-Gly-NH₂

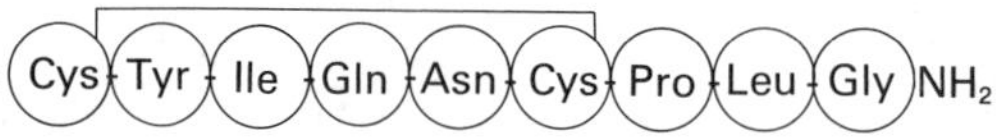

Placental hormones: Estriol

Progesterone

Human chorionic gonadotrophin, HCG (glycoprotein, LHlike)

Human placental lactogen, HPL; also called human chorionic somatomammotrophin, HCS (prolactinlike and growth hormonelike)

FIGURE 9.31 The principal hormones that influence reproduction in vertebrates.

Such a result, however, will be possible only if the environmental conditions are favorable. Biologically, the whole process can be deemed successful only when the progeny has itself come into reproductive condition. This cycle of events clearly will require a great deal of coordination. Several hormones play an important role in reproduction (Figure 9.31).

Briefly, hormones are involved in (1) the maturation maintenance, and preparation of the animal for breeding at an appropriate time and occasion, (2) the formation of the gametes, (3) the preparation of the gonoducts to receive and/or discharge the ova and sperm, (4) the appropriate arrangements, both spatial and temporal, for the union of the sperm and ova, and (5) the growth, development, and differentiation of the fertilized egg to a viable adult.

The reproductive apparatus consists essentially of the gonads (testes and ovaries), the pituitary gland and hypothalamus, the accessory sex organs, such as the oviducts and uterus (usually derived from the Mullerian ducts), the vas deferens (derived from the Wolffian ducts), and the various glands that produce secretions that assist the survival and delivery of the germ cells. Secondary sex characters are commonly involved in the preparation for the nuptials and include, to name only a few, integumental structures such as the pelage and plumage, the combs of cockerels, finlike crests in newts, and vocal sacs in frogs. Behavior appropriate to fertilization and including complex patterns of display and courtship are influenced by hormones. The care of the young is facilitated by the secretion of milk by the mammary glands in mammals, crop sac secretions in some birds, and the development of brood pouches (where the young can be protected) in some fishes and amphibians. A transient but important endocrine gland, the placenta, is developed during gestation in mammals. This produces several hormones involved in the maintenance of pregnancy.

11.1 Periodicity of Breeding and the Role of the Pituitary

Some animals can breed at any time of the year and during any season. This ability appears to be related to continually favorable environmental conditions, such as are experienced by some tropical and domesticated species, including man. However, most animals only come into breeding condition periodically, usually only once a year but sometimes more often. The timing of breeding may be critical to the ultimate success of the whole procedure since, to ensure survival of the young, it should coincide with favorable environmental conditions, such as those associated with the seasons. Thus in temperate areas the young are usually produced in the spring or summer or, in areas of seasonable or unpredictable rainfall,

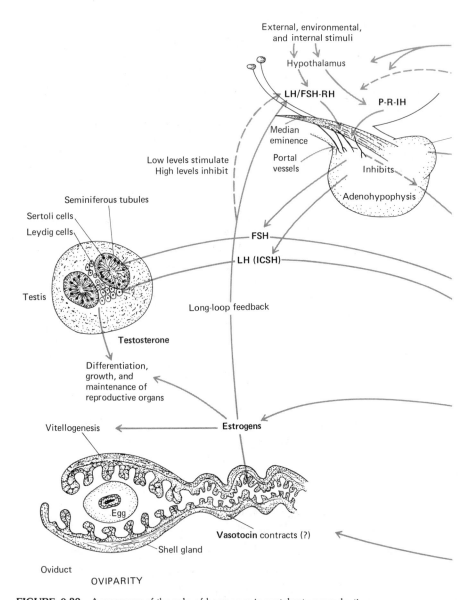

External, environmental,
and internal stimuli

Hypothalamus

LH/FSH-RH

P-R-IH

Median
eminence

Low levels stimulate
High levels inhibit

Portal
vessels

Inhibits

Adenohypophysis

Seminiferous tubules

Sertoli cells

Leydig cells

FSH

LH (ICSH)

Testis

Long-loop feedback

Testosterone

Differentiation,
growth, and
maintenance of
reproductive organs

Vitellogenesis

Estrogens

Egg

Vasotocin contracts (?)

Shell gland

Oviduct

OVIPARITY

FIGURE 9.32   A summary of the role of hormones in vertebrate reproduction.

during the time when water and food are most plentiful. As the preparations for breeding may be prolonged, various cues (or clues) are taken
from the environment and transformed into endocrine signals. These include variations in the duration and intensity of light (day length) and
sometimes temperature. Other factors, such as the animal's nutritional

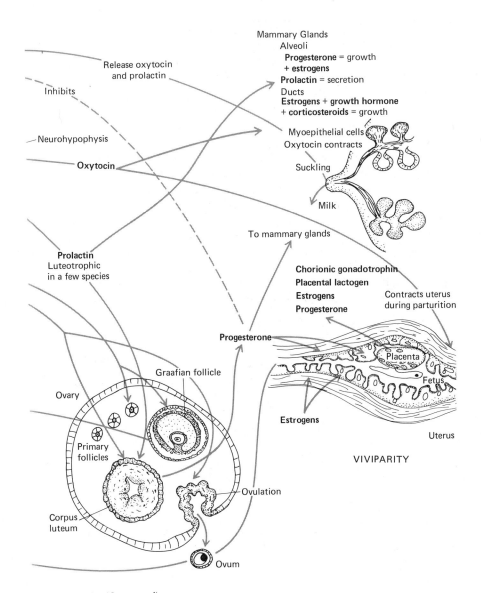

Mammary Glands
Alveoli
Progesterone = growth
+ estrogens
Prolactin = secretion
Ducts
Estrogens + growth hormone
+ corticosteroids = growth

Release oxytocin
and prolactin

Inhibits

Myoepithelial cells
Oxytocin contracts

Neurohypophysis

Suckling

Oxytocin

Milk

To mammary glands

Prolactin
Luteotrophic
in a few species

Chorionic gonadotrophin
Placental lactogen
Estrogens
Progesterone

Contracts uterus
during parturition

Progesterone

Placenta

Graafian follicle

Ovary

Fetus

Estrogens

Primary
follicles

Uterus

VIVIPARITY

Corpus
luteum

Ovulation

Ovum

FIGURE 9.32 (*Continued*)

state, the sight, sound, and smell of the opposite sex, courtship and display, and psychologically stimulating or disturbing conditions, such as captivity, may have profound influences.

These environmental circumstances act as *exteroceptive* stimuli that stimulate nerve transmission to the hypothalamus and median eminence, which controls the pituitary and hence the entire process of reproduction. There is evidence that in some species the pineal gland, like the median eminence, may also act as a neurotransducer and, by inhibiting the pituitary, help regulate the reproductive cycle. The adenohypophysis under the influence of a trophin LH/FSH-RH, from the median eminence, releases the hormones that control the reproductive cycle. There are two such *gonadotrophins* in mammals, birds, and at least some reptiles and amphibians: follicle stimulating hormone (FSH) and luteinizing hormone (LH). In many other vertebrates a single gonadotrophic hormone, incorporating both FSH and LH activity, probably exists. FSH stimulates the ovarian follicles to grow, mature, and, in conjunction with LH, to give forth their ova. The production of estrogens by the follicular cells is promoted by FSH and LH. During the maturation of the ovarian follicles, estrogens prepare and prime the oviducts or uterus for the reception of the ova. They also help to maintain the condition of the accessory and secondary sex organs and influence behavior. The levels of estrogen in the blood are controlled by a feedback to the hypothalamus that inhibits the release of the pituitrophin LH/FSH-RH and consequently FSH and LH from the

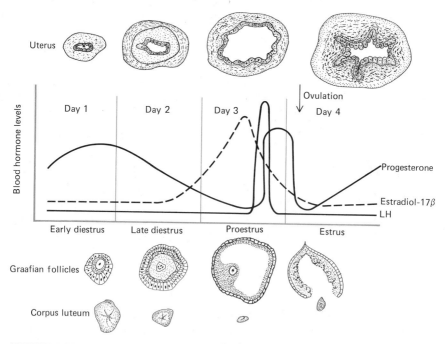

FIGURE 9.33   The estrus cycle of the laboratory rat. Note that in the rat the progesterone of the estrus cycle comes mainly from ovarian interstitial tissue. (From P. J. Bentley, *Comparative Vertebrate Endocrinology*. New York: Cambridge University Press, 1976.)

adenohypophysis. These processes and subsequent ones are summarized in Figure 9.32, and the changes that occur during the estrous cycle of the laboratory rat are shown in Figure 9.33.

11.2 Ovulation

Once the ovarian follicles are mature, the ova may then be extruded—a process called ovulation. In some teleost fishes, however, the eggs are fertilized within the follicle, and an ovoviviparous type of development may proceed in the ovary. Usually ovulation occurs prior to copulation and is initiated by a sudden increase (or "surge") in the level of LH in the blood. Most often ovulation takes place *spontaneously* following a series of well-ordered endocrine preparations, culminating in the stimulation of LH release by estrogens (and possibly progesterone in birds). It may also occur *reflexly* or be *induced*, following copulation and neural stimulation of the hypothalamus, as in the cat, rabbit, squirrel, and raccoon. In some birds, fishes, and reptiles, ovulation follows sexual play or stimulation, while in amphibians the presence of water can promote it. The egg clutch size in some birds is a well-regulated process, the number of eggs determining the subsequent ovulation (*indeterminate layers*). Thus if eggs are continually removed from the nest, further ovulation will be promoted, whereas if some are added, it will be inhibited. Such regulation is presumably mediated through the hypothalamus and pituitary. In other birds the number of eggs is determined by the number of mature follicles (*determinate layers*).

11.3 Fertilization

Occasionally in some teleosts the ova collect in a central cavity in the ovary, where fertilization and development occur. Usually, however, ova pass, via the body cavity or directly, into the oviduct. In cyclostomes there is no oviduct, and the ova pass through pores into the urinary ducts. In some aquatic species once they are in the uterus or oviduct the ova pass to the exterior where fertilization may take place. In terrestrial and some other aquatic species *internal fertilization* occurs in the gonoduct. The development of the egg then can proceed in different ways. (1) The fertilized egg, after the addition of a shell, nutrients and fluids, such as seen in reptiles and birds (amniotes), may be expelled to the exterior, where subsequent development occurs. This is the *oviparous* condition, which encompasses the anamniotes where external fertilization occurs. (2) In some vertebrates, including certain chondrichthyean and teleost fishes, amphibians, and reptiles, the eggs may be retained (*ovoviviparous*) in the oviduct or uterus, where their development proceeds. (3) In other species, which are also phyletically dispersed and include the mammals (except monotremes), the fetus directly derives sustenance from the mother across the uterine wall (the *viviparous* condition). The role of hormones varies somewhat in each of these situations in a manner consistent with the needs of the developing young.

In oviparous and ovoviviparous species the eggs must contain adequate nutrients for the development of the embryos, so the eggs are bigger and contain large quantities of yolk. In such species estrogens have a special role in promoting the formation of the yolk proteins, *vitellogenesis,* in the liver, whence they are transported in the blood and deposited in the developing follicles. This effect, not seen in viviparous species, is one of the main differences in the roles of estrogens among the vertebrates. In other respects their roles are remarkably similar.

11.4   Delayed Implantation

In a variety of marsupials (kangaroos and wallabies) and some placental mammals, such as the badger, pine marten, mink, and weasel, the fertilized egg is not always immediately implanted in the uterus. Instead, it may only develop to a stage where a blastocyst of 75 to 100 cells is formed, which remains quiescent for a period of time ranging from a few weeks to several months. This condition is called delayed implantation and may be dictated by unfavorable seasonal conditions (*obligatory delay*) such as the onset of winter or by the presence of suckling young (*facultative delay*). In placentals it seems that a shortening of the day length promotes this condition, but with an increased exposure to light, implantation and development may subsequently proceed. In kangaroos and wallabies with young in their pouches, suckling seems to provide the stimulus that inhibits further development of the egg. In both situations hormones appear to be involved. In the marsupials this may result from continual periodic release of oxytocin that occurs as a result of suckling. The onset of elevated progesterone levels, which occurs as suckling declines, may stimulate the development of the blastocyst, but the complete endocrine picture has not yet been drawn.

11.5   The Development of the Embryo: Gestation and Pregnancy

Oviparous species may or may not tend and protect the eggs. The latter is the more usual situation in birds but may also occur, though less often, in other vertebrates. Broodiness in birds (the desire to incubate the eggs) can be induced by suitable injections of prolactin, but neural mechanisms are also involved.

The viviparous or ovoviviparous retention of the developing embryo or egg in the uterus may occupy a period of time, *gestation,* ranging from a few days to several years. Such development takes four years in the amphibian *Salamandra atra* and as much as two years in some chondrichthyean fishes. During this time the uterus must remain in a suitable condition for the gestation, and simultaneous breeding behavior and ovulation are suppressed. This condition involves the maintenance of suitable hormone levels in the blood. In mammals the extrusion of the ovum from the Graafian follicle is followed by the growth of tissue in the evacuated cavity to produce the *corpus luteum.* This is promoted by LH (and in a few species also by prolactin) and secretes *progestins,* principally *progesterone,* that inhibit the secretion of LH. Progesterone has been termed "the hormone of pregnancy." It stimulates the development of tissues lining the uterus (the endometrium) into a suitable glandular condition to receive and to maintain the fertilized egg preparatory to its implantation. The muscle of the uterus (the myometrium) is also maintained in a suitable noncontractile condition by progesterone. The subsequent development of the placenta in eutherian mammals may reinforce the endocrine armentorium accompanying pregnancy by secreting progesterone, as well as estrogens and chorionic gonadotrophin. The latter exerts an LH-like action on the corpora lutea. The placenta also produces placental lactogen, or somatomammotrophin, which behaves like both the pituitary growth hormone and prolactin. Its function is not yet clear, but the fact that antibodies to it produce abortion in rats suggests that it has an important role. Such is the endocrine role of the placenta in some mammals that the ovaries can be removed and pregnancy continues as usual without the aid of its secretions. There are qualitative differences in the nature of the particular hormones produced by the placenta of different mammals. The placenta of marsupial mammals does not have an endocrine role, but no information is available about this potential function in viviparous nonmammals.

The ovarian follicles in many vertebrates also give rise to structures resembling the corpora lutea of mammals. This follows ovulation in the usual way. In addition, unovulated follicles may develop into structures called *corpora atretica.* It is not clear whether either of these structures also produces progesterone, though in some species of reptiles the corpora lutea undoubtedly do. Little is known about the corpora lutea in nonmammals, and it has been suggested that their presence phyletically antedates an endocrine function that is only seen in mammals.

11.6 Oviposition and Parturition

The expulsion of the egg or young—oviposition or parturition (labor), respectively—is a poorly understood process both with respect to timing and to its immediate mechanism. Fluctuating levels of the various hormones occur towards the end of pregnancy, and it has been suggested that hormonal changes, such as a decline in progesterone or an increase in estrogens, may be responsible. Removal of the ovaries and pituitary does not necessarily prevent oviposition or parturition, though it often makes the process more difficult. The oviducts and uterus of vertebrates contract in response to the neurohypophysial hormones, such as oxytocin (in mammals) and vasotocin (in other vertebrates). These hormones could aid this process, but only if the uterine muscle is in a suitably receptive condition. It is likely that oxytocin assists parturition in many mammals and possibly even in nonmammals.

11.7 Care of the Young

Many vertebrates tend, care, feed, and protect their young after birth or hatching. This is a response to hormones that, apart from suppressing further immediate reproduction, also supports the appropriate domestic and feeding activities. The crop sac of pigeons and doves secretes a fatty proteinaceous paste that is used to feed the young. This process is under the control of prolactin. All mammals produce milk (*lactation*), with which they feed the young. The physiology of lactation is a complex process involving several hormones. Estrogens promote the growth of the ducts in the mammae; progesterone, the secretory alveoli. Prolactin stimulates the latter to secrete milk. Other hormones, including growth hormone, insulin, thyroxin, and cortisol all are necessary for optimal production milk, so the mammae are the target of a multihormonal assault. The release, or "letdown," of milk is a reflex initiated by the young suckling on the nipple, producing a neural stimulus to the neurohypophysis that releases oxytocin. This hormone contracts the mammary myoepithelial cells, and milk is expelled.

11.8 The Male

The maturation of the sperm in the testes is a process comparable to that of the ova. The pituitary of the male contains the same adenohypophysial hormones as that of the female, but FSH in this instance promotes the maturation of the sperm. The testes are also the site of production of male sex hormones, or *androgens*, of which testosterone is the principal one. This steroid hormone is formed in tissue lying between the seminiferous tubular cells that contain the developing sperm. It is called the *interstitial tissue* and is composed of *Leydig cells*. Other cells, which are rich in lipid and called *Sertoli cells*, are associated with the basement membrane of the germinal epithelium. They are also probably a site of androgen secretion, especially of what is immediately necessary for the maturation of the sperm. The Sertoli cells are probably stimulated by FSH. The interstitial tissue, on the other hand, is stimulated by LH or, as it is called in males, *interstitial-cell stimulating hormone (ICSH)*. The development of the sperm proceeds under the influence of FSH and androgens.

In some species, particularly domesticated ones like man, rat, and feral pigeon, the active state of the testes is continually maintained. In most male vertebrates, however, a well-defined breeding season is observed that is consonant with that of the female. In homoiotherms the testes and the sperm usually ripen early in the spring, in the weeks immediately prior to breeding. This is termed *prenuptial spermatogenesis*. Most poikilotherms, on the other hand, undergo a *postnuptial spermatogenesis* in the summer, immediately after breeding and prior to winter hibernation. The sperm thus approach maturity early in preparation for breeding the next spring.

11.9 Growth and Differentiation

The sex hormones also play a role in the primary sexual differentiation of the fetus. The sex of the embryo is predetermined genetically, but the distinction is not immediately apparent during an early sexually indifferent period of embryonic development (see Figure 9.34). The potentiality for development of either sex is present, and the secretions of the embryonic gonads determine the ultimate result. The gonad consists of a cortex, or potential ovary, and a medulla, which may become the testis. Under the influence of testosterone, from the embryonic testes, the cortex regresses, and the Wolffian ducts differentiate to form the vas deferens and epididymis. The genital tubercle becomes the penis. In the female, under the influence of ovarian estrogens the medulla regresses, and the Mullerian ducts form the oviducts, uterus, and part of the vagina. The sex hormones also influence the development of the brain with respect to adult sexual behavior. If newborn female rats are injected with male sex hormones, the normal adult estrous cycle is suppressed. The sex hormones may even affect sex differentiation later in life, such as seen in some female fishes that may, under the influence of androgens, reverse their sex and become fully functioning fertile males.

In early life a number of other hormones, including growth hormone, insulin, and thyroxin, are necessary for adequate growth and development of the young. The thyroid hormones are particularly important since they play a vital role in the development of nervous tissue. The effects of their absence are seen in the condition of cretinism in man.

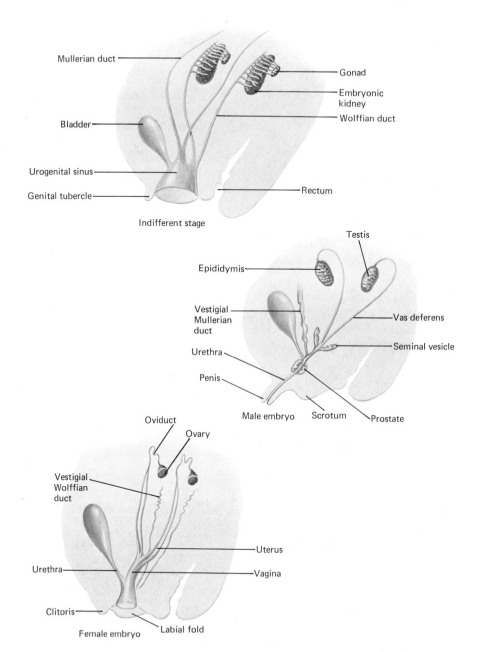

FIGURE 9.34 Differentiation of the sexual apparatus of the embryo in mammals. The correct development of the genetic male is dependent on the presence of androgens and the female estrogens. This pattern can be modified if the appropriate hormones are absent or if hormonal inhibitors of these or hormones appropriate to the genetically opposite sex are administered. (From B. E. Frye, *Hormonal Control in Vertebrates.* New York: Macmillan, 1967.)

11.10 Metamorphosis

A number of fishes and amphibians may undergo a major morphological and physiological change, such as during the transition from larvae to adults or from one adult form to another. This is most commonly observed in tadpoles; after a period of time, ranging from a few weeks to several years, they metamorphose into their adult form and become frogs. Similarly, fish larvae also metamorphose into other forms, such as those that may precede migrations between rivers and the sea, as well as the onset of a breeding condition. These changes are often interrelated with one another. In the eel there is a "first" metamorphosis from the larva into the elver (in the sea), and subsequently (in fresh water) a "second" metamorphosis from the yellow to the silver eel occurs. The latter migrates back into the sea to breed. Conversely, young salmon are hatched in fresh water and make a transition from the parr to the smolt, which is the form that travels to the sea and adapts to a marine life.

Although it is considered likely that hormones may trigger or assist such changes in fish, there is no information about this. Changes in thyroid hormones and TSH activity have been widely invoked, but the evidence that they are directly involved in fish metamorphosis remains equivocal. The activity of other glands such as the pituitary and adrenocortical tissue may change, but no causal relationship between this and metamorphosis has yet been established. This lack of precise information, however, should not lead to the conclusion that hormones are not involved.

The metamorphosis of larval amphibians is clearly related to the action of thytoid hormone. This is well known in the change of tadpoles into frogs. Thyroid hormone is also essential for the development of urodele amphibians. The axolotl is a neotenous salamander (*Ambystoma mexicanum*) that possesses prominent external gills and yet normally remains and breeds in the larval condition. However, if it is fed thyroid hormone, it metamorphoses to a normal adult salamander. Other urodeles, such as the mud puppy (*Necturus maculosus*) and the congo eel (*Amphiuma means*), are neotenous but do not metamorphose in response to thyroid hormone. Neotenous forms of newts (*Triturus helveticus*) that normally do not exist in this form in nature have been observed in ponds near fields of kale and turnips, which are plants that have a notorious antithyroid activity.

It is in the anuran tadpoles that the process of metamorphosis and the role of hormones has been particularly well studied. Tadpole metamorphosis can be quite short, less than a month in frogs from desert areas where free water is available only for a short time; on the other hand, it requires several years in some bullfrogs. During this time the animals grow and undergo periods of differentiation that include the disappearance of the external gills, appearance of the hindlimbs and then the forelimbs,

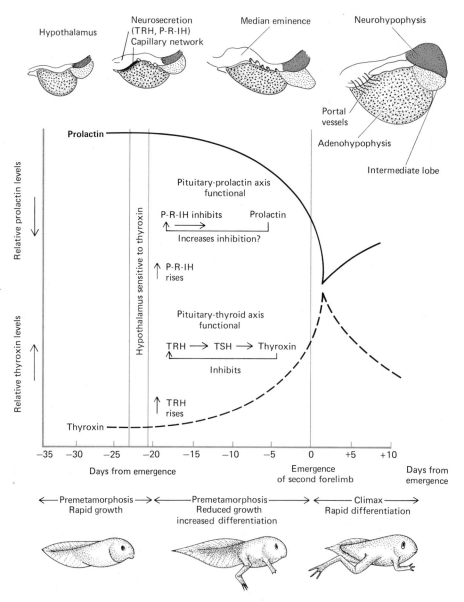

FIGURE 9.35 The role of endocrines in the growth and metamorphosis in the tadpole of a frog. (Upper) Development of the frog pituitary. (Lower) The changes that occur in the levels of hormones and the activity of the prolactin- and thyroxin-axes. TRH—thyrotrophin-releasing hormone; P-R-H—prolactin-release inhibiting hormone. (From P. J. Bentley, *Comparative Vertebrate Endocrinology.* New York: Cambridge University Press, 1976; based on W. Etkin, *Mem. Soc. Endocrin.* 18:137, 1970.)

and culminating in the absorption of the tail (see Figure 9.35) and the emergence of an adult frog that can live on dry land. Metamorphosis can be hastened by feeding the tadpoles thyroid hormones (*morphogenetic effect*) or delayed indefinitely by removing the pituitary or feeding them antithyroid drugs. Growth is facilitated, but metamorphosis is inhibited following the injection of prolactin, which also may have a physiological role antagonistic to thyroxin and so delays metamorphosis. These features apply almost equally to the larvae of newts and salamanders. The metamorphic cycle in relation to the development of the endocrine glands has been described by W. Etkin (see Figure 9.35). Following an initial period of growth when the levels of prolactin are high, and TSH and thyroxin low, the hypothalamus matures and produces TRH, which leads to the secretion of TSH and increased levels of thyroxin. There then follows a period of reduced growth and increased differentiation (*prometamorphosis*), during which the legs appear. On the emergence of the forelimbs the rate of differentiation accelerates (*climax*), the tail is reabsorbed, and the adult frog emerges. The relationship of this morphogenetic effect of thyroxin, along with prolactin, to the growth of young mammals is doubtful, and it may represent yet another use for these hormones.

REFERENCES

Austin, C. R., and Short, R. V., eds. *Reproduction in Mammals. Hormones in Reproduction.* Vol. III. New York: Cambridge University Press, 1972.

Axelrod, J. The Pineal Gland: A Neurochemical Transducer. *Science* 184:1341–1348, 1974.

Bagnara, J., and Hadley, M. E. *Chromatophores and Color Change.* Englewood Cliffs, N.J.: Prentice-Hall, 1972.

Barrington, E. J. W. *An Introduction to General and Comparative Endocrinology.* 2d. ed. New York: Oxford University Press, 1975.

Benson, G. K., and Phillips, J. G., eds. *Hormones and the Environment. Memoirs of the Society for Endocrinology.* Vol. 18. New York: Cambridge University Press, 1970.

Bentley, P. J. *Endocrines and Osmoregulation. A Comparative Account of the Regulation of Water and Salt in Vertebrates.* New York: Springer-Verlag, 1971.

_____. *Comparative Vertebrate Endocrinology.* New York: Cambridge University Press, 1976.

DeLuca, H. F. Vitamin D: The Vitamin and the Hormone. *Fed. Proc.* 33:2211–2219, 1974.

Gorbman, A., and Bern, H. A. *A Textbook of Comparative Endocrinology.* New York: Wiley, 1962.

Greep, R. O., and Astwood, E. B. (sect. eds.). *Handbook of Physiology, Section 7: Endocrinology.* Vols. 1–6. Bethesda, Md.: American Physiological Society, 1972–1976.

Guillemin, R., and Burgus, R. The Hormones of the Hypothalamus. *Sci. Am.* 227:24–33, Nov. 1972.

Holmes, R. L., and Ball, I. N. *The Pituitary Gland. A Comparative Account.* New York: Cambridge University Press, 1974.

Idler, D. R., ed. *Steroids in Non-mammalian Vertebrates.* New York: Academic Press, 1972.

Litwack, G., ed. *Biochemical Actions of Hormones.* Vols. 1–3. New York: Academic Press, 1970, 1972, 1975.

O'Malley, B., and Schrader, W. T. The Receptors of Steroid Hormones. *Sci. Am.* 234:32–43, 1976.

Schally, A. V., Arimura, A., and Kastin, A. J. Hypothalamic Regulatory Hormones. *Science* 179:341–350, 1973.

(Symposia) Proceedings of the Fifth, Sixth, and Seventh International Symposia on Comparative Endocrinology. *Gen. Comp. Endocr.* (*Suppl. 2 and 3*), 1969, 1972. Also published as Suppl. 1975. Thousand Oaks, Calif.: American Society of Zoologists.

(Symposium) Comparative Aspects of the Endocrine Pancreas. *Am. Zool.* 13:567–709, 1973.

(Symposium) Comparative Aspects of Ovarian Function. *Am. Zool.* 12:225–339, 1972.

(Symposium) The Current Status of Fish Endocrine Systems. *Am. Zool.* 13:711–936, 1973.

Waring, H. *Color Change Mechanism in Cold-Blooded Vertebrates.* New York: Academic Press, 1963.

10

TEMPERATURE REGULATION

John Kanwisher

Warm-bloodedness has been a fundamental factor in the ascendancy of higher organisms. It has allowed both an increased speed and versatility with which such animals could meet the competitive and unpredictable situations in nature. We will consider its part in the physiology of animals as varied as moths, hummingbirds, and whales. We place the subject at the end of the book because thermal physiology involves many aspects covered in earlier chapters (such as respiration, biochemistry, and nervous control).

An understanding of temperature regulation requires more physical considerations than other subjects in physiology. Our own senses give us an intimate, if qualitative, impression of heat as an important environmental parameter affecting our own well-being. But to appreciate the subject quantitatively, how heat is produced and the means by which it moves about, one should have a working knowledge of the basic processes, such as radiative transfer and conduction. The approach here has been to keep the explanations of this thermal physics as nontheoretical as possible.

In a physical sense, animals are messy subjects to treat experimentally in any refined, quantitative manner. We are unlikely, for instance, to know the total amount of heat produced in an animal with anything like the precision we learn to expect in most laboratory measurements. But this is largely unimportant. We should rather keep in mind that this quantity changes by ten- or twentyfold under different conditions of exercise or rest.

In the same way the structural details we consider cannot be defined as precise physical systems. As an example, we recognize countercurrent exchange between arterial and venous blood as an important feature in an animal's thermal physiology. But we can only loosely consider it through its basic details, such as blood flow and vessel diameter, in a manner that would allow us to compute its thermal effectiveness. We also loosely talk about the thermal conductivity of tissue, when in reality heat is transferred largely by flowing blood rather than thermal conduction.

But we should not become too distracted by this sloppiness of detail. A precise central control system regulates these essentially crude elements. And the varied life style of warm animals is thus made possible. Because they are warm they have literally inherited the earth.

1 • INTRODUCTION

Under the climatic stress of extreme heat or cold, a warm-blooded animal such as man will both consciously and unconsciously go to great lengths to achieve thermal comfort. In doing so he is fulfilling the requirements of mammals and birds, as well as some fishes, to keep the temperature deep within their bodies within narrow limits. These animals have managed this internal warmth by combining a high production of internal metabolic heat with various forms of insulation, such as fur, feathers, or blubber, to keep this heat within their bodies. This regulated body temperature is the most characteristic feature of these higher vertebrates.

Before them, other groups had found many ways to use both living and nonliving heat to keep themselves or their eggs warmer than the environment. But it was only with the development of the controllable internal heat source of the animal's own basal metabolism that animals were free in both time and space for a warm-blooded existence. This final ascent to thermal independence has been preceded by many strategies offering varying degrees of freedom from environmental temperatures.

In a general sense heat and life are inevitably related through the fundamental nature of an animal's biochemistry. The ordered chemical reactions of an organism's metabolism are described in Chapter 7, Section 3. It is by these reactions that tissue synthesis or external work is accomplished. Such reactions are uniformly inefficient in terms of energetics. Most of the energy in metabolic transformations appears as heat (Chapter 7, Section 3.5). For every unit of external work that an animal performs, there are about four units of metabolic heat produced. In this sense an animal can be considered thermally as a chemical furnace. The inevitable internal heat supply of any animal's metabolism always tends to make it warmer than its surroundings. Through most of evolution this warming has been of little consequence. Animals are not heat engines and so cannot extract work from the temperature differential.

Although this heat was originally something of an evolutionary accident, it now plays an important role in the physiology of mammals and birds, as well as some fishes. These animals are able to use it to produce a high constant body temperature. The details of this thermal physiology is our concern in this chapter. Throughout we will be concerned mostly with mammals, the most widespread and diverse of the homoiotherms.

1.1 Why Be Warm?

In terms of overall strategy, warm-bloodedness entails a considerable cost in both anatomical complexity and metabolic energy. But there are important advantages that compensate for this fact.

First, warm-bloodedness has freed mammals from restrictions of climate that limited earlier cold-blooded vertebrates. The homoiotherms are widely distributed from the tropics to the arctic, as well as over the oceans. They show a versatility in size that is unique; the weight ratio between a shrew, weighing a few grams, and a blue whale, the largest mammal, is 1 to nearly 100 million. Despite this phenomenal range, the basic workings of thermal physiology are essentially the same.

Internal warmth, represented by a higher body temperature, has also given homoiotherms access to larger amounts of power. This follows from the temperature dependence of metabolism. At higher temperatures reactions go faster, and more work becomes possible. This greater capacity for living becomes a reliable resource, provided that the temperature is kept constant as well as high. In a prey-predator world, life favors the swift. But the warm body in turn needs additional food to satisfy the greater metabolic needs.

Finally, temperature regulation is one of the many forms of homeostasis, discussed in Chapter 1, that work in the direction of keeping the internal environment constant. This buffering against the stresses of the environment permits greater complexity, along with reliability, in the internal workings of the animal. This progressive evolutionary sophistication has culminated in the central nervous system of man.

1.2 Metabolic Consequences of Size and Warmth

The evolutionary jump to homoiothermism involved, among other changes, a large increase in the rate of metabolic heat production. This has meant more energy from more food, with obvious implications to lifestyle. In a sense it is as though the entire process of living had been speeded up.

The metabolism in man when he is resting involves about 100 watts of power. About 20 watts of this goes to the brain and a roughly equal amount to the liver. During vigorous exercise the body's metabolism will amount to more than a kilowatt.

Such values are some 10 times higher than in a cold-blooded fish or reptile of equal size. For a more cosmic view, however, we should note that the rate of energy production in the human body is 10,000 times greater per unit of volume than the nuclear energy produced in the sun. With the latter, though, we are dealing with such a large volume to surface that the internal temperature is in the millions of degrees. Size is an important parameter when considering the effect of energy on temperature in both animals and stars.

Small mammals are, for their size, more active than large ones. This means there is an increasing metabolic intensity with decreasing animal size. It is an important aspect of homoiotherms when they are considered over their full size range. On a weight basis a shrew or mouse will have 50 times the resting (basal) metabolism of an elephant (Figure 10.1).

Thermal problems that follow from this purely geometrical effect of surface to volume account for part of this. In a large animal, heat produced throughout the great bulk must be lost to the environment through a rela-

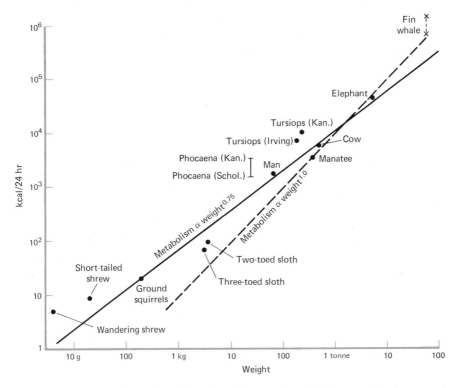

FIGURE 10.1 Logarithmic plot of the relation between the metabolism and size in mammals (mouse-to-elephant curve). The dashed line shows that the elephant and the whale are less metabolically, and therefore thermally, active than the much smaller, lethargic sloth. Small marine mammals fall consistently above the solid line that best represents terrestrial species.

tively smaller surface. Kleiber has pointed out that a cow with a mouse's specific metabolic rate would be at the boiling point of water, while a mouse with a cow's metabolic intensity would need a foot of insulating fur. Cold-blooded animals, where heat can hardly be involved, show an analogous variation of metabolism with size. Thus thermal considerations cannot be the only factor effecting size-dependent metabolism.

Although paleophysiology can only be guesswork, it would be interesting to know something about the thermal story of the extinct, really large reptiles. Great size most likely carried with it a low metabolic rate. But the geometrical factor, as noted for the sun, could have resulted in a heat disposal problem, particularly when the animals were out of the water. Many of the Permian dinosaurs had suggestive web structures along the back that may have functioned as a cooling mechanism.

The great energy need of a homoiotherm translates directly into food. An animal is always being consumed by its own metabolism. The higher this metabolism is the faster the animal will literally waste away in the absence of food. Thus specific metabolic rate puts a time scale on the frequency with which an animal must feed. If a homoiotherm has a metabolism that is 10 times higher than that of a cold-blooded animal, then it must eat 10 times more food and can only go one-tenth as long between meals. Since mammals are so successful, we must conclude that other advantages of being warm more than compensate.

Small mammals operate on a much shorter time scale than large ones because of this higher specific metabolism. Shrews may require food equal to their own weight every day. At the other end of the size scale, a whale can survive for many months on its blubber. Man is intermediate. One experimental subject survived 30 days of starvation during which 20 percent of his weight was metabolized away. A dog can withstand more severe attrition. With a roughly equal metabolic rate as in man, one survived 117 days without food, while losing 66 percent of its weight.

These starvation times are a direct consequence of the rate of metabolic heat production needed for homoiothermism. This more imminent threat of starvation in a small mammal has forced evolution in the direction of seasonal or daily stages of metabolic inactivity. The body temperature is temporarily lowered as in hibernation, and food stores are conserved.

Both the bat and the hummingbird do this on a daily basis. They only keep their body temperature up when they are active and feeding. Because metabolism decreases rapidly with lowered body temperature, the saving in stored reserves is considerable. The daily cycle of metabolism for the bat and hummingbird is shown in Figure 10.2, along with that of a shrew, a small homoiotherm that does not drop its temperature. These now become not very different functionally from the insects, which are

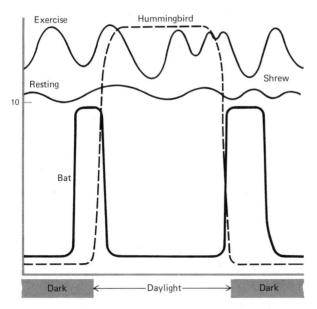

FIGURE 10.2 Idealized daily heat production curves of three small (3- to 5-g) homoiotherms. The hummingbird and the bat show a resting torpor which saves ten- to twentyfold in the amount of food stores they consume. They are only metabolically active during the part of the day when feeding. The shrew has a high metabolism even when resting, so it must eat much more.

warm only during flight. Hibernating mammals show an analogous saving on a seasonal time scale. This plasticity in the body temperature represents a temporary relaxing of the requirements of homoiothermy. It has evolved many times to relieve small birds and mammals from a constantly demanding search for food so that they can keep warm.

Prehomoiotherms

BEHAVIORAL THERMAL REGULATION. The probable evolutionary roots of homoiothermism can be seen in the many ways that earlier groups found to be warmer than their surroundings. Before the internal metabolic flame was bright enough for dependable warmth, use was made of external heat such as sunshine. Such solar heating is still important in poorly insulated, as well as in small, homoiotherms. People basking on a summer beach can be considered a behavioral adaptation in which the insulation of clothes is exchanged for the 1.5 cal/cm²/min of heat from sunshine. With this warmth added to the basal metabolism, thermal balance with a bare skin is possible. One must keep in mind that most of the true homoiotherms still retain a varying dependence on external heat.

But in reptiles and other poikilotherms nearly all of an animal's warmth is external. This can take unusual forms in special situations, such as hot springs. And it may show ingenious behavior sophistication, such as the incubation of eggs in the warmth of rotting leaf litter piled up for the purpose. Much of the time fishes and reptiles, when given the option, will move along a temperature gradient to a warm place. They will move down the gradient as well if the temperature is too high for their liking. The needs of this behavioral thermal regulation place a restriction on the movements of such an animal.

Basking in the sun, which is so effective in warming reptiles, allows the animal to move about more. A lizard or snake in the sun can be 15°C or more warmer than the air. To avoid overheating it is only necessary to stay partly in the shade. Or a variable skin pigmentation may control the fraction of the incident solar heating that gets into the body. When shade is not possible, some animals line up in a way that minimizes the amount of sunshine that strikes them. If none of these alternatives suffice, a lizard will open its mouth and use the cooling from the evaporation of saliva to avoid thermally unpalatable body temperatures. Behavioral temperature regulation may use varying amounts of these stratagems to effect a body temperature that is constant within a few degrees.

INSECT THERMAL REGULATION.   When warmth is critical but not readily available, an animal must resort to its own metabolism. The python hatches its eggs by lying on them and rhythmically contracting its muscles. The price for the higher temperature is a 10 times higher metabolism and the food necessary to support it. But the eggs hatch more reliably and quickly. This use of muscle contraction for warmth was fully developed earlier in the insects. We are only now discovering how effective this can be.

Insects initially developed flight as an escape from predation. In this early time, 350 million years ago, the air was an unoccupied refuge from terrestrial enemies. To remain in the air insects needed a sustained high rate of metabolism. The inevitable inefficiency of muscle biochemistry requires that there be a parallel high rate of heat production. In a flying insect this may equal 5 to 10 cal/g weight/min. Such a metabolic intensity is many times the basal rate in man.

Such a high muscle metabolism is possible only in tissue that is already warm. Thus a cold insect must raise its temperature before it can take off. Many species manage this by absorbing sunshine with a pigmented absorbing body surface. They are unable to fly in cold, cloudy weather.

But it is also possible to use the flight muscles to warm themselves in a thermal bootstrap mode much like shivering. Such an insect will make small, aerodynamically ineffective wing strokes until the body temperature reaches some critical value.

As the muscles warm up, they are increasingly capable of a higher metabolism and thus more heat production. But this is balanced by a more rapid heat flow along the increasing outward thermal gradient. The result is a linear rise in temperature toward the value where flight is possible. Once the insect is in the air, the metabolism of flight keeps the insect warm.

In a nonregulated system this warming would continue, under conditions of strenuous flight in warm air, to temperatures that might cause thermal injury. It is unlikely that flight in larger insects ever occurs without an active means to dampen such a continuing temperature rise. We can only wonder about the thermal functioning in the 60-cm dragonflies of the past. But temperature recording of the internal temperature in a moth does show an active regulation during flight. The means of this regulation is now understood.

Structurally an insect consists of three relatively separate parts: the head, the thorax, and the abdomen. The flight muscles are in the thorax. During flight they account for nearly all of the metabolic heat production. Much of this heat is retained in the thorax. The retention may be aided by the insulation of surface scales.

As the temperature rises, it may become necessary to dispose of heat. This is accomplished by pumping cool blood from the adjacent, cooler abdomen, where the heart is under the control of the temperature that is sensed in the thorax. The abdomen is relatively less insulated and can be used as a thermal window. By this means the temperature in a moth may be actively maintained within a degree or two in spite of its being as much as 20°C higher than that of the air (Figure 10.3).

The cooling of flight may be too great for the muscle heat to keep the temperature high enough for flight. This is the case with a bumblebee in cold air conditions of the arctic. They must periodically land and warm up with no flying muscle activity.

Thus the much smaller insects use the same physiological mechanisms as mammals and birds to realize the analogous end result of a high, regulated body temperature. Although these mechanisms operate only when the animal is active, the same is true for some small mammals and birds. The similarity between them shows the universality of the physical laws governing the production and transport of heat in living systems.

Even some plants, particularly those of the arum family, have evolved high metabolic rates in order to be warm. Tropical members, such as the voodoo lily, have such a high rate of metabolism in the floral parts during the first day of their flowering that the local temperature may be 15°C higher than that of the surrounding air. This heat is a pollination

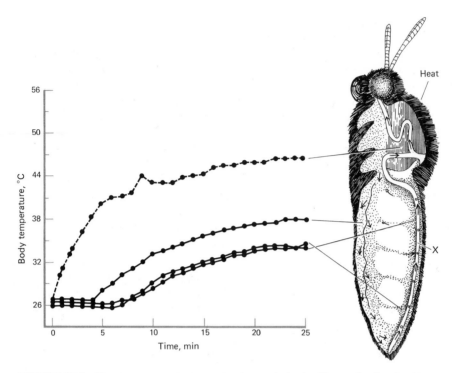

FIGURE 10.3   Time course temperature curves in a restrained sphinx moth when heat is applied to the thorax. When its temperature reaches 40°C, the beating of the dorsal vessel (X) increases. This allows blood flow to transfer excess heat to the abdomen, which then rises in temperature. (From B. Heinrich, *Science* 185:747–775, 1974.)

mechanism. It volatizes amines and ammonia, which smell like carrion and attract insects.

In colder conditions the heat allows skunk cabbage to bloom through the snow in the early spring. This heat is the product of sugar metabolism in the flower. The sugar is supplied from starch in the root. Since the rate increases with lower air temperatures, one can consider this a crude form of temperature regulation.

2 • NATURE OF THE PHYSICAL PROBLEM

2.1   Heat Production and Loss

In its simplest sense a homoiotherm can be considered as a heat source surrounded by a layer of insulation. A central thermostat senses body temperature and makes necessary physiological adjustments. These vary heat production and its loss to the surroundings such that deep body tem-

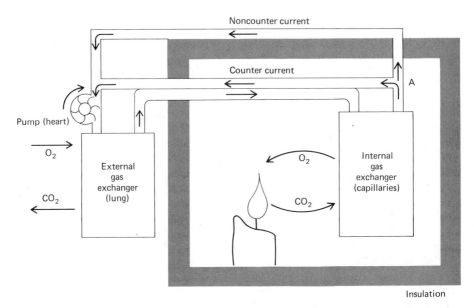

FIGURE 10.4 Schematic representation of the way in which blood flow can provide the respiratory needs of the body as well as control heat loss across an insulation barrier. The latter is accomplished at A, where the circulation is partitioned between countercurrent and noncountercurrent flow. In this way the static insulation can be bypassed to prevent overheating.

perature stays within narrow limits. For any extended period heat loss must equal heat gained. Otherwise the animal would heat up or cool down (Figure 10.4).

This continuous thermal adjustment between supply and loss is necessary since the size of the animal's internal heat source, as well as the cooling or warming power of the animal's environment, can vary over wide ranges. Thus to keep a thermal balance the effective insulation must be even more variable. This is illustrated by the two extremes of exercising in a warm climate and resting in a cold one.

Physical exertion in an animal can raise metabolic heat production at least tenfold. If it is not to overheat, the animal must dispose of this heat with what is in effect a 10 times smaller thermal barrier between it and the environment. Likewise, the cooling power of a cold climate can be 10 times greater than a warm one. This requires additional insulative variability. To meet the above extreme needs of varying external temperature and also a changing internal heat production simultaneously requires a hundredfold change in the effectiveness of an animal's thermal barrier or insulation. The purpose here is to point out the great range necessary for physiological processes to meet the variety of situations in nature. In a homoiotherm few factors relating to temperature are unchanging for long and most vary by large amounts. Such an animal must be viewed as a dynamic system of interacting parts whose interdependencies function to maintain an overall thermal integrity. Figure 10.4 is a physical model that aids in visualizing some of the above features of heat production and loss.

A variety of functional and behavioral adjustments are made to accomplish thermal balance. Increased heat loss is effected by sweating, with its evaporative cooling, increased blood flow (vasodilation), which carries more warm blood to the surface, and increased breathing (panting), which can be considered as internal sweating. Heat loss is reduced during cold stress by decreasing circulation (vasoconstriction). Added heat is produced by shivering and sometimes by direct thermogenesis. These unconscious reactions (autonomic) must always be aided in extremes by behavioral responses.

Getting out of the sun may be necessary for an animal to avoid overheating. Putting on more clothes is a behavioral response to protect man against the cold. This is thermally equivalent to the ruffling of feathers on a bird or the fluffing of fur on a mammal. To understand the overall thermal performance of a warm-blooded animal we must know how and under what conditions these individual tactics are brought into play.

It should be kept in mind that no control system, as described in Chapter 1, works without an error signal. In this case it is some finite temperature difference between the set point of the thermostat and the actual body temperature. It provides the necessary trigger for the ensuing actions that attempt to move the temperature back toward the set point. This difference is proportional to the thermal stress. A healthy runner finishing a long-distance race may have a body temperature a few degrees above normal. Such a value, which would have grave implications in a clinical situation, results from his intense heat production. The body can only be forced to a correspondingly high heat loss by this large error signal.

As the opposite extreme, the body temperature of a naked aborigine sleeping through the cold desert night will drop a couple of degrees below the control point. This maximally stimulates the physiological and behavioral adjustments that minimize heat loss. Such a person will shiver violently to raise his body temperature on awakening. His adaptation to this stress seems to be primarily behavioral. He can sleep because he tolerates the discomfort of a lowered body temperature that would keep most people awake. In most normal situations in which an animal is less stressed, the error signal is a small fraction of a degree.

2.2   Properties of Heat

Knowledge of some basic properties of heat as a form of energy will be necessary to deal fluently with the various details of thermal physiology. First, the more heat in a body the higher will be its temperature. This heat always flows or diffuses downhill. It moves from a region of higher temperature to one of lower temperature. Such a flow has the effect of erasing temperature differences, either in an animal or in inanimate nature. An animal is warmer than its surroundings by the continual production of internal heat from stored chemical energy in its food and also because it receives external heat such as sunshine. In the tropical situation where the air is warmer than the body, heat will conduct inward.

As heat passes from inside an animal to the outside, the temperature does not change abruptly. Instead, at the surface of the body it shows a continuous variation between that of the deep body and the outside environment (Figure 10.5). This change is found mostly within the insulation, which, because it impedes heat flow, creates the gradient in that region.

Heat can be considered in an analogous manner to Ohm's law in electricity. The heat that flows (current) through an insulating layer (resistance) depends on the driving force of the temperature difference (voltage) across this layer. Thus the quantity of heat that flows through a unit area of the animal's surface is determined by the steepness of the temperature gradient and the thermal conductivity of this insulation. Fur and feathers, as well as clothes, ideally approach the low thermal conductivity value of air. They work mostly by trapping air in small spaces. Blubber, the only form suitable to marine mammals, is a relatively much poorer insulation. Surface body fat, such as that on man and pigs, can be an important bar-

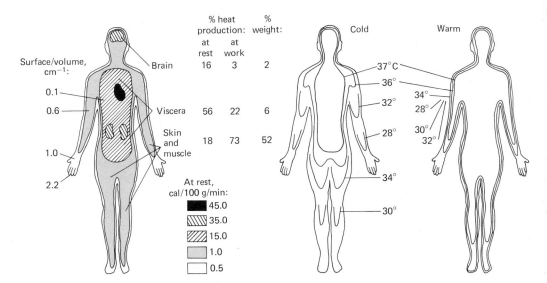

FIGURE 10.5 Varying rate of heat production within a quiet human and the interval isotherms in a warm and a cool environment. (From J. Aschoff and R. Wever, *Naturwissenschaften* 20:483, 1958.)

rier to the outflow of body heat in land animals. For a given rate of heat flow, a greater temperature difference is developed across a layer of greater thermal resistance or lower conductivity—that is, better insulation.

Such a flow of heat is really the transfer of the kinetic energy of molecular vibrations being passed along from molecule to molecule. It is inherently a much faster process than the diffusion of a dissolved substance through a liquid medium. This latter is a random-walk type of molecular process, in which the direction of movement of an individual diffusing molecule can instantaneously be in any direction. The large difference in rate of these two types of diffusion, roughly a hundredfold, has important anatomical consequences.

2.3   Heat in the Animal

The thermal situation in the interior of an animal can best be described as untidy, in the sense that one cannot describe it with physical precision. The heat is generated biochemically at the cellular level at rates that vary greatly between the different organs. In a resting human nearly half of this metabolism is in the brain, kidney, and liver, as shown in Figure 10.5. At maximum exercise, when the brain has the same rate of heat production as when resting, that in the skeletal muscles may have gone up 50 times.

When there is aerobic metabolism, there must also be blood circulating to provide the necessary respiratory gas exchange. This is dictated by the slowness of diffusion as a transport mechanism in getting O_2 and CO_2 in and out of tissues. As the blood moves through the capillary bed, it thermally equilibrates with the tissue. It will thus pick up heat at more active sites and leave it where the rate is lower. Such a mass transfer by circulating blood is the means by which most heat is moved within the body core. Circulation evolved originally for transport. It has since undergone extensive secondary changes for thermal purposes.

Because of blood circulation, the internal temperature differences that would result from large local variations in heat production are equalized to a large extent. We call the resulting average value the *core temperature*. But temperature differences between individual tissues are not completely erased. One can frequently indicate metabolically more active regions by slightly higher local temperatures. In part, though, the system is self-regulating. More metabolism generates more heat. But this in turn requires more local blood flow, which tends to transfer this heat away.

2.4   Processes of Heat Exchange

Thermal exchange between an animal and its environment involves the three individual processes of conduction, radiation, and evaporation (Fig-

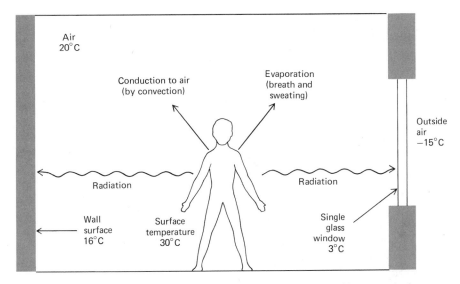

FIGURE 10.6 Pathways of heat loss from a man in a heated house. About two-thirds goes as thermal radiation to the colder surrounding surfaces. One-tenth is convected away as conduction to the air. The remainder is accounted for as evaporation in the expired breath and in sweating. The person would likely sense the cold glass surface of the window since the net heat flux away from him would be greater in that direction.

ure 10.6). We must consider the sum of all three to find the net flow into or out of the animal, but this is an extremely complicated physical measuring problem. Such complete determinations have been possible only on a subject peddling a bicycle ergometer in a highly instrumented laboratory setup. In most situations we have only qualitative data, or we must work by difference. This involves measuring two variables such as heat flow and temperature gradient. We can then estimate the third quantity of thermal resistance, which we cannot determine directly. We can frequently make use of the long-term equality that must exist between heat production in an animal and its exchange with the environment.

Any of the three processes can, under different conditions, involve heat loss or gain by the animal. Consider an animal standing in the sun. It is receiving heat from sunshine at the same time as it may have a heat loss due to the evaporative cooling of sweating or panting. To assess the overall thermal picture, one must understand the separate processes.

Conduction.   Heat flow by conduction has been described earlier as energy transfer by vibrating molecules bumping against adjacent molecules. The more energetic molecules are always tending to impart kinetic energy to their colder, less active neighbors. In this way heat inevitably flows down a gradient from warm to cold. Without new thermal energy produced by nuclear processes in the sun, nature would run toward an iso-thermal death.

A cold object feels cold because heat flows from the hand into it. This locally lowers the skin temperature. Cold receptors are activated, and this signal, by way of the central nervous system, elicits the sensation of coolness. Warmth is sensed in the same way from another set of sensors (Figure 10.7).

Air at the body surface has an insulating capability because its thermal conductivity is so low that it cannot carry body heat away very rapidly. A temperature gradient is established in the surrounding air layer, and a naked skin surface can be many degrees warmer than the main mass of the surrounding air. With high winds this shell of air is increasingly torn away, destroying the insulating properties. The skin temperature is lowered, and more heat is carried away. This cooling power of wind is an important thermal stress in cold air. Like all turbulent flow phenomena, it can be treated only empirically. Fur and feathers are adaptations to contain the insulating air next to the skin. By erecting the

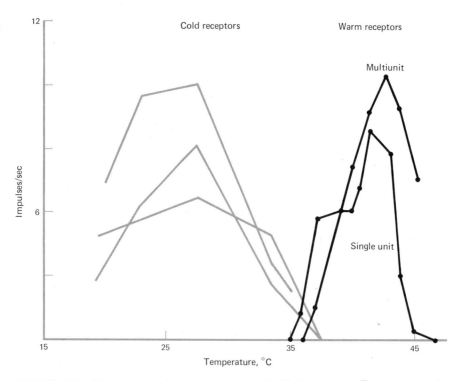

FIGURE 10.7   Nervous activity from primate warm and cold skin receptors. The two types act as error detectors to signal how far the skin is above or below a neutral point of about 36°C. (From A. J. Iggo, *Am. J. Physiol.* 200:403, 1969.)

hair of fur and also by ruffling the feathers, the thickness of the insulating air layer can be varied.

Radiation. Animals also lose and gain heat by radiation. An object of any temperature radiates electromagnetic energy at a rate proportional to the fourth power of its absolute (Kelvin) temperature. This occurs mostly at invisible infrared wavelengths. The radiation from a really hot body, such as the sun or a light bulb, has an appreciable visible component. If an object is next to some of different temperatures, there is a net difference in the energy radiated back and forth between them.

At a given temperature the ability of a surface to radiate, and also absorb, can vary. A perfect radiator is spoken of as a "blackbody," and the radiative temperature determined by measuring the infrared energy coming from this surface is the blackbody temperature. These terms have their roots in the laboratory practice of preparing a good absorbing and radiating surface by coating it with carbon soot. The opposite of such a surface is one with a silvery metallic luster.

The visual appearance of an animal's fur or feathers is a poor guide for estimating the thermal radiative properties of such a surface. At environmental temperatures the radiated energy is all in the infrared, and it is the property of the surface at these wavelengths that determines the radiative thermal interchange. Most animal surfaces can be considered fairly good blackbodies. This means that the surface temperature measured radiatively will be close to that determined with a contact thermometer. A black and a white mouse appear identical at environmental thermal wavelengths.

The net outgoing radiative flux from an animal is the sum of the energy radiated from each element of body surface. These same surface elements receive thermal radiation from all lines of sight directions in the surroundings. The net flux, which the animal must deal with thermally, is the difference of these two sums.

The sky, which makes up a large part of the field of view, has the properties of a thermal sink. It is partially transparent since only CO_2 and water vapor (both minor constituents of air) absorb in the infrared. Thus much of the thermal energy radiated at the sky is lost to outer space. This is particularly true with a clear sky and low humidity. Since the sky absorbs poorly, it follows that it must also radiate back poorly. A radiation thermometer pointed at the sky will register as much as 25°C lower than the actual air temperature.

This radiative loss to the nighttime sky will act as a serious thermal deficit to a naked man sleeping in the desert. It can be partially avoided by sleeping under a tree so that the body "sees" leaves rather than sky in an upward direction. This is an example of using behavior to obtain comfort.

Evaporation.   Conduction can no longer be used to get rid of heat when the surrounding air is warmer than the body surface. At higher temperatures when this point is approached, an animal must increase its evaporative cooling through sweating and panting. This makes use of the large amount of heat (540 cal/g) needed to change water from a liquid to a vapor. Sweating imposes a parallel stress of salt loss, which the animal must eventually make up in the food. And panting involves respiratory difficulties. The excessive ventilation draws CO_2 from the body and upsets the internal acid-base chemistry of the blood, which can have secondary complications.

Nature has broadly used both forms of evaporative cooling. Sweating is not adapted to an animal with fur because the air cannot freely get to the skin to carry off the water vapor. Evaporation on the surface of the fur is only of a little help in keeping cool because the body heat would still have to flow out through the insulation. This accounts for the predominance of panting in furred animals, particularly those that live in the desert, where keeping warm at night may be as important as keeping cool during the day.

2.5 Circulatory Adaptations

The thermal needs of a homoiotherm require moving variable amounts of heat through different tissues. This is accomplished mostly by changes from the initial evolutionary plan of circulating blood only for the purpose of local tissue metabolic needs. A variable blood flow in excess of such needs can be used to carry a correspondingly variable amount of heat. Such changes in heat flow are produced by opening up (*vasodilation*) and closing down (*vasoconstriction*) local blood vessels. This is the primary means of controlling heat flow to the body surface, increasing it when warm and decreasing it when cool. In the latter case, heat from some region of the body is moved to another where the cooling heat loss is more than the local metabolism can make up.

In a cold situation the restriction of blood flow near the surface can be almost complete. The skin will cool, and a temperature gradient will exist in the underlying tissues. Such shell cooling allows the surface layers of an animal to function as part of its thermal barrier between the warm interior and the outside air. This is a particularly important part of thermal control in animals, such as pigs and man, in which the surface insulation is poor (Figure 10.5).

A different situation arises when an animal cannot stand the thermal cost of keeping some of its appendages warm against a severe cooling stress. The feet of a duck or the tail of a muskrat in cold water would both take large amounts of heat to keep them warm. To save heat, the entire appendage is operated at close to ambient temperature. The foot of a duck may be near 0°C, whereas a few centimeters away the body is 40°C.

The living tissue in such appendages must still have a blood flow. To maintain the steep temperature gradients between the warm core and the cool limb it is necessary to have thermal isolation in spite of blood flow. This is accomplished anatomically by having the arterial supply from the warm interior run alongside the venous return in a manner such as that shown in Figure 10.8A, in which heat is allowed to diffuse between them. Such a parallel countercurrent flow allows temperature differences between different parts of an animal to be maintained in spite of the mass transfer of nutrients and gases by the blood. These materials must not be diffusible along the same path taken by the heat. Since heat diffuses about 100 times more rapidly than respiratory gases, it is easy to satisfy this condition by proper vessel sizes and rates of blood flow. Countercurrent thermal exchange is a common circulatory strategy in homoiothermy. Such secondary adaptations of the circulation form some of the most important elements of temperature regulation (Figure 10.4).

The human forearm provides an easily observable example of how circulatory control is used to effect a variable heat loss. Two separate venous returns are involved. One, on the surface, effects a greater loss of

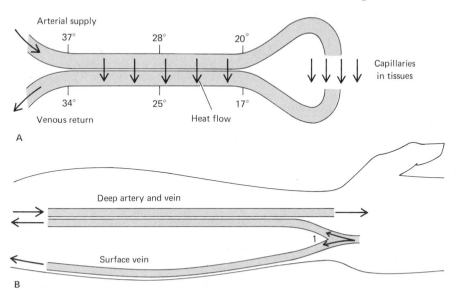

FIGURE 10.8 (**A**) Countercurrent heat exchange takes place between arteries and veins when they are adjacent to each other. (**B**) In the human arm a separate venous return allows this to be bypassed when heat conservation is unnecessary. Central nervous control between the two paths at 1 can then adjust heat loss.

heat to the air. The other lies deep within the arm alongside the arterial supply, where countercurrent heat exchange is effective.

Consider a person who is initially warm and then goes into the cold. Surface veins will first be in use and can be seen as a varicose pattern on the back of one's hand. Eventually the slightly lower body temperature that results from the cold will operate on the central thermostat in the hypothalamus. The response is an altered circulation in which the venous return is increasingly switched to the alternate path lying deep within the arm. This allows countercurrent exchange to operate. The hand will now drop appreciably below that of the deep body. The cooled venous blood from the hand will be warmed by the outgoing arterial supply as it flows up the arm, as shown schematically in Figure 10.8B. In ways such as this the deep body temperature is maintained. The thermally less critical hands cool down (Figure 10.5).

3 • NATURE OF THE THERMOSTAT

3.1 Temperature Variations within the Body

Any homoiotherm under different types of temperature stress can have considerable thermal inhomogeneity throughout its body. This reflects, as we have already said, the different local rates of heat production and the varying degrees of temperature control exerted at different locations. The hands and feet of a man in the cold may be allowed to drop below 10°C (with the painful sensation of cold). This represents a 30°C change from the other extreme when hot. Even greater changes are seen in the feet of a duck or the leg of a caribou.

Although such temperature variations decrease as we consider sites deeper within the body, the core is not truly isothermal either. These variations indicate the nature and priorities of the regulating system. Not surprisingly, the toes are considered thermally less critical than the brain, where a 1°C change is physiologically important.

In understanding the whole animal as a temperature-regulating system, we must take into account multiple temperature sensors throughout the body affecting the various control mechanisms, such as blood flow, shivering, and sweating. These act in concert to keep local temperature variations within the safe physiological limits for that tissue. Some sensors act locally. But most of their data are processed and acted upon centrally, where all of the animal's other needs, in addition to those of temperature, are balanced against one another.

Since temperature variations in the brain are the smallest, we can conclude that thermal physiology places the greatest emphasis on its value. It follows that the most sensitive sensor must also be in the brain. Physiologists were misled by an early preoccupation with rectal temperatures. More recent work has shown that thermally sensitive functions such as sweating and shivering, are more closely correlated with the temperature measured with a sensor pushed against the eardrum. This indication that the brain is the site of the central thermostat is hardly unexpected. It is substantiated by a variety of direct neural evidence that the actual location is the anterior hypothalamus (Figure 10.9). In retrospect there is little to support the dedication to rectal values other than their convenience. The more varying values they produced were mistaken as indicating a poorer master thermostat than actually existed.

The sensitivity to brain temperature of the physiological functions that together accomplish thermal regulatory is very marked. Small changes about the "set point" strongly influence the appropriate functions. Variations of less than 0.1°C above the set point cause perceptible sweating. And shivering may begin with an equally small drop in temperature below that value. This indicates the operation in homoiotherms of a very precise thermostat.

The value of the temperature set point in the hypothalamus should be viewed as the result of a complex physiological compromise between the internal needs of the homiotherm and the thermal realities of the external world. Most mammals have a temperature of 37°C, while birds operate closer to 40°C. Some primitive homoiotherms have temperatures in the low 30s, whereas a bluefin tuna keeps its muscle in the range of 24°C to 27°C.

3.2   Hypothalamus

The hypothalamus processes the temperature information flowing to it from sensors distributed throughout the body. This is weighted with the temperature of the hypothalamus itself. The composite thermal picture is continually adjusted by control of blood flow, shivering, sweating, and various behavioral actions. The temperature in the brain is most closely controlled, and we call the temperature sensor in this region the thermostat. We should keep in mind, though, that temperatures at other places in the body exert varying degrees of control.

This thermostat in the hypothalamus is composed of neural networks that have a genetically predetermined set point close to 37°C in most species. Individual cells in this region respond separately to either warmth or cold. Their response is conditioned also by their chemical environment. Eventually they control the details of thermal physiology throughout the body. Experiments and anatomical studies illuminate many details concerning how the neural mechanisms function.

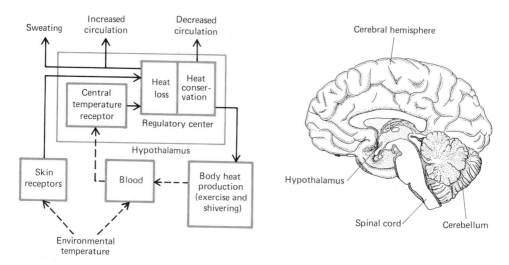

FIGURE 10.9 The temperature control center of man is in the hypothalamus deep in the brain. It senses the temperature of the blood and also receives neural information from temperature sensors over the body. Separate centers control heat loss and heat saving by nervous control of sweating, shivering, and altering blood flow. Dashed lines—heat flow; solid lines—nervous information.

It was early realized that separate posterior and anterior centers in the hypothalamus were concerned with temperature regulation (Figure 10.9). The posterior is temperature-insensitive itself. It receives information from temperature sensors throughout the body and processes this for the anterior to act upon.

The anterior responds mostly to its own temperature at this point deep within the brain. Primarily from this and secondarily from posterior inputs it judges what effective physiological action should be taken. Since this central temperature takes precedence, it follows that the hypothalamus itself is the point of the least temperature variation in the body. Changes of hypothalamic temperatures of less than a tenth of a degree from the set point can be seen to trigger responses such as a change in circulation.

The chemical sensitivity of the anterior hypothalamus is unique. Microinjections of fever-forming agents (*pyrogens*) only at this point cause an increase in body temperature. In the reverse reaction, morphine and anesthetics produce a lower body temperature (*hypothermia*) by acting at the same site. The concentration of inorganic ions at this point can also change the temperature set point up or down.

There appears to be a balance between two antagonistic classes of physiologically active compounds in the hypothalamus that tend to set the temperature in opposite directions. These compounds are largely monoamines such as epinephrine, norepinephrine, histamine, and dopamine.

Fluorescent microscopy, in which a tissue slice is observed under ultraviolet light, shows that individual cells contain one or the other of these amines. The function of such a cell is probably thus indicated. In addition, there are bacterially produced toxins that act analogously in diseases such as typhoid fever and malaria. These alternately change the temperature set point up and down. The body responds by shivering and sweating in its attempts to obey these changing central thermal commands.

We have mentioned the more secondary role that peripheral temperature sensors play in body temperature control. An example of how they work is provided by the local variations in blood circulating to the body surface. The flow to an area can be as much as doubled by heating the skin with a lamp at that point. This activates warm receptors, which signal for greater blood flow to cool the region. With an increase of only 0.5°C in the hypothalamic temperature, circulation to the same surface area can increase as much as sevenfold. The flushed skin of an overheated person is a visible sign of such a blood flow. The control factor represented by such physiological feedback, expressed as amount per °C, may be 20 times greater in the brain than locally in the skin. This is necessary since the brain must in some situations override the surface control. Otherwise an internally overheated animal in a cold environment would let a cold skin decrease blood flow. The more serious condition of the overheated interior must have priority.

It is possible to find a combination of temperature parameters that will cause the system to make a poor judgment. Consider a diver emerging from cold water with a seriously lowered body temperature. A natural instinct would be to warm up with a hot shower. The hot water would stimulate the warm receptors in the skin. This increases circulation between the already cooled core and the even colder surface layers. The thermal averaging from more blood flow would lower the core still further. To avoid this temporary further temperature drop, it is better to warm by shivering. This produces much of the heat deep in the body where it is needed most.

Receptors that send data on temperature at the local tissue level can be directly studied. Action potentials have been recorded from individual nerves in the skin of man which respond with increased firing rate to lowered surface temperature. The analogous warm receptors increase in firing rate above the local set point. Figure 10.7 shows data from the skin receptors of a primate. Information from both types goes to the posterior hypothalamus.

Electrical recording can also be done from individual nerves in the anterior hypothalamus. Again, separate elements that increase in firing rate both above and below the set point have been found. These are the elements that eventually control shivering and sweating.

3.3   Thermogenesis

Much of metabolism is concerned with the production of energy-rich compounds, such as ATP. These are necessary for muscle contraction when an animal performs external work. Shivering in a cold animal provides the added heat necessary to keep up the body temperature. The muscular work is wasted.

A cold, naked aborigine after a night asleep under the desert sky may have a body temperature as low as 35°C. With violent shivering his metabolic heat production can be three or four times higher than basal. This will restore the body temperature in less than an hour.

In some animals the shivering is not externally obvious. A pigeon below its critical temperature shows little movement even though the oxygen consumption is increased. But an electrode in the musculature picks up electrical activity characteristic of contraction. The individual fibers apparently are working in an uncoordinated fashion and do not seem to make obvious movements. Electrical activity of muscle is also linearly related to cold-induced heat production in small pigs.

It would be convenient if biochemistry could be arranged for direct heat production without muscle activity. Many animals show some increase in basal metabolism after prolonged cold exposure. Other species can produce excess metabolic heat on a shorter time scale. This direct thermogenesis without muscle contraction is not widespread, but it is found in animals as varied as rats, infant humans, and guinea pigs. It usually occurs in special brown adipose tissue (Chapter 7, Section 3.5), sometimes called brown fat. The metabolism at such sites is responsive to control by hormones in the circulating blood. The fat is strategically located in the body for local heat generation. This strategy has probably not been more commonly used because shivering has been sufficient.

It is not clear what happens to any ATP that might be produced by nonshivering thermogenesis biochemistry. The phosphorus may be directly uncoupled, producing ADP, but this has not been demonstrated. Another suggestion has been that nerve membranes may become more leaky through some form of central control. The ATP would then be used up in pumping the sodium back out of the nerves.

Brown fat, and the accompanying hormone-mediated thermogenesis, is most common in the newborn. As the animal develops its musculature and is able to shiver, the brown fat disappears. Its thermal role is largely limited to this early phase in mammals. Birds have large mobile fat reserves that can be burned quickly (Chapter 7, Section 3.1) because of the high energy needs of flight. One over-water migratory flight can consume half the bird's weight in the form of this fat. These migratory birds can readily burn this fat for thermal purposes also. Game birds, which typically fly only briefly for escape, have less brown fat.

4 • CLIMATIC STRESS

4.1 Critical Temperature and the Cold

Animals typical of the cold, such as the arctic fox and polar bear, are different in appearance from animals in the tropics. Their obviously greater insulation has led to experimental determinations of how efficient it is. This is usually done by stressing such an animal in an increasingly cold environment while measuring its metabolism (heat production). The animal will increasingly bring into play the various circulatory and postural means to control its heat loss. At some temperature these have been exhausted, and lower temperatures can be tolerated only by increasing heat production above the basal rate. Such a critical temperature provides a test of how good its insulation is and therefore how cold-adapted a species is.

This approach to cold adaptation considers only the amount of metabolic heat in combination with the total insulation available to the animal. At the critical temperature the basal amount of heat flowing through the maximum insulation produces a temperature difference that just allows the deep body to keep in the vicinity of 37°C. For lower environmental temperatures the insulation cannot be increased, so a greater internal heat production from increased metabolism is needed.

At very high temperatures the metabolism of a homoiotherm also increases due to the work of keeping cool (panting and so on) as well as the direct effect of tissue temperature on the local biochemistry. Between this temperature range the variable thermal barrier between the inside of the animal and its external thermal environment is able to actively keep heat loss equal to the resting rate of heat production.

The metabolic data in Figure 10.10 provide a simple way of viewing the degree of cold adaptation of a species. With a basal rate of heat production the fox is comfortable at −20°C. But below 20°C a tropical species, such as the coati, will start to shiver for additional heat.

The rate of increase in metabolism below the critical temperature reflects the added thermal flux necessary to produce the increasing temperature difference across the animal's maximum insulation. When this difference is doubled, the metabolism must be twice as great. Since shivering can produce such a doubling of heat output, a fox could withstand −80°C, whereas tropical species might have trouble at 10°C.

This view of temperature adaptation, called *Newton's law of cooling*, tells nothing about the physiological details involved in the active control of heat production and loss. The static view of insulation as a thermal barrier must frequently be qualified in real situations. Wind can increase heat loss by pushing away the air in fur and feathers. An exposed animal in nature might be much worse off thermally than measurements in the quiet of a physiologist's metabolic chamber would indicate. Conversely, some an-

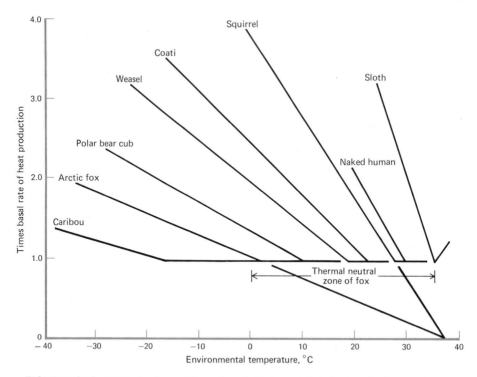

FIGURE 10.10   Within the range of thermal neutrality the metabolic heat production remains constant as the environmental temperature varies. Below some critical temperature the heat-saving mechanisms can no longer keep the body temperature up. Greater heat production is necessary. The slope of this increase extrapolates back to the core temperature.

imals huddle together and are collectively more cold-protected than a single individual.

Man is basically a naked, tropical animal. His critical temperature is above 20°C. Through cultural adaptations he carries a warm climate inside his clothes. This has allowed him to spread even into polar regions. Most of an Eskimo's skin is therefore warm. Cold stress largely comes through the hands and face. These are protected from cold damage by being supplied with an increased blood flow from the warm core. This is sufficient to keep tissue temperatures above some critical value. This physiological capability can develop in anyone during a period of some weeks of cold exposure. It is not a racial trait of arctic inhabitants. The additional heat loss by this circulation must be paid for by increased metabolism and the associated food intake.

An incident I observed in the arctic while doing cold research shows the general nature of some of the thermal adaptations. While traveling by dog team, we built an igloo each night for temporary occupancy.

The dogs slept exposed to the wind, which frequently drifted snow over them. Yet they were easily comfortable under such conditions in spite of air temperatures as low as −30°C. Their low critical temperature allows them to keep a thermal balance under all arctic conditions with little more than their basal heat production.

One morning a dog broke loose during the harnessing. An Eskimo and I chased it through deep snow. The sun was shining, but the air was still down to −20°C. When we finally caught the dog, all three of us were overheated from the exertion. The dog immediately splayed out its legs and pressed its lightly furred belly against the snow. Body heat conducted rapidly through this thermal window and melted snow. In a few minutes the dog was comfortable.

The Eskimo picked up handfuls of snow, and soon melted water was streaming from his hands. He was likewise using enhanced blood flow to a thermal window to get rid of his excess body heat. Such blood flow can serve equally well to avoid overheating as well as aiding in keeping hands warm by the transfer of extra body heat. Increased circulation is a fundamental part of adaptation to both heat and cold. Man can greatly increase his own circulation capability during exposure to either of these. When I picked up snow, my hands remained cold even though I was sweating inside my parka. But after a month or two in the cold, I found that I could cool off the same as the Eskimo.

4.2 The Dry, Warm Desert

The unique condition of dry heat in the desert has produced as extensive a set of adaptations as those related to arctic cold. When the air temperature is higher than that of the body, heat will flow down the gradient into the animal. This provides a thermal load, in addition to the animal's own metabolic heat, that the animal must deal with.

In addition, the desert typically offers little shade. This exposes the animal to a potentially large heat input from sunshine on the body surface. The ground is heated by the sun, so it radiates additional heat to an animal from below.

Among the strategies to moderate such thermal stresses, the most direct is to avoid them. Small desert animals do this by coming out of their burrows mostly at night when it is cool. Large animals must deal with the heat physiologically.

When the sun shines on fur or feathers, the surface may be as hot as 70°C. This represents a reverse gradient of more than 30°C across the insulation. We usually think the purpose of fur or feathers is to keep body heat in. It is equally efficient in keeping the sun's heat out. The back of the camel and ostrich are heavily insulated for this purpose.

Since very little solar heat penetrates, the body has to only deal with its own metabolic heat. Such an animal can rarely afford extensive exercise in the hot daytime. This might increase the heat load tenfold and tax the evaporative cooling capabilities and also the water resources of the animal. In the desert the lost water may be difficult to replace except at long intervals. Large desert animals are typically more resistant to dehydration. They can lose as much as a fourth of their body water and make it up in a few minutes if they have an opportunity to drink.

Size is an advantage in meeting the daily thermal stress in a warm climate. A large animal has a lower specific metabolism than a small one, and it therefore heats up more slowly with time. If all of the resting human heat production were kept in the body, the temperature rise would be a little more than a degree per hour. In an elephant the warming is less than half this amount, whereas in a mouse it is five times as great. Moreover, a large species has more mass in which to store external thermal inputs, such as sunshine, which enter proportionally to the body surface area. Together these factors can be thought of as a kind of thermal inertia that draws out thermal crises to a longer time scale. If this time is long enough, the large animal can make it to the dependable cool refuge of the approaching evening.

An important additional strategy in large tropical animals is to sacrifice some constancy in deep body temperature. In essence such species heat up all day and cool down at night. This temporarily reduces the need for evaporative cooling. Even under moist conditions this is a help because sweating is less efficient since water does not evaporate as readily. Thus we find diurnally labile body temperatures in the camel, ostrich, hippopotamus, rhinoceros, and probably the elephant.

Some tropical species can allow their body temperatures to rise considerably above the usual mammalian range, particularly when running. This reduces the temperature differential with the environment and slows heat conduction into the animal. A cape hunting dog may reach 42°C. The higher core temperature reduces the heat inflow from the still warmer environment. This means less water must be lost by evaporative cooling. Such a saving gives it the advantage of being able to hunt in the middle of the day.

Even higher temperatures are tolerated by some large African antelopes, such as the eland and the oryx. They can tolerate extended periods in midday with a body temperature of 45°C. This so reduces the need for cooling, because of the lower heat inflow, that these animals survive without drinking.

But the entire animal does not warm up. The brain is selectively cooled by evaporation in the nostrils. The venous return from the latter cools the arterial supply to the former. This is another example of how vascular countercurrent operating as a heat exchanger allows an animal to maintain temperature differentials in its body. The sheep uses the same stratagem of nasal cooling for the brain.

An ostrich, which lives in semiarid areas, illustrates many details of specialized hot desert physiology. Its size prevents it from taking advantage of the microclimate in burrows. Because it is a bird it has no sweat glands. Under heat stress it must rely on increased evaporation from the respiratory system as the major avenue for heat dissipation. Water loss from the skin is a negligible fraction of the bird's total.

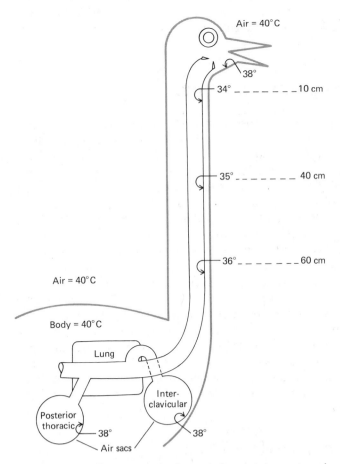

FIGURE 10.11 Temperature distribution in the respiratory system of an ostrich when the air and body temperature are both 40°C. No heat is conducted in or out of the bird since no temperature difference exists. The bird's own heat production is disposed of by evaporative cooling in the air sacs and the trachea.

Surface temperatures within the respiratory system gives information about the site of evaporation. Figure 10.11 shows the situation in which the ambient temperature (40°C) is the same as that inside the bird. No heat exchange can occur through conduction or radiation, and the entire metabolic heat production must be dissipated by evaporation of water. Temperatures lower than ambient indicate sites of evaporation. Unfortunately, they do not tell the amounts of heat transferred. The surface temperature is lowered by increased evaporation and raised by increased blood flow. A low temperature is therefore not a precise measure of the actual amount of cooling taking place. Evaporative cooling is sufficient to maintain the ostrich in thermal balance for many hours at temperatures even above 50°C. The resulting water loss was found to be more than 15 g/min.

5 • AQUATIC HOMOIOTHERMS

Water represents a particularly hostile thermal environment. It has 10 to 100 times the cooling power of air when it lies next to the body surface. In spite of this fact, a variety of aquatic mammals and some fishes as well have a body temperature considerably above that of the surrounding water. Animals such as whales and seals manage this by substituting a layer of blubber in place of the fur used on land. These remarkable products of evolution regulate the deep body temperature as carefully as any terrestrial species. In comparison, man is a poor thermal performer. He has a survival time measured in minutes when he is in cold water.

The thermal success of whales and porpoises (cetaceans), the most impressive marine group, is shown by their existence in all seas from the tropics to the edge of the polar ice. They pass their entire life cycle in the open sea without ever leaving the water. This includes the thermally difficult time of the birth of the young. The seals and sea lions (pinnipeds), the other major marine group, come onto land at this time. The pups are born with fur to meet the thermally less difficult life in air. They keep this until they put on a layer of blubber in a few weeks of nursing on a mother's milk which is 40 percent fat. This provides the thermal equipment needed to return to the water.

5.1 Small Porpoise in Cold Water

Probably nowhere in nature is the cold stress as severe as that on a small mammal, such as a porpoise, in the winter ocean. The porpoise fits rather well our physical model of a warm core surrounded by a well-defined layer of insulation. One can more readily analyze the heat loss from such a marine mammal than is possible to do on a terrestrial species. Heat is lost mostly by simple conduction through a blubber layer to the surrounding water. Skin temperature is close to that of the water because of the latter's high thermal conductivity. Physically this amounts to a known temperature difference existing across a known thermal barrier. The computation of the total outgoing heat involves only known temperatures, the area and thickness of the blubber, and also its thermal conductive properties.

I have worked with one of the smallest cetaceans, the common harbor porpoise, *Phocaena phocaena*. It is found off Norway in water below 10°C. This means it has a temperature difference of about 30°C across its 2-cm-thick blubber layer.

When the thermal flux through this layer is estimated for the animal in such cold water, it is found to be surprisingly high. Most terrestrial animals show a close correlation between size and metabolic heat production. The resulting plot (see Figure 10.1) is known as Benedict's "mouse-to-elephant curve." Our computation for the small porpoise shows a heat flux roughly three times greater than for a terrestrial animal of comparable size.

So many animals lie close to Benedict's curve that we are reluctant to allow such an exception. But two possibilities present themselves: either the porpoise is "running" all the time to keep warm in cold water, or it is violating Benedict's rule. Fortunately, one can make proper measurements on live specimens to decide which is the case.

Metabolic heat production in an animal is usually inferred from a determination of oxygen consumed in respiratory gas exchange. This is called *indirect calorimetry*, and it is a measure of the total oxidative chemistry going on in the animal. With an aquatic animal such as a porpoise, it is a simple matter to determine the heat by direct calorimetry as well. The porpoise is placed in a box of water, and the heat flowing out of the animal is determined by the rate of rise of the water temperature.

We have said that homoiothermy requires that heat production equal heat loss. We can measure both production and loss in a porpoise. The animal in the calorimeter box is connected to breathing tubes that allow its oxygen consumption to be measured. Figure 10.12 shows such results. The animal was overheated slightly during capture in the large aquarium and subsequent transfer to the calorimeter box. Its muscle or deep body temperature dropped during the first hour. This cooling means that the animal was losing heat to the water faster than it was producing heat metabolically. The measurements of these quantities showed this independently. Apparently the animal was coming into thermal equilibrium with the 8°C water. It was metabolizing close to a basal rate, since it showed only a slight amount of shivering and little other movement.

The oxygen consumption of the animal was about 380 ml/min,

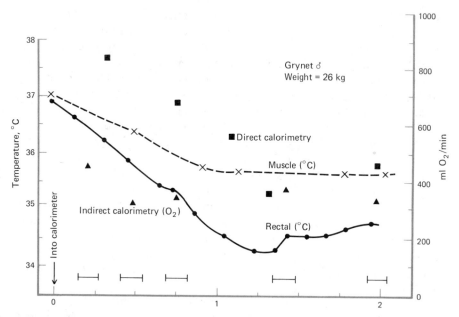

FIGURE 10.12 Thermal measurements on an initially overheated porpoise in a box of water (calorimeter). During the first hour the animal cools down. This is verified by the greater direct heat loss (squares) than the heat loss produced metabolically (triangles). In the second hour the animal's temperature levels off, and the two heat losses are nearly equal.

which is nearly half again greater than that for a man or other land animal weighing twice as much. On an overall weight basis the porpoise is metabolically at least three times as active as a man. The blubber accounts for 40 percent of the animal's weight but contains little of the metabolism. So the difference in specific metabolic activity of the tissues is even greater than indicated.

This basal metabolism, then, shows that the animal has increased its heat production about threefold in order to keep warm as a small species in cold water. It hardly seems likely that porpoises would thrive in a situation in nature where they would be in danger of freezing if they stopped swimming! Analogously all the terrestrial mammals in the arctic also have an insulation combined with a heat production which makes them able to meet the coldest situation with only a slight rise in basal metabolism.

The unusually high cetacean metabolic rate represents the severe cost of maintaining a high mammalian body temperature in cold water. Instead of swimming all the time to keep warm, small marine mammals have increased their heat production to bring it into line with their needs. In porpoises, and also seals, an increase in metabolic rate can be regarded as an evolutionary adaptation to the stress of thermal regulation in cold water.

The same is also true of terrestrial homoiotherms, where a parallel is found in the smallest land mammals. The little shrew weighing a few grams is the terrestrial equivalent to the porpoise. The thermal barrier of air practically limits these animals from smaller size. But retaining heat with air at the body surface is easier, so they are 10^4 times smaller than the porpoise. Thermally we can consider the porpoise a "marine shrew."

We find that the whole physiology of the porpoise is exaggerated in order to allow the animal to keep warm. It has pushed far up the amount of insulation carried and the amount of heat produced. To satisfy the oxygen demand of heat production, the porpoise sacrifices diving ability, and it must devote a great deal of its body to the needs of gas exchange; that is, it must breathe more air, pump more blood, and so on. All these points contribute to the impression that evolution in porpoises has reached a steep barrier limiting them from smaller size.

5.2 Comparison of Porpoise and Whale

Even with its higher heat production, the porpoise has been able to keep warm only by devoting 40 percent of its weight to the surface blubber. One can hardly expect more from a predator, which depends on rapid movement. The big whales, which are nearly 10,000 times larger, are an interesting comparison.

The greater volume-to-surface ratio in the whale means that nearly 10 times more heat for each unit of area is flowing out through the side of the animal. If a porpoise can keep warm with a 2-cm layer of blubber, the whale could thermally make do with 2 mm. Instead it has 20 cm, which is 10 times more rather than 10 times less. The comparative values for the two are shown in Table 10.1. It is clear that the whale has at least 50 times

TABLE 10.1   Thermal Comparison of the Porpoise and Whale

Characteristic	Porpoise	Whale
Area, m²	0.75	137
Metabolic heat, kcal/24 hr	3.5×10^3	1.5×10^6
Heat/area, kcal/m²/hr	175	400
Conduction heat		
(using ΔT and t), kcal/m²/hr	200	20
ΔT across blubber	30°C	40°C
Blubber thickness t, cm	2	20
Percent blubber	40–45	20–25
Percent muscle	20–25	40–45

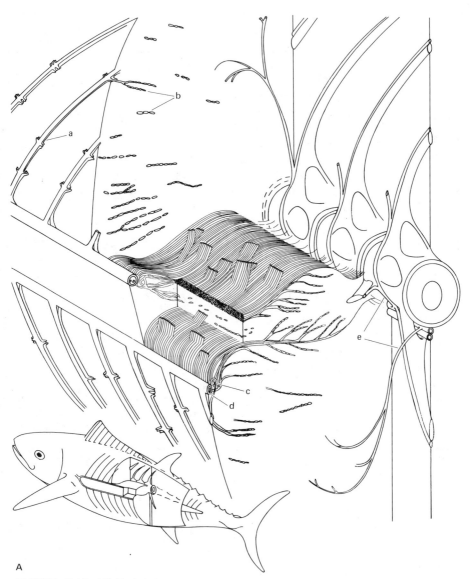

A

FIGURE 10.13 (A) Exploded view of tuna fish circulation showing the lateral arteries and veins (c and d) branching into intermixed combinations that provide a countercurrent supply to the red muscle (shaded). The supply to the white muscle (a and b) is also countercurrent. The more conventional blood flow along the backbone (e), which is a small part of the total, is not.

more blubber than it needs thermally. By this token the blubber on a whale cannot be considered as insulation. It may well be a food store or serve for hydrostatic buoyancy. Or it may be an evolutionary hangover from small

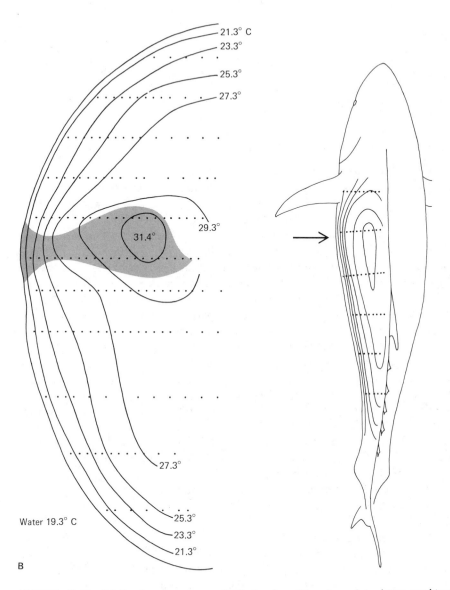

21.3° C
23.3°
25.3°
27.3°

29.3°
31.4°

27.3°

Water 19.3° C

25.3°
23.3°
21.3°

B

FIGURE 10.13 (**B**) Resultant temperature distribution from this pattern of circulation, combined with the tuna's high metabolism.

species that were probably the first to invade the water.

Table 10.1 lists the percentages of body constituents and shows how the general architecture of the porpoise has had to be varied from that of the whale to meet thermal requirements: the amounts of muscle and blubber in the two species are roughly interchanged. The porpoise is forced to carry along so much blubber and metabolically supporting cardiovascular equipment that there is not as much muscle to drive him through the water. I was never able to frighten *Phocaena* into swimming faster than 9 knots.

5.3   Warm-Blooded Fishes

Larger cetaceans, with their clean external form and great power resources, can swim at speeds up to 20 knots. But a large tuna fish can go through the water at twice this speed. It appears the advantages of the mammalian approach are not as great as one might have thought. The answer lies in an unexpected thermal story provided by these far-ranging fishes. They have managed to be as much as 15°C warmer than the surrounding water, in spite of the acute thermal loss represented by having blood aerated in a gill rather than an air-breathing lung. Their high muscle temperature, with the consequently greater muscle power it makes available, combined with some of the most exquisite streamlining in nature, give the tuna its impressive performance. The explanation is found in a circulatory system radically altered from that of other fishes.

Fishes obtain their oxygen by a gill, which brings the blood in intimate diffusive contact with the water. Heat diffuses at least an order of magnitude faster than the dissolved oxygen. It follows then that the thermal equilibration for blood passing through the gill will be much more complete than will that for respiratory gas exchange. The presence of hemoglobin to increase the oxygen capacity of the blood accentuates the difference. This unavoidable physical chemical fact of diffusion determines that the blood leaving a gill must be very close to the temperature of the water. For this reason fishes were previously thought to be always at a temperature close to that of the water.

We can compare the heat loss in a lung with that in a gill. The lung gets its oxygen from air, which is made up of 21 percent oxygen. The oxygen dissolved in water amounts to only 1 part in 10^5. Since thermal equilibrium must be nearly complete with both in order to get reasonable oxygen exchange, it follows that a 10,000 times greater weight of water than of air is exposed to the blood. The much greater viscosity of water over air also means that more physical work is necessary to bring the water to the exchanging surface. These features of gill ventilation, described in Chapter 5, Section 2.1, would appear to discourage the evolution of both warm-bloodedness and a related high metabolic rate in any fish. As in the temperature regulation of a small porpoise, evolution has apparently found a way around such difficulties.

Once the blood has been aerated (and cooled) in the gill, it can be pumped through a countercurrent system that is dimensioned for heat and not for the dissolved oxygen. The latter is transported through the exchanger to the muscle on the far side. Here it is used to support metabolism that produces heat. This would ordinarily be transported by the venous return back to the gill and lost to the water. But the heat exchanger prevents its passage. The heat therefore accumulates in the muscle, and the temperature of the tissue rises.

The circulatory system of the tuna fish has been drastically altered for this thermal purpose. This has been accomplished without the separation of the systemic and pulmonary circulation (Chapter 5, Section 4.2) that birds and mammals have used. The muscles of the tuna are supplied by a pair of lateral arteries running from the gills down the sides along the surface of the fish rather than more conventionally by way of the dorsal aorta. The important anatomical element is the countercurrent heat exchanger. This is formed from the lateral arteries and veins, which break up into many branches as they run inward to the muscle. These are intermixed in an alternating array to allow maximum thermal contact between the two flows. The arrangement in the fish is shown in Figure 10.13. The temperature gradient in such a system occurs along the axis of the exchanger, and the center of the muscle becomes the warmest place in the fish.

The greater speed the tuna obtains from its warmth gives it a competitive edge. These fishes will probably turn out to be the fastest animals in the sea, and they can pursue or escape from anything they desire. Two bluefin tuna were tagged at Bimini, off Miami, Florida, and 40 days later they were caught off Norway. This wide-ranging movement is partly made possible by the interior warmth of the fish. Other large, cold fishes, such as marlin and swordfish, do not seem to range into cold surface water. But they have thermal structures that may keep the eyeball or brain warm, for reasons which are not clear. Although the tuna requires a larger food supply to maintain itself, its increased freedom of movement opens up new sources.

REFERENCES

Aschoff, J., and Wever, R. Kern und Schole im Warmehaushalt des Menschen. *Die Naturwissenschaften* 20:483, 1958.

Benzinger, T. H. Heat Regulation: Homeostasis of Central Temperature in Man. *Physiol. Rev.* 49:671, 1969.

Carey, F. G., Teal, J. M., Kanwisher, J. W., Lawson, K. D., and Beckett, J. S. Warm-bodied Fish. *Am. Zool.* 11:135, 1971.

Eisenman, J. S. Pyrogen-induced Changes in Thermoresponsiveness of Septal, Preoptic and Hypothalamic Neurons. *Physiologist* 10:160, 1967.

Heinrich, B. Thermal Regulation in Endothermic Insects. *Science* 185:747–750, 1974.

Hemmingsen, A. M. Energy Metabolism as Related to Body Size. *Reports of the Steno Memorial Hospital (Copenhagen)* 9:1, 1960.

Iggo, A. J. Cutaneous Thermal Receptors in Mammals. *J. Physiol.* 200:403, 1969.

Irving, L. Ecology and Thermoregulation. *Les Concepts de Claude Bernard sur le Milieu Intérieur.* Paris: Masson, 1967.

Kanwisher, J., and Sundnes, G. Physiology of a Small Cetacean. *Hvalradets Skrifter* 48:45, 1968.

Kleiber, M. *The Fire of Life.* New York: Wiley, 1961.

Lasiewski, R. C., and Dawson, W. R. A Reexamination of the Relation between Standard Metabolic Rate and Body Weight in Birds. *Condor* 69:13, 1967.

Meyers, R. D. Thermal Regulation: Neurochemical Systems in the Hypothalamus. *The Hypothalamus and Pituitary in Health and Disease.* William Locke, ed. Springfield, Ill: Thomas, 1972.

Prosser, C. L. Temperature. *Comparative Animal Physiology.* Philadelphia: Saunders, 1973.

Scholander, P. F. Countercurrent Exchange. A Principle in Biology. *Hvalradets Skrifter* 44:1–24, 1958.

Scholander, P. F., Hock, R., Walters, V., and Irving, L. Adaptations to Cold in Arctic and Tropical Mammals and Birds. *Biol. Bull.* 99:259, 1950.

Schmidt-Nielsen, K. *How Animals Work.* New York: Cambridge University Press, 1972.

Schmidt-Nielsen, K., Kanwisher, J., Lasiewski, R. C., Cohn, J. E., and Bretz, W. L. Temperature Regulation and Respiration in the Ostrich. *Condor* 71:341–352, 1969.

Taylor, C. R. Dehydration and Heat: Effects on Temperature Regulation of East African Ungulates. *Am. J. Physiol.* 219:1136–1139, 1970.

INDEX